UNITEXT for Physics

UNITEXT for Physics series, formerly UNITEXT Collana di Fisica e Astronomia, publishes textbooks and monographs in Physics and Astronomy, mainly in English language, characterized of a didactic style and comprehensiveness. The books published in UNITEXT for Physics series are addressed to graduate and advanced graduate students, but also to scientists and researchers as important resources for their education, knowledge and teaching.

More information about this series at http://www.springer.com/series/13351

Carlos Maña

Probability and Statistics for Particle Physics

Springer

Carlos Maña
Departamento de Investigación Básica
Centro de Investigaciones Energéticas,
Medioambientales y Tecnológicas
Madrid
Spain

ISSN 2198-7882 ISSN 2198-7890 (electronic)
UNITEXT for Physics
ISBN 978-3-319-55737-3 ISBN 978-3-319-55738-0 (eBook)
DOI 10.1007/978-3-319-55738-0

Library of Congress Control Number: 2017936885

Printed on acid-free paper

This Springer imprint is published by Springer Nature
The registered company is Springer International Publishing AG
The registered company address is: Gewerbestrasse 11, 6330 Cham, Switzerland

Contents

Introduction

> *They say that understanding ought to work by the rules of right reason. These rules are, or ought to be, contained in Logic; but the actual science of logic is conversant at present only with things either certain, impossible, or entirely doubtful, none of which (fortunately) we have to reason on. Therefore the true logic of this world is the calculus of Probabilities, which takes account of the magnitude of the probability which is, or ought to be, in a reasonable man's mind.*
>
> J.C. Maxwell

These notes, based on a one-semester course on probability and statistics given in the former *Doctoral Program* of the *Department of Theoretical Physics* at the *Universidad Complutense* in Madrid, are a more elaborated version of a series of lectures given at different places to advanced graduate and Ph.D. students. Although they certainly have to be tailored for undergraduate students, they contain a humble overview of the basic concepts and ideas one should have in mind before getting involved in data analysis and I believe they will be a useful reference for both students and researchers.

I feel, maybe wrongly, that there is a recent tendency in a subset of the particle physics community to consider statistics as a collection of prescriptions *written in some holy references* that are used blindly with the only arguments that either "*everybody does it that way*" or that "*it has always been done this way*." In the lectures, I have tried to demystify the "*how-to*" recipes not because they are not useful but because, on the one hand, they are applicable under some conditions that tend to be forgotten and, on the other, because if the concepts are clear so will be the way to proceed ("at least formally") for the problems that come across in particle physics. At the end, the quote from Laplace given at the beginning of the first lecture is what it is all about.

There is a countable set of books on probability and statistics and a sizable subset of them are very good, out of which I would recommend the following ones (a personal choice function). Chapter 1 deals with probability and this is just a measure, a finite nonnegative measure, so it will be very useful to read some sections of *Measure Theory* (2006; Springer) by V.I. Bogachev, in particular Chaps. 1 and 2 of the first volume. However, for those students who are not yet familiar with measure theory, there is an appendix to this chapter with a short digression on some basic concepts. A large fraction of the material presented in this lecture can be found in more depth, together with other interesting subjects, in the book *Probability: A Graduate Course* (2013; Springer Texts in Statistics) by A. Gut. Chapter 2 is about statistical inference, Bayesian inference in fact, and a must for this topic is the *Bayesian Theory* (1994; John Wiley & Sons) by J.M. Bernardo and A.F.M. Smith that contains also an enlightening discussion about the Bayesian and frequentist approaches in the Appendix B. It is beyond question that in any worthwhile course on statistics the ubiquitous frequentist methodology has to be taught as well and there are excellent references on the subject. Students are encouraged to look, for instance, at *Statistical Methods in Experimental Physics* (2006; World Scientific) by F. James, *Statistics for Nuclear and Particle Physicists* (1989; Cambridge University Press) by L. Lyons, or *Statistical Data Analysis* (1997; Oxford Science Pub.) by G. Cowan. Last, Chap. 3 is devoted to Monte Carlo simulation, an essential tool in statistics and particle physics, and Chap. 4 to information theory, and, like for the first chapters, both have interesting references given along the text.

"*Time is short, my strength is limited,…*", Kafka dixit, so many interesting subjects that deserve a whole lecture by themselves are left aside. To mention some: an historical development of probability and statistics, Bayesian networks, generalized distributions (a different approach to probability distributions), decision theory (games theory), and Markov chains for which we shall state only the relevant properties without further explanation.

I am grateful to Drs. J. Berdugo, J. Casaus, C. Delgado, and J. Rodriguez for their suggestions and a careful reading of the text and much indebted to Dr. Hisako Niko. Were not for her interest, this notes would still be in the drawer. My gratitude goes also to Mieke van der Fluit for her assistance with the edition.

Chapter 1
Probability

The Theory of Probabilities is basically nothing else but common sense reduced to calculus

P.S. Laplace

1.1 The Elements of Probability: $(\Omega, \mathcal{B}, \mu)$

The axiomatic definition of probability was introduced by A.N. Kolmogorov in 1933 and starts with the concepts of *sample space* (Ω) and *space of events* $(\mathcal{B}_\Omega)$ with structure of σ-algebra. When the pair $(\Omega, \mathcal{B}_\Omega)$ is equipped with a *measure* μ we have a *measure space* $(E, \mathcal{B}, \mu)$ and, if the measure is a *probability measure* P we talk about a *probability space* $(\Omega, \mathcal{B}_\Omega, P)$. Lets discuss all these elements.

1.1.1 Events and Sample Space: (Ω)

To learn about the state of nature, we do experiments and observations of the natural world and ask ourselves questions about the outcomes. In a general way, the *object* of questions we may ask about the result of an experiment such that the possible answers are *it occurs* or *it does not occur* are called **events**. There are different kinds of events and among them we have the **elementary events**; that is, those results of the random experiment that **can not** be decomposed in others of lesser entity. The **sample space** (Ω) is the set of **all** the possible **elementary outcomes (events)** of a random experiment and they have to be:

(i) **exhaustive**: any possible outcome of the experiment has to be included in Ω;
(ii) **exclusive**: there is no overlap of elementary results.

C. Maña, *Probability and Statistics for Particle Physics*,
UNITEXT for Physics, DOI 10.1007/978-3-319-55738-0_1

To study random phenomena we start by specifying the *sample space* and, therefore, we have to have a clear idea of what are the possible results of the experiment. To center the ideas, consider the simple experiment of rolling a die with 6 faces numbered from 1 to 6. We consider as *elementary events*

$$e_i = \{\text{get the number } i \text{ on the upper face}\}; \quad i = 1, \ldots, 6$$

so $\Omega = \{e_1, \ldots, e_6\}$. Note that any possible outcome of the roll is included in Ω and we can not have two or more elementary results simultaneously. But there are other types of events besides the elementary ones. We may be interested for instance in the parity of the number so we would like to consider also the possible results[1]

$$A = \{\text{get an even number}\} \qquad \text{and} \qquad A^c = \{\text{get an odd number}\}$$

They are not *elementary* since the result $A = \{e_2, e_4, e_6\}$ is equivalent to get e_2, e_4 or e_6 and $A^c = \Omega \setminus A$ to get e_1, e_3 or e_5. In general, an **event** is any subset[2] of the sample space and we shall distinguish between:

elementary events: any element of the *sample space* Ω;

events: any subset of the *sample space*;

and two extreme events:

sure events: $S_S = \{\text{get any result contained in } \Omega\} \equiv \Omega$

impossible events: $S_I = \{\text{get any result not contained in } \Omega\} \equiv \emptyset$

Any event that is neither *sure* nor *impossible* is called **random event**. Going back to the rolling of the die, sure events are

$$S_S = \{\text{get a number } n \mid 1 \leq n \leq 6\} = \Omega \qquad \text{or}$$
$$S_S = \{\text{get a number that is even or odd}\} = \Omega$$

impossible events are

$$S_I = \{\text{get an odd number that is not prime}\} = \emptyset \qquad \text{or}$$
$$S_I = \{\text{get the number } 7\} = \emptyset$$

---

[1]Given two sets $A, B \subset \Omega$, we shall denote by A^c the *complement* of A (that is, the set of all elements of Ω that are not in A) and by $A \setminus B \equiv A \cap B^c$ the *set difference* or *relative complement of B in A* (that is, the set of elements that are in A but not in B). It is clear that $A^c = \Omega \setminus A$.

[2]This is not completely true if the sample space is non-denumerable since there are subsets that can not be considered as events. It is however true for the subsets of $\mathcal{R}^n$ we shall be interested in. We shall talk about that in Sect. 1.1.2.2.

and random events are any of the e_i or, for instance,

$$S_r = \{\text{get an even number}\} = \{e_2, e_4, e_6\}$$

Depending on the number of possible outcomes of the experiment, the the sample space can be:

finite: if the number of elementary events is finite;

Example: In the rolling of a die, $\Omega = \{e_i; i = 1, \ldots, 6\}$ so $\dim(\Omega) = 6$.

countable: when there is a one-to-one correspondence between the elements of Ω and $\mathcal{N}$;

Example: Consider the experiment of flipping a coin and stopping when we get H. Then $\Omega = \{H, TH, TTH, TTTH, \ldots\}$.

non-denumerable: if it is neither of the previous;

Example: For the decay time of an unstable particle $\Omega = \{t \in R | t \geq 0\} = [0, \infty)$ and for the production polar angle of a particle $\Omega = \{\theta \in R | 0 \leq \theta \leq \pi\} = [0, \pi]$.

It is important to note that the *events* are not necessarily numerical entities. We could have for instance the die with colored faces instead of numbers. We shall deal with that when discussing *random quantities*. Last, given a sample space Ω we shall talk quite frequently about a *partition* (or a *complete system of events*); that is, a sequence $\{S_i\}$ of events, finite or countable, such that

$$\Omega = \bigcup_i S_i \quad \text{(complete system)} \quad \text{and} \quad S_i \bigcap_{\forall i,j} S_j = \emptyset; \quad i \neq j \quad \text{(disjoint events)}.$$

1.1.2 σ-algebras ($\mathcal{B}_\Omega$) and Measurable Spaces ($\Omega, \mathcal{B}_\Omega$)

As we have mentioned, in most cases we are interested in events other than the elementary ones. We single out them in a class of events that contains all the possible results of the experiment we are interested in such that when we ask about the union, intersection and complements of events we obtain elements that belong the same class. A non-empty family $\mathcal{B}_\Omega = \{S_i\}_{i=1}^n$ of subsets of the sample space Ω that is *closed* (or stable) under the operations of *union* and *complement*;

that is

$$S_i \cup S_j \in \mathcal{B}; \quad \forall S_i, S_j \in \mathcal{B} \qquad \text{and} \qquad S_i{}^c \in \mathcal{B}; \quad \forall S_i \in \mathcal{B}$$

is an **algebra** (*Boole algebra*) if Ω is finite. It is easy to see that if it is closed under unions and complements it is also closed under intersections and the following properties hold for all $S_i, S_j \in \mathcal{B}_\Omega$:

$$\Omega \in \mathcal{B}_\Omega \qquad \emptyset \in \mathcal{B}_\Omega \qquad S_i \cap S_j \in \mathcal{B}_\Omega$$

$$S_i{}^c \cup S_j{}^c \in \mathcal{B}_\Omega \qquad (S_i{}^c \cup S_j{}^c)^c \in \mathcal{B}_\Omega \qquad S_i \setminus S_j \in \mathcal{B}_\Omega$$

$$\cup_{i=1}^m S_i \in \mathcal{B}_\Omega \qquad \cap_{i=1}^m S_i \in \mathcal{B}_\Omega$$

Given a sample space Ω we can construct different Boole algebras depending on the events of interest. The smaller one is $\mathcal{B}_m = \{\emptyset, \Omega\}$, the minimum algebra that contains the event $A \subset \Omega$ has 4 elements: $\mathcal{B} = \{\emptyset, \Omega, A, A^c\}$ and the largest one, $\mathcal{B}_M = \{\emptyset, \Omega, \text{all possible subsets of } \Omega\}$ will have $2^{\dim(\Omega)}$ elements. From $\mathcal{B}_M$ we can engender any other algebra by a finite number of unions and intersections of its elements.

1.1.2.1 σ-algebras

If the sample space is countable, we have to generalize the Boole algebra such that the unions and intersections can be done a countable number of times getting always events that belong to the same class; that is:

$$\bigcup_{i=1}^{\infty} S_i \in \mathcal{B} \qquad \text{and} \qquad \bigcap_{i=1}^{\infty} S_i \in \mathcal{B}$$

with $\{S_i\}_{i=1}^{\infty} \in \mathcal{B}$. These algebras are called σ-**algebras**. Not all the Boole algebras satisfy these properties but the *σ-algebras* are always Boole algebras (closed under finite union).

Consider for instance a finite set E and the class $\mathcal{A}$ of subsets of E that are either finite or have finite complements. The finite union of subsets of $\mathcal{A}$ belongs to $\mathcal{A}$ because the finite union of finite sets is a finite set and the finite union of sets that have finite complements has finite complement. However, the countable union of finite sets is countable and its complement will be an infinite set so it does not belong to $\mathcal{A}$. Thus, $\mathcal{A}$ is a Boole algebra but not a σ-algebra.

Let now E be any infinite set and $\mathcal{B}$ the class of subsets of E that are either countable or have countable complements. The finite or countable union of countable sets is countable and therefore belongs to $\mathcal{B}$. The finite or countable union of sets whose complement is countable has a countable complement and also belongs to $\mathcal{B}$. Thus, $\mathcal{B}$ is a Boole algebra and σ-algebra.

1.1.2.2 Borel σ-algebras

Eventually, we are going to assign a probability to the events of interest that belong to the algebra and, anticipating concepts, probability is just a bounded measure so we need a class of *measurable sets* with structure of a *σ-algebra*. Now, it turns out that when the sample space Ω is a non-denumerable topological space there exist non-measurable subsets that obviously can not be considered as events.[3] We are particularly interested in $\mathcal{R}$ (or, in general, in $\mathcal{R}^n$) so we have to construct a family $\mathcal{B}_{\mathcal{R}}$ of measurable subsets of $\mathcal{R}$ that is

(i) closed under countable number of intersections: $\{B_i\}_{i=1}^{\infty} \in \mathcal{B}_{\mathcal{R}} \longrightarrow \cap_{i=1}^{\infty} B_i \in \mathcal{B}_{\mathcal{R}}$
(ii) closed under complements: $B \in \mathcal{B}_{\mathcal{R}} \rightarrow B^c = \mathcal{R} \backslash B \in \mathcal{B}_{\mathcal{R}}$

Observe that, for instance, the family of all subsets of $\mathcal{R}$ satisfies the conditions (i) and (ii) and the intersection of any collection of families that satisfy them is a family that also fulfills this conditions but not all are measurable. Measurably is the key condition. Let's start identifying what we shall considered the *basic set* in $\mathcal{R}$ to engerder an algebra. The sample space $\mathcal{R}$ is a linear set of points and, among it subsets, we have the **intervals**. In particular, if $a \leq b$ are any two points of $\mathcal{R}$ we have:

- open intervals: $(a, b) = \{x \in \mathcal{R} \mid a < x < b\}$
- closed intervals: $[a, b] = \{x \in \mathcal{R} \mid a \leq x \leq b\}$
- half-open intervals on the right: $[a, b) = \{x \in \mathcal{R} \mid a \leq x < b\}$
- half-open intervals on the left: $(a, b] = \{x \in \mathcal{R} \mid a < x \leq b\}$

When $a = b$ the closed interval reduces to a point $\{x = a\}$ (*degenerated interval*) and the other three to the null set and, when $a \to -\infty$ or $b \to \infty$ we have the *infinite intervals* $(-\infty, b)$, $(-\infty, b]$, (a, ∞) and $[a, \infty)$. The whole space $\mathcal{R}$ can be considered as the interval $(-\infty, \infty)$ and any interval will be a subset of $\mathcal{R}$. Now, consider the class of all intervals of $\mathcal{R}$ of any of the aforementioned types. It is clear that the intersection of a finite or countable number of intervals is an interval but the union is not necessarily an interval; for instance $[a_1, b_1] \cup [a_2, b_2]$ with $a_2 > b_1$ is not an interval. Thus, this class is not additive and therefore not a closed family. However,

[3]Is not difficult to show the existence of Lebesgue non-measurable sets in $\mathcal{R}$. One simple example is the *Vitali set* constructed by G. Vitali in 1905 although there are other interesting examples (Hausdorff, Banach–Tarsky) and they all assume the Axiom of Choice. In fact, the work of R.M. Solovay around the 70s shows that one can not prove the existence of Lebesgue non-measurable sets without it. However, one can not specify the choice function so one can prove their existence but can not make an explicit construction in the sense Set Theorists would like. In Probability Theory, we are interested only in Lebesgue measurable sets so those which are not have nothing to do in this business and Borel's algebra contains only measurable sets.

it is possible to construct an additive class including, along with the intervals, other measurable sets so that any set formed by countably many operations of unions, intersections and complements of intervals is included in the family. Suppose, for instance, that we take the half-open intervals on the right $[a, b)$, $b > a$ as the initial class of sets[4] to generate the algebra $\mathcal{B}_{\mathcal{R}}$ so they are in the bag to start with. The open, close and degenerate intervals are

$$(a, b) = \bigcup_{n=1}^{\infty} [a - 1/n, b); \quad [a, b] = \bigcap_{n=1}^{\infty} [a, b + 1/n) \quad \text{and} \quad a = \{x \in \mathcal{R} | x = a\} = [a, a]$$

so they go also to the bag as well as the half-open intervals $(a, b] = (a, b) \cup [b, b]$ and the countable union of unitary sets and their complements. Thus, countable sets like $\mathcal{N}$, $\mathcal{Z}$ or $\mathcal{Q}$ are in the bag too. Those are the sets we shall deal with.

The smallest family $\mathcal{B}_{\mathcal{R}}$ (or simply $\mathcal{B}$) of measurable subsets of $\mathcal{R}$ that contains all intervals and is closed under complements and countable number of intersections has the structure of a *σ-algebra*, is called **Borel's algebra** and its elements are generically called **Borel's sets** or *borelians* Last, recall that half-open sets are Lebesgue measurable ($\lambda((a, b]) = b - a$) and so is any set built up from a countable number of unions, intersections and complements so all Borel sets are Lebesgue measurable and every Lebesgue measurable set differs from a Borel set by at most a set of measure zero. Whatever has been said about $\mathcal{R}$ is applicable to the n-dimensional euclidean space $\mathcal{R}^n$.

The pair $(\Omega, \mathcal{B}_\Omega)$ is called **measurable space** and in the next section it will be equipped with a *measure* and "upgraded" to a *measure space* and eventually to a *probability space*.

1.1.3 Set Functions and Measure Space: $(\boldsymbol{\Omega}, \boldsymbol{\mathcal{B}_\Omega}, \boldsymbol{\mu})$

A function $f : A \in \mathcal{B}_\Omega \longrightarrow \mathcal{R}$ that assigns to each set $A \in \mathcal{B}_\Omega$ one, and only one real number, finite or not, is called a **set function**. Given a sequence $\{A_i\}_{i=1}^{n}$ of subset of $\mathcal{B}_\Omega$ pair-wise disjoint, $(A_i \cap A_j = \emptyset;\ i, j = 1, \ldots, n;\ i \neq j)$ we say that the *set function* is **additive** (*finitely additive*) if:

$$f\left(\bigcup_{i=1}^{n} A_i\right) = \sum_{i=1}^{n} f(A_i)$$

or σ-**additive** if, for a countable the sequence $\{A_i\}_{i=1}^{\infty}$ of pair-wise disjoint sets of $\mathcal{B}$,

$$f\left(\bigcup_{i=1}^{\infty} A_i\right) = \sum_{i=1}^{\infty} f(A_i)$$

[4] The same algebra is obtained if one starts with (a, b), $(a, b]$ or $[a, b]$.

It is clear that any σ-additive set function is additive but the converse is not true. A *countably additive set function* is a **measure on** the algebra $\mathcal{B}_\Omega$, a *signed measure* in fact. If the σ-additive set function is $\mu : A \in \mathcal{B}_\Omega \longrightarrow [0, \infty)$ (i.e., $\mu(A) \geq 0$) for all $A \in \mathcal{B}_\Omega$ it is a **non-negative measure**. In what follows, whenever we talk about *measures* $\mu, \nu, \ldots$ on a σ-algebra we shall assume that they are always non-negative measures without further specification. If $\mu(A) = 0$ we say that A is a set of *zero measure*.

The "*trio*" $(\Omega, \mathcal{B}_\Omega, \mu)$, with Ω a non-empty set, $\mathcal{B}_\Omega$ a σ-algebra of the sets of Ω and μ a measure over $\mathcal{B}_\Omega$ is called **measure space** and the elements of $\mathcal{B}_\Omega$ *measurable sets*.

In the particular case of the n-dimensional euclidean space $\Omega = \mathcal{R}^n$, the σ-algebra is the Borel algebra and **all** the Borel sets are measurable. Thus, the intervals I of any kind are *measurable sets* and satisfy that

(i) If $I \in \mathcal{R}$ is measurable $\longrightarrow$ $I^c = \mathcal{R} - I$ is measurable;
(ii) If $\{I\}_{i=1}^{\infty} \in \mathcal{R}$ are measurable $\longrightarrow$ $\cup_{i=1}^{\infty} I_i$ is measurable;

Countable sets are Borel sets of zero measure for, if μ is the Lebesgue measure (see Appendix 2), we have that $\mu([a, b)) = b - a$ and therefore:

$$\mu(\{a\}) = \lim_{n\to\infty} \mu([a, a + 1/n)) = \lim_{n\to\infty} \frac{1}{n} = 0$$

Thus, *any point* is a Borel set with *zero Lebesgue measure* and, being μ a σ-additive function, any countable set has zero measure. The converse is not true since there are borelians with zero measure that are not countable (i.e. Cantor's ternary set).

In general, a measure μ over $\mathcal{B}$ satisfies that, for any $A, B \in \mathcal{B}$ not necessarily disjoint:

(m.1) $\mu(A \cup B) = \mu(A) + \mu(B \backslash A)$
(m.2) $\mu(A \cup B) = \mu(A) + \mu(B) - \mu(A \cap B)$ $\quad (\mu(A \cup B) \leq \mu(A) + \mu(B))$
(m.3) If $A \subseteq B$, then $\mu(B \backslash A) = \mu(B) - \mu(A)$ $\quad (\geq 0$ since $\mu(B) \geq \mu(A))$
(m.4) $\mu(\emptyset) = 0$

(m.1) $A \cup B$ is the union of two disjoint sets A and $B \backslash A$ and the measure is an additive set function;
(m.2) $A \cap B^c$ and B are disjoint and its union is $A \cup B$ so $\mu(A \cup B) = \mu(A \cap B^c) + \mu(B)$. On the other hand $A \cap B^c$ and $A \cap B$ are disjoint at its union is A so $\mu(A \cap B^c) + \mu(A \cap B) = \mu(A)$. It is enough to substitute $\mu(A \cap B^c)$ in the previous expression;
(m.3) from (m.1) and considering that, if $A \subseteq B$, then $A \cup B = B$
(m.4) from (m.3) with $B = A$.

A measure μ over a measurable space $(\Omega, \mathcal{B}_\Omega)$ is **finite** if $\mu(\Omega) < \infty$ and **σ-finite** if $\Omega = \cup_{i=1}^{\infty} A_i$, with $A_i \in \mathcal{B}_\Omega$ and $\mu(A_i) < \infty$. Clearly, any finite measure is σ-finite but the converse is not necessarily true. For instance, the Lebesgue measure λ in $(\mathcal{R}^n, \mathcal{B}_{\mathcal{R}^n})$ is not finite because $\lambda(\mathcal{R}^n) = \infty$ but is σ-finite because

$$\mathcal{R}^n = \bigcup_{k \in \mathcal{N}} [-k, k]^n$$

and $\lambda([-k, k]^n) = (2k)^n$ is finite. As we shall see in Chap. 2, in some circumstances we shall be interested in the limiting behaviour of σ-finite measures over a sequence of compact sets. As a second example, consider the measurable space $(\mathcal{R}, \mathcal{B})$ and μ such that for $A \subset \mathcal{B}$ is $\mu(A) = \text{card}(A)$ if A is finite and ∞ otherwise. Since $\mathcal{R}$ is an uncountable union of finite sets, μ is not σ-finite in $\mathcal{R}$. However, it is σ-finite in $(\mathcal{N}, \mathcal{B}_{\mathcal{N}})$.

1.1.3.1 Probability Measure

Let $(\Omega, \mathcal{B}_\Omega)$ be a measurable space. A measure P over $\mathcal{B}_\Omega$ (that is, with domain in $\mathcal{B}_\Omega$), image in the closed interval $[0, 1] \in \mathcal{R}$ and such that $P(\Omega) = 1$ (finite) is called a **probability measure** and its properties a just those of finite (non-negative) measures. Expliciting the axioms, a *probability measure* is a *set function* with *domain* in $\mathcal{B}_\Omega$ and *image* in the closed interval $[0, 1] \in \mathcal{R}$ that satisfies three *axioms*:

(i) **additive**:	is an additive set function;
(ii) **no negativity**:	is a measure;
(iii) **certainty**:	$P(\Omega) = 1$.

These properties coincide obviously with those of the *frequency and combinatorial probability* (see Note 1). All probability measures are finite ($P(\Omega) = 1$) and any bounded measure can be converted in a *probability measure* by proper normalization. The *measurable space* $(\Omega, \mathcal{B}_\Omega)$ provided with and probability measure P is called the **probability space** $(\Omega, \mathcal{B}_\Omega, P)$. It is straight forward to see that if $A, B \in \mathcal{B}$, then:

(p.1) $P(A^c) = 1 - P(A)$
(p.2) $P(\emptyset) = 0$
(p.3) $P(A \cup B) = P(A) + P(B \backslash A) = P(A) + P(B) - P(A \cap B) \leq P(A) + P(B))$

The property (p.3) can be extended by recurrence to an arbitrary number of events $\{A_i\}_{i=1}^n \in \mathcal{B}$ for if $S_k = \cup_{j=1}^k A_j$, then $S_k = A_k \cup S_{k-1}$ and $P(S_n) = P(A_n) + P(S_{n-1}) - P(A_n \cap S_{n-1})$.

Last, note that in the probability space $(\mathcal{R}, \mathcal{B}, P)$ (or in $(\mathcal{R}^n, \mathcal{B}_n, P)$), the set of points $W = \{\forall x \in \mathcal{R} \mid P(x) > 0\}$ is countable. Consider the partition

$$W = \bigcup_{k=1}^{\infty} W_k \qquad \text{where} \qquad W_k = \{\forall x \in \mathcal{R} \mid 1/(k+1) < P(x) \leq 1/k\}$$

If $x \in W$ then it belongs to one W_k and, conversely, if x belongs to one W_k then it belongs to W. Each set W_k has at most k points for otherwise the sum of probabilities of its elements is $P(W_k) > 1$. Thus, the sets W_k are finite and since W is a countable union of finite sets is a countable set. In consequence, *we can assign finite probabilities on at most a countable subset of* $\mathcal{R}$.

NOTE 1: What is probability?

It is very interesting to see how along the 500 years of history of probability many people (Galileo, Fermat, Pascal, Huygens, Bernoulli, Gauss, De Moivre, Poisson,...) have approached different problems and developed concepts and theorems (Laws of Large Numbers, Central Limit, Expectation, Conditional Probability,...) and a proper definition of probability has been so elusive. Certainly there is a before and after Kolmogorov's *"General Theory of Measure and Probability Theory"* and *"Grundbegriffe der Wahrscheinlichkeitsrechnung"* so from the mathematical point of view the question is clear after 1930s. But, as Poincare said in 1912: *"It is very difficult to give a satisfactory definition of Probability"*. Intuitively, What is probability?

The first *"definition"* of probability was the *Combinatorial Probability* (~1650). This is an objective concept (i.e., independent of the individual) and is based on Bernoulli's *Principle of Symmetry or Insufficient Reason*: all the possible outcomes of the experiment equally likely. For its evaluation we have to know the cardinal ($\nu(\cdot)$) of all possible results of the experiment ($\nu(\Omega)$) and the probability for an event $A \subset \Omega$ is *"defined"* by the Laplace's rule: $P(A) = \nu(A)/\nu(\Omega)$. This concept of probability, implicitly admitted by Pascal and Fermat and explicitly stated by Laplace, is an *a priory probability* in the sense that can be evaluated before or even without doing the experiment. It is however meaningless if Ω is a countable set ($\nu(\Omega) = \infty$) and one has to justify the validity of the Principle of Symmetry that not always holds originating some interesting debates. For instance, in a problem attributed to D'Alembert, a player A tosses a coin twice and wins if H appears in at least one toss. According to Fermat, one can get $\{(TT), (TH), (HT), (HH)\}$ and A will loose only in the first case so being the four cases equally likely, the probability for A to win is $P = 3/4$. Pascal gave the same result. However, for Roberval one should consider only $\{(TT), (TH), (H\cdot)\}$ because if A has won already if H appears at the first toss so $P = 2/3$. Obviously, Fermat and Pascal were right because, in this last case, the three possibilities are not all equally likely and the *Principle of Symmetry* does not apply.

The second interpretation of probability is the *Frequentist Probability*, and is based on the idea of *frequency of occurrence* of an event. If we repeat the experiment n times and a particular event A_i appears n_i times, the relative frequency of occurrence is $f(A_i) = n_i/n$. As n grows, it is observed (experimental fact) that this number stabilizes around a certain value and in consequence the probability of occurrence of A_i is defined as $P(A_i) \equiv \lim_{n\to\infty}^{exp} f(A_i)$. This is an objective concept inasmuch it is independent of the observer and is a posteriori since it is based on what has been observed after the experiment has been done through an *experimental limit* that obviously is not attainable. In this sense, it is more a practical rule than a definition. It was also implicitly assumed by Pascal and Fermat (letters of de Mere to Pascal: *I have observed in my die games...*), by Bernoulli in his *Ars Conjectandi* of 1705 (*Law of Large Numbers*) and finally was clearly explicited at the beginning of the XX century (Fisher and Von Mises).

Both interpretations of probability are restricted to observable quantities. What happen for instance if they are not directly observable? What if we can not repeat the experiment a large number of times and/or under the same conditions? Suppose that you jump from the third floor down to ground (imaginary experiment). Certainly, we can talk about the probability that you break your leg but, how many times can we repeat the experiment under the same conditions?

During the XX century several people tried to pin down the concept of probability. Pierce and, mainly, Popper argumented that probability represents the *propensity* of Nature to give a particular result in a **single** trial without any need to appeal at *"large numbers"*. This assumes that the *propensity*, and therefore the probability, exists in an *objective* way even though the *causes* may be difficult to understand. Others, like Knight, proposed that randomness is not a measurable property but just a problem of knowledge. If we toss a coin and know precisely its shape, mass, acting forces, environmental conditions,... we should be able to determine with certainty if the result will be head or tail but since we lack the necessary information we can not predict the outcome with certainty so we are lead to consider that as a random process and use the Theory of Probability. Physics suggests that it is not only a question of knowledge but randomness is deeply in the way Nature behaves.

The idea that probability *is a quantification of the degree of belief that we have in the occurrence of an event* was used, in a more inductive manner, by Bayes and, as we shall see, Bayes's theorem and the idea of information play an essential role in its axiomatization. To quote again Poincare, "*... the probability of the causes, the most important from the point of view of scientific applications.*". It was still an open the question whether this quantification is *subjective* or not. In the 20s, Keynes argumented that it is not because, if we know all the elements and factors of the experiment, what is likely to occur or not is determined in an objective sense regardless what is our opinion. On the contrary, Ramsey and de Finetti argued that the probability that is to be assigned to a particular event depends on the degree of knowledge we have (*personal beliefs*) and those do not have to be shared by everybody so it is *subjective*. Furthermore they started the way towards a mathematical formulation of this concept of probability consistent with Kolmogorov's axiomatic theory. Thus, within the Bayesian spirit, it is logical and natural to consider that **probability is a measure of the degree of belief we have in the occurrence of an event** that characterizes the random phenomena and we shall assign probabilities to events based on the prior knowledge we have. In fact, to some extent, all statistical procedures used for the analysis of natural phenomena are subjective inasmuch they all are based on a mathematical idealizations of Nature and all require a priory judgments and hypothesis that have to be assumed.

1.1.4 Random Quantities

In many circumstances, the possible outcomes of the experiments are not numeric (a die with colored faces, a person may be sick or healthy, a particle may decay in different modes,...) and, even in the case they are, the possible outcomes of the experiment may form a non-denumerable set. Ultimately, we would like to deal with numeric values and benefit from the algebraic structures of the real numbers and the theory behind measurable functions and for this, given a measurable space $(\Omega, \mathcal{B}_\Omega)$, we define a function $X(w) : w \in \Omega \longrightarrow \mathcal{R}$ that assigns to each event w of the sample space Ω *one and only one real number*.

In a more formal way, consider two measurable spaces $(\Omega, \mathcal{B}_\Omega)$ and $(\Omega', \mathcal{B}'_\Omega)$ and a function

$$X(w) : \; w \in \Omega \longrightarrow X(w) \in \Omega'$$

Obviously, since we are interested in the events that conform the σ-algebra $\mathcal{B}_\Omega$, the same structure has to be maintained in $(\Omega', \mathcal{B}'_\Omega)$ by the application $X(w)$ for otherwise we wont be able to answer the questions of interest. Therefore, we require the function $X(w)$ to be *Lebesgue measurable with respect to the σ-algebra* $\mathcal{B}_\Omega$; i.e.:

$$X^{-1}(B') = B \subseteq \mathcal{B}_\Omega \qquad\qquad \forall \; B' \in \mathcal{B}'_\Omega$$

so we can ultimately identify $P(B')$ with $P(B)$. Usually, we are interested in the case that $\Omega' = \mathcal{R}$ (or $\mathcal{R}^n$) so $\mathcal{B}'_\Omega$ is the Borel σ-algebra and, since we have generated the Borel algebra $\mathcal{B}$ from half-open intervals on the left $I_x = (-\infty, x]$ with $x \in \mathcal{R}$, we have that $X(w)$ will be a Lebesgue measurable function over the Borel algebra (*Borel measurable*) if, and only if:

$$X^{-1}(I_x) = \{w \in \Omega \mid X(w) \leq x\} \in \mathcal{B}_\Omega \qquad\qquad \forall \; x \in \mathcal{R}$$

We could have generated as well the Borel algebra from open, closed or half-open intervals on the right so any of the following relations, all equivalent, serve to define a Borel measurable function $X(w)$:

(1) $\{w | X(w) > c\} \in \mathcal{B}_\Omega \; \forall c \in \mathcal{R}$;
(2) $\{w | X(w) \geq c\} \in \mathcal{B}_\Omega \; \forall c \in \mathcal{R}$;
(3) $\{w | X(w) < c\} \in \mathcal{B}_\Omega \; \forall c \in \mathcal{R}$;
(4) $\{w | X(w) \leq c\} \in \mathcal{B}_\Omega \; \forall c \in \mathcal{R}$

To summarize:

- Given a probability space $(\Omega, \mathcal{B}_\Omega, Q)$, a **random variable** is a function $X(w) : \Omega \rightarrow \mathcal{R}$, Borel measurable over the σ-algebra $\mathcal{B}_\Omega$, that allows us to work with the induced probability space $(\mathcal{R}, \mathcal{B}, \mathcal{P})$.[5]

Form this definition, it is clear that the name *"random variable"* is quite unfortunate inasmuch it is a univoque function, neither random nor variable. Thus, at least to get rid of *variable*, the term "**random quantity**" it is frequently used to design

[5]It is important to note that a random variable $X(w) : \Omega \longrightarrow \mathcal{R}$ is measurable **with respect to** the σ-algebra $\mathcal{B}_\Omega$.

a numerical entity associated to the outcome of an experiment; outcome that is uncertain before we actually do the experiment and observe the result, and distinguish between the random quantity $X(w)$, that we shall write in upper cases and usually as X assuming understood the w dependence, and the value x (lower case) taken in a particular realization of the experiment. If the function X takes values in $\Omega_X \subseteq \mathcal{R}$ it will be a *one dimensional random quantity* and, if the image is $\Omega_X \subseteq \mathcal{R}^n$, it will be an ordered n-tuple of real numbers $(X_1, X_2, \ldots, X_n)$. Furthermore, attending to the cardinality of Ω_X, we shall talk about *discrete random quantities* if it is finite or countable and about *continuous random quantities* if it is uncountable. This will be explained in more depth in Sect. 1.1.3.1. Last, if for each $w \in \Omega$ is $|X(w)| < k$ with k finite, we shall talk about a *bounded random quantity*.

The properties of random quantities are those of the measurable functions. In particular, if $X(w) : \Omega \rightarrow \Omega'$ is measurable with respect to $\mathcal{B}_\Omega$ and $Y(x) : \Omega' \rightarrow \Omega''$ is measurable with respect to $\mathcal{B}_{\Omega'}$, the function $Y(X(w)) : \Omega \rightarrow \Omega''$ is measurable with respect to $\mathcal{B}_\Omega$ and therefore is a random quantity. We have then that

$$P(Y \leq y) = P(Y(X) \leq y) = P(X \in Y^{-1}(I_X))$$

where $Y^{-1}(I_X)$ is the set $\{x | x \in \Omega'\}$ such that $Y(x) \leq y$.

Example 1.1 Consider the measurable space $(\Omega, \mathcal{B}_\Omega)$ and $X(w) : \Omega \rightarrow \mathcal{R}$. Then:

- $X(w) = k$, constant in $\mathcal{R}$. Denoting by $A = \{w \in \Omega | X(w) > c\}$ we have that if $c \geq k$ then $A = \emptyset$ and if $c < k$ then $A = \Omega$. Since $\{\emptyset, E\} \in \mathcal{B}_\Omega$ we conclude that $X(w)$ is a measurable function. In fact, it is left as an exercise to show that for the minimal algebra $\mathcal{B}_\Omega^{min} = \{\emptyset, \Omega\}$, the only functions that are measurable are $X(w) = $ constant.
- Let $G \in \mathcal{B}_\Omega$ and $X(w) = \mathbf{1}_G(w)$ (see Appendix 1.1). We have that if $I_a = (-\infty, a]$ with $a \in \mathcal{R}$, then $a \in (-\infty, 0) \rightarrow X^{-1}(I_a) = \emptyset$, $a \in [0, 1) \rightarrow X^{-1}(I_a) = G^c$, and $a \in [1, \infty) \rightarrow X^{-1}(I_a) = \Omega$ so $X(w)$ is a measurable function with respect to $\mathcal{B}_\Omega$. A simple function

$$X(w) = \sum_{k=1}^{n} a_k \, \mathbf{1}_{A_k}(w)$$

  where $a_k \in \mathcal{R}$ and $\{A_k\}_{k=1}^n$ is a partition of Ω is Borel measurable and any random quantity that takes a finite number of values can be expressed in this way.
- Let $\Omega = [0, 1]$. It is obvious that if G is a non-measurable Lebesgue subset of $[0, 1]$, the function $X(w) = \mathbf{1}_{G^c}(w)$ is not measurable over $\mathcal{B}_{[0,1]}$ because $a \in [0, 1) \rightarrow X^{-1}(I_a) = G \notin \mathcal{B}_{[0,1]}$.
- Consider a coin tossing, the elementary events

$$e_1 = \{H\}, \quad \text{and} \quad e_2 = \{T\} \quad \longrightarrow \quad \Omega = \{e_1, e_2\}$$

the algebra $\mathcal{B}_\Omega = \{\emptyset, \Omega, \{e_1\}, \{e_2\}\}$ and the function $X : \Omega \longrightarrow \mathcal{R}$ that denotes the number of heads

$$X(e_1) = 1 \quad \text{and} \quad X(e_2) = 0$$

Then, for $I_a = (-\infty, a]$ with $a \in \mathcal{R}$ we have that:

$$\begin{aligned} a \in (-\infty, 0) &\longrightarrow \quad X^{-1}(I_a) = \emptyset \in \mathcal{B}_\Omega \\ a \in [0, 1) &\longrightarrow \quad X^{-1}(I_a) = e_2 \in \mathcal{B}_\Omega \\ a \in [1, \infty) &\longrightarrow \quad X^{-1}(I_a) = \{e_1, e_2\} = \Omega \in \mathcal{B}_E \end{aligned}$$

so $X(w)$ is measurable in $(\Omega, \mathcal{B}_\Omega, P)$ and therefore an admissible random quantity with $P(X = 1) = P(e_1)$ and $P(X = 0) = P(e_2)$. It will not be an admissible random quantity for the trivial minimum algebra $\mathcal{B}_\Omega^{min} = \{\emptyset, \Omega\}$ since $e_2 \notin \mathcal{B}_\Omega^{min}$.

Example 1.2 Let $\Omega = [0, 1]$ and consider the sequence of functions $X_n(w) = 2^n \mathbf{1}_{\Omega_n}(w)$ where $w \in \Omega$, $\Omega_n = [1/2^n, 1/2^{n-1}]$ and $n \in \mathcal{N}$. Is each $X_n(w)$ is measurable iff $\forall r \in \mathcal{R}$, $A = \{w \in \Omega \,|\, X_n(w) > r\}$ is a Borel set of $\mathcal{B}_\Omega$. Then:

(1) $r \in (2^n, \infty) \longrightarrow A = \emptyset \in \mathcal{B}_\Omega$ with $\lambda(A) = 0$;
(2) $r \in [0, 2^n] \longrightarrow A = [1/2^n, 1/2^{n-1}] \in \mathcal{B}_\Omega$ with $\lambda(A) = 2/2^n - 1/2^n = 1/2^n$.
(3) $r \in (-\infty, 0) \longrightarrow A = [0, 1] = \Omega$ with $\lambda(\Omega) = 1$.

Thus, each $X_n(w)$ is a measurable function.

Problem 1.1 Consider the experiment of tossing two coins, the elementary events

$$e_1 = \{H, H\}\,, \qquad e_2 = \{H, T\}\,, \qquad e_3 = \{T, H\}\,, \qquad e_4 = \{T, T\}$$

the sample space $\Omega = \{e_1, e_2, e_3, e_4\}$ and the two algebras

$$\begin{aligned} \mathcal{B}_1 &= \{\emptyset, \Omega, \{e_1\}, \{e_4\}, \{e_1, e_2, e_3\}, \{e_2, e_3, e_4\}, \{e_1, e_4\}, \{e_2, e_3\}\} \\ \mathcal{B}_2 &= \{\emptyset, \Omega, \{e_1, e_2\}, \{e_3, e_4\}\} \end{aligned}$$

The functions $X(w) : \Omega \longrightarrow \mathcal{R}$ such that $X(e_1) = 2$; $X(e_2) = X(e_3) = 1$; $X(e_4) = 0$ (number of heads) and $Y(w) : \Omega \longrightarrow \mathcal{R}$ such that $Y(e_1) = Y(e_2) = 1$; $Y(e_3) = Y(e_4) = 0$, with respect to which algebras are admissible random quantities? (sol.: X wrt $\mathcal{B}_1$; Y wrt $\mathcal{B}_2$)

Problem 1.2 Let $X_i(w) : \mathcal{R} \longrightarrow \mathcal{R}$ with $i = 1, \ldots, n$ be random quantities. Show that

$$Y = \max\{X_1, X_2\}, \qquad Y = \min\{X_1, X_2\}, \qquad Y = \sup\{X_k\}_{k=1}^n \quad \text{and} \quad Y = \inf\{X_k\}_{k=1}^n$$

are admissible random quantities.

Hint: It is enough to observe that

$$
\begin{aligned}
\{w|\max\{X_1, X_2\} \le x\} &= \{w|X_1(w) \le x\}\cap\{w|X_2(w) \le x\} \in \mathcal{B} \\
\{w|\min\{X_1, X_2\} \le x\} &= \{w|X_1(w) \le x\}\cup\{w|X_2(w) \le x\} \in \mathcal{B} \\
\{w|\sup_n X_n(w) \le x\} &= \cap_n\{w|X_n(w) \le x\} \in \mathcal{B} \\
\{w|\inf_n X_n(w) < x\} &= \cup_n\{w|X_n(w) < x\} \in \mathcal{B}.
\end{aligned}
$$

1.2 Conditional Probability and Bayes Theorem

Suppose and experiment that consists on rolling a die with faces numbered from one to six and the event $e_2 =$\{*get the number two on the upper face*\}. If the die is fair, based on the Principle of Insufficient Reason you and your friend would consider reasonable to assign equal chances to any of the possible outcomes and therefore a probability of $P_1(e_2) = 1/6$. Now, if I look at the die and tell you, and only you, that the outcome of the roll is an even number, you will change your beliefs on the occurrence of event e_2 and assign the new value $P_2(e_2) = 1/3$. Both of you assign different probabilities because you do not share the same knowledge so it may be a truism but it is clear that *the probability we assign to an event is subjective and is conditioned by the information we have about the random process*. In one way or another, probabilities are always conditional degrees of belief since there is always some state of information (even before we do the experiment we know that whatever number we shall get is not less than one and not greater than six) and we always assume some hypothesis (the die is fair so we can rely on the Principle of Symmetry).

Consider a probability space (Ω, B_Ω, P) and two events $A, B \subset B_\Omega$ that are not disjoint so $A \cap B \neq \emptyset$. The probability for both A **and** B to happen is $P(A \cap B) \equiv P(A, B)$. Since $\Omega = B \cup B^c$ and $B \cap B^c = \emptyset$ we have that:

$$
P(A) \equiv P(A \bigcap \Omega) = \underbrace{P(A \bigcap B)}_{\substack{\text{probability for A} \\ \textbf{and} \text{ B to occur}}} + \underbrace{P(A \bigcap B^c) = P(A \backslash B)}_{\substack{P(A\backslash B)\ :\ \text{probability for} \\ \text{A to happen and } \textbf{not} \text{ B}}}
$$

What is the probability for A to happen if we know that B has occurred? The probability of A *conditioned* to the occurrence of B is called **conditional probability of** A **given** B and is expressed as $P(A|B)$. This is equivalent to calculate the probability for A to happen in the probability space (Ω', B'_Ω, P') with Ω' the reduced sample space where B has already occurred and B'_Ω the corresponding sub-algebra that does not contain B^c. We can set $P(A|B) \propto P(A \cap B)$ and define (Kolmogorov) the conditional probability for A to happen once B has occurred as:

$$
P(A|B) \overset{def.}{=} \frac{P(A \bigcap B)}{P(B)} = \frac{P(A, B)}{P(B)}
$$

provided that $P(B) \neq 0$ for otherwise the conditional probability is not defined. This normalization factor ensures that $P(B|B) = P(B \cap B)/P(B) = 1$. Conditional probabilities satisfy the basic axioms of probability:

(i) **non-negative** since $(A \bigcap B) \subset B \rightarrow 0 \leq P(A|B) \leq 1$
(ii) **unit measure** (*certainty*) since $P(\Omega|B) = \dfrac{P(\Omega \bigcap B)}{P(B)} = \dfrac{P(B)}{P(B)} = 1$
(iii) **σ-additive**: For a countable sequence of disjoint set $\{A_i\}_{i=1}^{\infty}$

$$P\left(\bigcup_{i=1}^{\infty} A_i|B\right) = \frac{P\left((\bigcup_{i=1}^{\infty} A_i) \bigcap B\right)}{P(B)} = \frac{\sum_{i=1}^{\infty} P(A_i \bigcap B)}{P(B)} = \sum_{i=1}^{\infty} P(A_i|B)$$

Generalizing, for n events $\{A_i\}_{i=1}^{n}$ we have, with $j = 0, \ldots, n-1$ that

$$\begin{aligned} P(A_1, \ldots, A_n) &= P(A_n, \ldots, A_{n-j}|A_j, \ldots, A_1)P(A_j, \ldots, A_1) = \\ &= P(A_n|A_1, \ldots, A_{n-1})P(A_3|A_2, A_1)P(A_2|A_1)P(A_1). \end{aligned}$$

1.2.1 Statistically Independent Events

Two events $A, B \in B_{\Omega}$ are **statistically independent** when the occurrence of one does not give any information about the occurrence of the other[6]; that is, when

$$P(A, B) = P(A)P(B)$$

A necessary and sufficient condition for A and B to be independent is that $P(A|B) = P(A)$ (which implies $P(B|A) = P(B)$). Necessary because

$$P(A, B) = P(A)P(B) \longrightarrow P(A|B) = \frac{P(A, B)}{P(B)} = \frac{P(A)P(B)}{P(B)} = P(A)$$

[6]In fact for the events $A, B \in B_{\Omega}$ we should talk about *conditional independence* for it is true that if $C \in B_{\Omega}$, it may happen that $P(A, B) = P(A)P(B)$ but conditioned on C, $P(A, B|C) \neq P(A|C)P(B|C)$ so A and B are related through the event C. On the other hand, that $P(A|B) \neq P(A)$ does not imply that B has a *"direct"* effect on A. Whether this is the case or not has to be determined by reasoning on the process and/or additional evidences. Bernard Shaw said that we all should buy an umbrella because there is statistical evidence that doing so you have a higher life expectancy. And this is certainly true. However, it is more reasonable to suppose that instead of the umbrellas having any mysterious influence on our health, in London, at the beginning of the XX century, if you can afford to buy an umbrella you have most likely a well-off status, healthy living conditions, access to medical care,…

Sufficient because

$$P(A|B) = P(A) \longrightarrow P(A, B) = P(A|B)P(B) = P(A)P(B)$$

If this is not the case, we say that they are *statistically dependent* or **correlated**. In general, we have that:

$P(A|B) > P(A) \rightarrow$ the events A and B are **positively correlated**; that is, that B has already occurred increases the chances for A to happen;

$P(A|B) < P(A) \rightarrow$ the events A and B are **negatively correlated**; that is, that B has already occurred reduces the chances for A to happen;

$P(A|B) = P(A) \rightarrow$ the events A and B are **not correlated** so the occurrence of B does not modify the chances for A to happen.

Given a finite collection of events $\mathcal{A} = \{A_i\}_{i=1}^{n}$ with $A_{\forall i} \subset B_{\Omega}$, they are statistically independent if

$$P(A_1, \ldots, A_m) = P(A_1) \cdots P(A_m)$$

for any finite subsequence $\{A_k\}_{k=j}^{m}$; $1 \leq j < m \leq n$ of events. Thus, for instance, for a sequence of 3 events $\{A_1, A_2, A_3\}$ the condition of independence requires that:

$$P(A_1, A_2) = P(A_1)P(A_2); \quad P(A_1, A_3) = P(A_1)P(A_3); \quad P(A_2, A_3) = P(A_2)P(A_3)$$
$$\textbf{and} \quad P(A_1, A_2, A_3) = P(A_1)P(A_2)P(A_3)$$

so the events $\{A_1, A_2, A_3\}$ may be statistically dependent and be pairwise independent.

Example 1.3 In four cards (C_1, C_2, C_3 and C_4) we write the numbers 1 (C_1), 2 (C_2), 3 (C_3) and 123 (C_4) and make a fair random extraction. Let be the events

$$A_i = \{\text{the chosen card has the number } i\}$$

with $i = 1, 2, 3$. Since the extraction is fair we have that:

$$P(A_i) = P(C_i) + P(C_4) = 1/2$$

Now, I look at the card and tell you that it has number j. Since you know that A_j has happened, you know that the extracted card was either C_j or C_4 and the only possibility to have $A_i \neq A_j$ is that the extracted card was C_4 so the conditional probabilities are

$$P(A_i|A_j) = 1/2; \quad i, j = 1, 2, 3; \quad i \neq j$$

The, since

$$P(A_i|A_j) = P(A_i); \quad i, j = 1, 2, 3; \quad i \neq j$$

any two events (A_i, A_j) are (pairwise) independent. However:

$$P(A_1, A_2, A_3) = P(A_1|A_2, A_3)\, P(A_2|A_3)\, P(A_3)$$

and if I tell you that events A_2 and A_3 have occurred then you are certain that chosen card is C_4 and therefore A_1 has happened too so $P(A_1|A_2, A_3) = 1$. But

$$P(A_1, A_2, A_3) = 1\,\frac{1}{2}\,\frac{1}{2} \neq P(A_1)P(A_2)P(A_3) = \frac{1}{8}$$

so the events $\{A_1, A_2, A_3\}$ are **not** independent even though they are pairwise independent.

Example 1.4 (*Bonferroni's Inequality*) Given a finite collection $A = \{A_1, \ldots, A_n\} \subset \mathcal{B}$ of events, Bonferroni's inequality states that:

$$P(A_1 \bigcap \cdots \bigcap A_n) \equiv P(A_1, \ldots, A_n) \geq P(A_1) + \cdots + P(A_n) - (n-1)$$

and gives a lover bound for the joint probability $P(A_1, \ldots, A_n)$. For $n = 1$ it is trivially true since $P(A_1) \geq P(A_1)$. For $n = 2$ we have that

$$P(A_1 \bigcup A_2) = P(A_1) + P(A_2) - P(A_1 \bigcap A_2) \leq 1 \longrightarrow P(A_1 \bigcap A_2) \geq P(A_1) + P(A_2) - 1$$

Proceed then by induction. Assume the statement is true for $n-1$ and see if it is so for n. If $B_{n-1} = A_1 \cap \cdots \cap A_{n-1}$ and apply the result we got for $n = 2$ we have that

$$P(A_1 \bigcap \cdots \bigcap A_n) = P(B_{n-1} \bigcap A_n) \geq P(B_{n-1}) + P(A_n) - 1$$

but

$$P(B_{n-1}) = P(B_{n-2} \bigcap A_{n-1}) \geq P(B_{n-2}) + P(A_{n-1}) - 1$$

so

$$P(A_1 \bigcap \cdots \bigcap A_n) \geq P(B_{n-2}) + P(A_{n-1}) + P(A_n) - 2$$

and therefore the inequality is demonstrated.

1.2.2 Theorem of Total Probability

Consider a probability space (Ω, B_Ω, P) and a partition $\mathcal{S} = \{S_i\}_{i=1}^n$ of the sample space. Then, for any event $A \in B_\Omega$ we have that $A = A \bigcap \Omega = A \bigcap (\bigcup_{i=1}^n S_i)$ and therefore:

$$P(A) = P\left(A \bigcap \left[\bigcup_{i=1}^n S_i\right]\right) = P\left(\bigcup_{i=1}^n \left[A \bigcap S_i\right]\right) = \sum_{i=1}^n P(A \bigcap S_i) = \sum_{i=1}^n P(A|S_i) \cdot P(S_i)$$

Consider now a second different partition of the sample space $\{B_k\}_{k=1}^m$. Then, for each set B_k we have

$$P(B_k) = \sum_{i=1}^n P(B_k|S_i)P(S_i); \quad k = 1, \ldots, m$$

and

$$\sum_{k=1}^m P(B_k) = \sum_{i=1}^n P(S_i)\left[\sum_{k=1}^m P(B_k|S_i)\right] = \sum_{i=1}^n P(S_i) = 1$$

Last, a similar expression can be written for conditional probabilities. Since

$$P(A, B, \mathcal{S}) = P(A|B, \mathcal{S})P(B, \mathcal{S}) = P(A|B, \mathcal{S})P(\mathcal{S}|B)P(B)$$

and

$$P(A, B) = \sum_{i=1}^n P(A, B, S_i)$$

we have that

$$P(A|B) = \frac{P(A, B)}{P(B)} = \frac{1}{P(B)} \sum_{i=1}^n P(A, B, S_i) = \sum_{i=1}^n P(A|B, S_i)P(S_i|B).$$

Example 1.5 We have two indistinguishable urns: U_1 with three white and two black balls and U_2 with two white balls and three black ones. What is the probability that in a random extraction we get a white ball?

Consider the events:

$$A_1 = \{\text{choose urn } U_1\}; \quad A_2 = \{\text{choose urn } U_2\} \quad \text{and} \quad B = \{\text{get a white ball}\}$$

It is clear that $A_1 \cap A_2 = \emptyset$ and that $A_1 \cup A_2 = \Omega$. Now:

$$P(B|A_1) = \frac{3}{5}; \qquad P(B|A_2) = \frac{2}{5} \qquad \text{and} \qquad P(A_1) = P(A_2) = \frac{1}{2}$$

so we have that

$$P(B) = \sum_{i=1}^{2} P(B|A_i) \cdot P(A_i) = \frac{3}{5}\frac{1}{2} + \frac{2}{5}\frac{1}{2} = \frac{1}{2}$$

as expected since out of 10 balls, 5 are white.

1.2.3 Bayes Theorem

Given a probability space (Ω, B_Ω, P) we have seen that the joint probability for for two events $A, B \in B_\Omega$ can be expressed in terms of conditional probabilities as:

$$P(A, B) = P(A|B)P(B) = P(B|A)P(A)$$

The Bayes Theorem (Bayes ∼1770s and independently Laplace ∼1770s) states that if $P(B) \neq 0$, then

$$P(A|B) = \frac{P(B|A)P(A)}{P(B)}$$

apparently a trivial statement but with profound consequences. Let's see other expressions of the theorem. If $\mathcal{H} = \{H_i\}_{i=1}^{n}$ is a partition of the sample space then

$$P(A, H_i) = P(A|H_i)P(H_i) = P(H_i|A)P(A)$$

and from the Total Probability Theorem

$$P(A) = \sum_{k=1}^{n} P(A|H_k)P(H_k)$$

so we have a different expression for Bayes's Theorem:

$$P(H_i|A) = \frac{P(A|H_i)P(H_i)}{P(A)} = \frac{P(A|H_i)P(H_i)}{\sum_{k=1}^{n} P(A|H_k)P(H_k)}$$

Let's summarize the meaning of these terms[7]:

$P(H_i)$: is the probability of occurrence of the event H_i before we know if event A has happened or not; that is, the degree of confidence we have in the occurrence of the event H_i before we do the experiment so it is called **prior probability**;

$P(A|H_i)$: is the probability for event A to happen given that event H_i has occurred. This may be different depending on $i = 1, 2, \ldots, n$ and when considered as function of H_i is usually called **likelihood**;

$P(H_i|A)$: is the degree of confidence we have in the occurrence of event H_i given that the event A has happened. The knowledge that the event A has occurred provides information about the random process and modifies the beliefs we had in H_i before the experiment was done (expressed by $P(H_i)$) so it is called **posterior probability**;

$P(A)$: is simply the normalizing factor.

Clearly, if the events A and H_i are independent, the occurrence of A does not provide any information on the chances for H_i to happen. Whether it has occurred or not does not modify our beliefs about H_i and therefore $P(H_i|A) = P(H_i)$.

In first place, it is interesting to note that the occurrence of A restricts the sample space for $\mathcal{H}$ and modifies the prior chances $P(H_i)$ for H_i in the same proportion as the occurrence of H_i modifies the probability for A because

$$P(A|H_i)P(H_i) = P(H_i|A)P(A) \quad \longrightarrow \quad \frac{P(H_i|A)}{P(H_i)} = \frac{P(A|H_i)}{P(A)}$$

Second, from Bayes Theorem we can obtain *relative posterior probabilities* (in the case, for instance, that $P(A)$ is unknown) because

$$\frac{P(H_i|A)}{P(H_j|A)} = \frac{P(A|H_i)}{P(A|H_j)} \frac{P(H_i)}{P(H_j)}$$

Last, conditioning all the probabilities to H_0 (maybe some conditions that are assumed) we get a third expression of Bayes Theorem

[7] Although is usually the case, the terms *prior* and *posterior* do not necessarily imply a temporal ordering.

$$P(H_i|A, H_0) = \frac{P(A|H_i, H_0)P(H_i|H_0)}{P(A|H_0)} = \frac{P(A|H_i, H_0)P(H_i|H_0)}{\sum_{k=1}^{n} P(A|H_k, H_0)P(H_k|H_0)}$$

where H_0 represents to some initial state of information or some conditions that are assumed. The *posterior degree of credibility* we have on H_i is certainly meaningful when we have an initial degree of information and therefore is relative to our *prior* beliefs. And those are subjective inasmuch different people may assign a different prior degree of credibility based on their previous knowledge and experiences. Think for instance in soccer pools. Different people will assign different prior probabilities to one or other team depending on what they know before the match and this information may not be shared by all of them. However, to the extent that they share common prior knowledge they will arrive to the same conclusions.

Bayes's rule provides a natural way to include new information and update our beliefs in a sequential way. After the event (data) D_1 has been observed, we have

$$P(H_i) \quad\longrightarrow\quad P(H_i|D_1) = \frac{P(D_1|H_i)}{P(D_1)} P(H_i) \propto P(D_1|H_i)P(H_i)$$

Now, if we get additional information provided by the observation of D_2 (new data) we *"update"* or beliefs on H_i as:

$$P(H_i|D_1) \longrightarrow P(H_i|D_2, D_1) = \frac{P(D_2|H_i, D_1)}{P(D_2|D_1)} P(H_i|D_1) = \frac{P(D_2|H_i, D_1)P(D_1|H_i)P(H_i)}{P(D_2, D_1)}$$

and so on with further evidences.

Example 1.6 An important interpretation of Bayes Theorem is that based on the relation **cause-effect**. Suppose that the event A (*effect*) has been produced by a certain *cause* H_i. We consider all possible causes (so $\mathcal{H}$ is a complete set) and among them we have interest in those that seem more plausible to explain the observation of the event A. Under this scope, we interpret the terms appearing in Bayes's formula as:

$P(A|H_i, H_0)$: is the probability that the **effect** A is produced by the **cause** (or hypothesis) H_i;

$P(H_i, H_0)$: is the prior degree of credibility we assign to the **cause** H_i before we know that A has occurred;

$P(H_i|A, H_0)$: is the posterior probability we have for H_i being the cause of the event (effect) A that has already been observed.

Let's see an example of a clinical diagnosis just because the problem is general enough and conclusions may be more disturbing. If you want, replace *individuals* by *events* and for instance (*sick, healthy*) by (*signal, background*). Now, the incidence of certain rare disease is of 1 every 10,000 people and there is an efficient diagnostic test such that:

(1) If a person is sick, the tests gives positive in 99% of the cases;
(2) If a person is healthy, the tests may fail and give positive (false positive) in 0.5% of the cases;

In this case, the **effect** is to give positive (T) in the test and the exclusive and exhaustive hypothesis for the **cause** are:

$$H_1: \text{ be sick} \qquad \text{and} \qquad H_2: \text{ be healthy}$$

with $H_2 = H_1{}^c$. A person, say you, is chosen **randomly** (H_0) among the population to go under the test and give positive. Then you are scared when they tell you: "*The probability of giving positive being healthy is* 0.5%, *very small*" (p-value). There is nothing wrong with the statement but it has to be correctly interpreted and usually it is not. It means no more and no less than what the expression $P(T|H_2)$ says: "*under the assumption that you are healthy* (H_2) *the chances of giving positive are* 0.5%" and this is nothing else but a feature of the test. It doesn't say anything about $P(H_1|T)$, the chances you have to be sick giving positive in the test that, in the end, is what you are really interested in. The two probabilities are related by an additional piece of information that appears in Bayes's formula: $P(H_1|H_0)$; that is, *under the hypothesis that you have been chosen at random* (H_0)*, What are the prior chances to be sick?* From the prior knowledge we have, the degree of credibility we assign to both hypothesis is

$$P(H_1|H_0) = \frac{1}{10000} \qquad \text{and} \qquad P(H_2|H_0) = 1 - P(H_1) = \frac{9999}{10000}$$

On the other hand, if T denotes the event *give positive in the test* we know that:

$$P(T|H_1) = \frac{99}{100} \qquad \text{and} \qquad P(T|H_2) = \frac{5}{1000}$$

Therefore, Bayes's Theorem tells that the probability to be sick giving positive in the test is

$$P(H_1|T) = \frac{P(T|H_1)\cdot P(H_1|H_0)}{\sum_{i=1}^{2} P(T|H_i)\cdot P(H_i|H_0)} = \frac{\frac{99}{100}\frac{1}{10000}}{\frac{99}{100}\frac{1}{10000} + \frac{5}{1000}\frac{9999}{10000}} \simeq 0.02$$

Thus, even if the test looks very efficient and you gave positive, the fact that you were chosen at random and that the incidence of the disease in the population is very small, reduces dramatically the degree of belief you assign to be sick. Clearly, if you were not chosen randomly but because there is a suspicion from to other symptoms that you are sic, prior probabilities change.

1.3 Distribution Function

A one-dimensional Distribution Function is a real function $F : \mathcal{R} \rightarrow \mathcal{R}$ that:

(p.1) is monotonous non-decreasing: $F(x_1) \leq F(x_2) \ \ \forall x_1 < x_2 \in \mathcal{R}$
(p.2) is everywhere continuous on the right: $\lim_{\epsilon \to 0^+} F(x+\epsilon) = F(x) \ \forall x \in \mathcal{R}$
(p.3) $F(-\infty) \equiv \lim_{x \to -\infty} F(x) = 0$ and $F(\infty) \equiv \lim_{x \to +\infty} F(x) = 1$.

and there is [1] a unique Borel measure μ on $\mathcal{R}$ that satisfies $\mu((-\infty, x]) = F(x)$ for all $x \in \mathcal{R}$. In the Theory of Probability, we define the Probability Distribution Function[8] of the random quantity $X(w) : \Omega \rightarrow \mathcal{R}$ as:

$$F(x) \overset{def.}{=} P(X \leq x) = P\left(X \in (-\infty, x]\right); \quad \forall x \in \mathcal{R}$$

Note that the Distribution Function $F(x)$ is defined for all $x \in \mathcal{R}$ so if $\text{supp}\{P(X)\} = [a, b]$, then $F(x) = 0 \ \forall x < a$ and $F(x) = 1 \ \forall x \geq b$. From the definition, it is easy to show the following important properties:

(a) $\forall x \in \mathcal{R}$ we have that:

$$\begin{aligned}
&\text{(a.1)} \ \ P(X \leq x) \overset{def.}{=} F(x) \\
&\text{(a.2)} \ \ P(X < x) = F(x-\epsilon) \ ; \\
&\text{(a.3)} \ \ P(X > x) = 1 - P(X \leq x) = 1 - F(x) \ ; \\
&\text{(a.4)} \ \ P(X \geq x) = 1 - P(X < x) = 1 - F(x-\epsilon) \ ;
\end{aligned}$$

(b) $\forall x_1 < x_2 \in \mathcal{R}$ we have that:

$$\begin{aligned}
&\text{(b.1)} \ \ P(x_1 < X \leq x_2) = P(X \in (x_1, x_2]) = F(x_2) - F(x_1) \ ; \\
&\text{(b.2)} \ \ P(x_1 \leq X \leq x_2) = P(X \in [x_1, x_2]) = F(x_2) - F(x_1 - \epsilon) \\
&\qquad\quad \text{(thus, if } x_1 = x_2 \text{ then } P(X = x_1) = F(x_1) - F(x_1 - \epsilon)); \\
&\text{(b.3)} \ \ P(x_1 < X < x_2) = P(X \in (x_1, x_2)) = F(x_2 - \epsilon) - F(x_1) = \\
&\qquad\qquad\qquad\qquad\quad = F(x_2) - F(x_1) - P(X = x_2) \ ; \\
&\text{(b.4)} \ \ P(x_1 \leq X < x_2) = P(X \in [x_1, x_2)) = F(x_2 - \epsilon) - F(x_1 - \epsilon) = \\
&\qquad\qquad\qquad\qquad\quad = F(x_2) - F(x_1) - P(X = x_2) + P(X = x_1) \ .
\end{aligned}$$

The Distribution Function is discontinuous at all $x \in \mathcal{R}$ where $F(x-\epsilon) \neq F(x+\epsilon)$. Let D be the set of all points of discontinuity. If $x \in D$, then $F(x-\epsilon) < F(x+\epsilon)$ since it is monotonous non-decreasing. Thus, we can associate to each $x \in D$ a rational number $r(x) \in \mathcal{Q}$ such that $F(x-\epsilon) < r(x) < F(x+\epsilon)$ and all will be different because if $x_1 < x_2 \in D$ then $F(x_1+\epsilon) \leq F(x_2-\epsilon)$. Then, since $\mathcal{Q}$ is a countable set, we have that the set of points of discontinuity of $F(x)$ is either **finite** or **countable**.

---

[8]The condition $P(X \leq x)$ is due to the requirement that $F(x)$ be continuous on the right. This is not essential in the sense that any non-decreasing function $G(x)$, defined on $\mathcal{R}$, bounded between 0 and 1 and continuous on the left ($G(x) = \lim_{\epsilon \to 0^+} G(x-\epsilon)$) determines a distribution function defined as $F(x)$ for all x where $G(x)$ is continuous and as $F(x+\epsilon)$ where $G(x)$ is discontinuous. In fact, in the general theory of measure it is more common to consider continuity on the left.

At each of them the distribution function has a "jump" of amplitude (property b.2):

$$F(x) - F(x - \epsilon) = P(X = x)$$

and will be continuous on the right (condition p.2).

Last, for each Distribution Function there is a **unique** probability measure P defined over the Borel sets of $\mathcal{R}$ that assigns the probability $F(x_2) - F(x_1)$ to each half-open interval $(x_1, x_2]$ and, conversely, to any probability measure defined on a measurable space $(\mathcal{R}, \mathcal{B})$ corresponds one Distribution Function. Thus, the Distribution Function of a random quantity contains all the information needed to describe the properties of the random process.

1.3.1 Discrete and Continuous Distribution Functions

Consider the probability space $(\Omega, \mathcal{F}, Q)$, the random quantity $X(w) : w \in \Omega \to X(w) \in \mathcal{R}$ and the induced probability space $(\mathcal{R}, \mathcal{B}, P)$. The function $X(w)$ is a **discrete random quantity** if its range (image) $D = \{x_1, \ldots, x_i, \ldots\}$, with $x_i \in \mathcal{R}$, $i = 1, 2, \ldots$ is a finite or countable set; that is, if $\{A_k; k = 1, 2, \ldots\}$ is a finite or countable partition of Ω, the function $X(w)$ is either:

simple: $X(w) = \sum_{k=1}^{n} x_k \mathbf{1}_{A_k}(w)$ or **elementary**: $X(w) = \sum_{k=1}^{\infty} x_k \mathbf{1}_{A_k}(w)$

Then, $P(X = x_k) = Q(A_k)$ and the corresponding Distribution Function, defined for all $x \in \mathcal{R}$, will be

$$F(x) = P(X \leq x) = \sum_{\forall x_k \in D} P(X = x_k)\, \mathbf{1}_A(x_k) = \sum_{\forall x_k \leq x} P(X = x_k)$$

with $A = (-\infty, x] \cap D$ and satisfies:

(i) $F(-\infty) = 0$ and $F(+\infty) = 1$;
(ii) is a monotonous non decreasing step-function;
(iii) continuous on the right $(F(x + \epsilon) = F(x))$ and therefore constant but on the finite or countable set of points of discontinuity $D = \{x_1, \ldots\}$ where

$$F(x_k) - F(x_k - \epsilon) = P(X = x_k)$$

Familiar examples of discrete Distribution Functions are Poisson, Binomial, Multinomial,...

The random quantity $X(w) : \Omega \longrightarrow \mathcal{R}$ is **continuous** if its range is a non-denumerable set; that is, if for all $x \in \mathcal{R}$ we have that $P(X = x) = 0$. In this case, the Distribution Function $F(x) = P(X \leq x)$ is continuous for all $x \in \mathcal{R}$ because

(i) from condition (p.2): $F(x + \epsilon) = F(x)$;
(ii) from property (b.2): $F(x - \epsilon) = F(x) - P(X = x) = F(x)$

Now, consider the measure space $(\Omega, \mathcal{B}_\Omega, \mu)$ with μ countably additive. If $f : \Omega \rightarrow [0, \infty)$ is integrable with respect to μ, it is clear that $\nu(A) = \int_A f d\mu$ for $A \in \mathcal{B}_\Omega$ is also a non-negative countably additive set function. (see Appendix 1.2). More generally, we have:

• **Radon–Nikodym Theorem** (Radon(1913), Nikodym(1930)): If ν and μ are two σ-additive measures on the measurable space $(\Omega, \mathcal{B}_\Omega)$ such that ν is **absolutely continuous** with respect to μ ($\nu << \mu$; that is, for every set $A \in \mathcal{B}_\Omega$ for which $\mu(A) = 0$ it is $\nu(A) = 0$), then there exists a μ-integrable function $p(x)$ such that

$$\nu(A) = \int_A d\nu(w) = \int_A \frac{d\nu(w)}{d\mu(w)} d\mu(w) = \int_A p(w) d\mu(w)$$

and, conversely, if such a function exists then $\nu << \mu$ (see Appendix 1.3 for the main properties).
The function $p(w) = d\nu(w)/d\mu(w)$ is called *Radon density* and is unique up to at most a set of measure zero; that is, if

$$\nu(A) = \int_A p(w) d\mu(w) = \int_A f(w) d\mu(w)$$

then $\mu\{x | p(x) \neq f(x)\} = 0$. Furthermore, if ν and μ are *equivalent* ($\nu \sim \mu$; $\mu << \nu$ and $\nu << \mu$) then $d\nu/d\mu > 0$ almost everywhere. In consequence, if we have a probability space $(\mathcal{R}, \mathcal{B}, P)$ with P equivalent to the Lebesgue measure, there exists a non-negative Lebesgue integrable function (see Appendix 2) $p : \mathcal{R} \longrightarrow [0, \infty)$, unique a.e., such that

$$P(A) \equiv P(X \in A) = \int_A p(x)\, dx; \qquad \forall A \in \mathcal{B}$$

The function $p(x)$ is called **probability density function** and satisfies:

(i) $p(x) \geq 0 \; \forall x \in \mathcal{R}$;
(ii) at any bounded interval of $\mathcal{R}$, $p(x)$ is bounded and is Riemann-integrable;
(iii) $\int_{-\infty}^{+\infty} p(x)\, dx = 1$.

Thus, for an **absolutely continuous random quantity** X, the Distribution Function $F(x)$ can be expressed as

$$F(x) = P(X \leq x) = \int_{-\infty}^{x} p(w)\, dw$$

Usually we shall be interested in random quantities that take values in a subset $D \subset \mathcal{R}$. It will then be understood that $p(x)$ is $p(x)\mathbf{1}_D(x)$ so it is defined for all $x \in \mathcal{R}$. Thus, for instance, if $\text{supp}\{p(x)\} = [a, b]$ then

$$\int_{-\infty}^{+\infty} p(x)\,dx \equiv \int_{-\infty}^{+\infty} p(x)\,\mathbf{1}_{[a,b]}(x)\,dx = \int_{a}^{b} p(x)\,dx = 1$$

and therefore

$$F(x) = P(X \leq x) = \mathbf{1}_{(a,\infty)}(x)\mathbf{1}_{[b,\infty)}(x) + \mathbf{1}_{[a,b)}(x) \int_{a}^{x} p(u)\,du$$

Note that from the previous considerations, the value of the integral will not be affected if we modify the integrand on a countable set of points. In fact, what we actually integrate is an equivalence class of functions that differ only in a set of measure zero. Therefore, a probability density function $p(x)$ has to be continuous for all $x \in \mathcal{R}$ but, at most, on a countable set of points. If $F(x)$ is not differentiable at a particular point, $p(x)$ is not defined on it but the set of those points is of zero measure. However, if $p(x)$ is continuous in $\mathcal{R}$ then $F'(x) = p(x)$ and the value of $p(x)$ is univocally determined by $F(x)$. We also have that

$$P(X \leq x) = F(x) = \int_{-\infty}^{x} p(w)dw \longrightarrow P(X > x) = 1 - F(x) = \int_{x}^{+\infty} p(w)dw$$

and therefore:

$$P(x_1 < X \leq x_2) = F(x_2) - F(x_1) = \int_{x_1}^{x_2} p(w)\,dw$$

Thus, since $F(x)$ is continuous at all $x \in \mathcal{R}$:

$$P(x_1 < X \leq x_2) = P(x_1 < X < x_2) = P(x_1 \leq X < x_2) = P(x_1 \leq X \leq x_2)$$

and therefore $P(X = x) = 0\ \forall x \in \mathcal{R}$ $(\lambda([x]) = 0)$ even though $X = x$ is a possible outcome of the experiment so, in this sense, unlike discrete random quantities "*probability*" 0 does not correspond necessarily to impossible events. Well known examples absolutely continuous Distribution Functions are the Normal, Gamma, Beta, Student, Dirichlet, Pareto, …

Last, if the continuous probability measure P is not absolutely continuous with respect to the Lebesgue measure λ in $\mathcal{R}$, then the probability density function does not exist. Those are called **singular** random quantities for which $F(x)$ is continuous but $F'(x) = 0$ almost everywhere. A well known example is the Dirac's singular measure $\delta_{x_0}(A) = \mathbf{1}_A(x_0)$ that assigns a measure 1 to a set $A \in \mathcal{B}$ if $x_0 \in A$ and 0 otherwise. As we shall see in the Examples 1.9 and 1.20, dealing with these cases is no problem because the Distribution Function always exists. The Lebesgue's *General*

Decomposition Theorem establishes that any Distribution Function can be expressed as a convex combination:

$$F(x) = \sum_{i=1}^{N_d} a_i \, F_d(x) + \sum_{j=1}^{N_{ac}} b_j \, F_{ac}(x) + \sum_{k=1}^{N_s} c_k \, F_s(x)$$

of a discrete Distribution Functions ($F_d(x)$), absolutely continuous ones ($F_{ac}(x)$ with derivative at every point so $F'(x) = p(x)$) and singular ones ($F_s(x)$). For the cases we shall deal with, $c_k = 0$.

Example 1.7 Consider a real parameter $\mu > 0$ and a discrete random quantity X that can take values $\{0, 1, 2, \ldots\}$ with a Poisson probability law:

$$P(X = k|\mu) = e^{-\mu} \, \frac{\mu^k}{\Gamma(k+1)}; \qquad k = 0, 1, 2, \ldots$$

The Distribution Function will be

$$F(x|\mu) = P(X \leq x|\mu) = e^{-\mu} \sum_{k=0}^{m=[x]} \frac{\mu^k}{\Gamma(k+1)}$$

where $m = [x]$ is the largest integer less or equal to x. Clearly, for $\epsilon \to 0^+$:

$$F(x + \epsilon|\mu) = F([x + \epsilon]|\mu) = F([x]|\mu) = F(x|\mu)$$

so it is continuous on the right and for $k = 0, 1, 2, \ldots$

$$F(k|\mu) - F(k-1|\mu) = P(X = k|\mu) = e^{-\mu} \, \frac{\mu^k}{\Gamma(k+1)}$$

Therefore, for reals $x_2 > x_1 > 0$ such that $x_2 - x_1 < 1$, $P(x_1 < X \leq x_2) = F(x_2) - F(x_1) = 0$.

Example 1.8 Consider the function $g(x) = e^{-ax}$ with $a > 0$ real and support in $(0, \infty)$. It is non-negative and Riemann integrable in $\mathcal{R}^+$ so we can define a probability density

$$p(x|a) = \frac{e^{-ax}}{\int_0^\infty e^{-\alpha x} \, dx} \, \mathbf{1}_{(0,\infty)}(x) = a \, e^{-ax} \, \mathbf{1}_{(0,\infty)}(x)$$

and the Distribution Function

$$F(x) = P(X \leq x) = \int_{-\infty}^{x} p(u|a) du = \begin{cases} 0 & x < 0 \\ 1 - e^{-ax} & x \geq 0 \end{cases}$$

Clearly, $F(-\infty) = 0$ and $F(+\infty) = 1$. Thus, for an absolutely continuous random quantity $X \sim p(x|a)$ we have that for reals $x_2 > x_1 > 0$:

$$\begin{aligned} P(X \leq x_1) &= F(x_1) &&= 1 - e^{-ax_1} \\ P(X > x_1) &= 1 - F(x_1) &&= e^{-ax_1} \\ P(x_1 < X \leq x_2) &= F(x_2) - F(x_1) &&= e^{-ax_1} - e^{-ax_2} \end{aligned}$$

Example 1.9 The ternary Cantor Set $Cs(0, 1)$ is constructed iteratively. Starting with the interval $Cs_0 = [0, 1]$, at each step one removes the open middle third of each of the remaining segments. That is; at step one the interval $(1/3, 2/3)$ is removed so $Cs_1 = [0, 1/3] \cup [2/3, 1]$ and so on. If we denote by D_n the union of the 2^{n-1} disjoint open intervals removed at step n, each of length $1/3^n$, the Cantor set is defined as $Cs(0, 1) = [0, 1] \setminus \cup_{n=1}^{\infty} D_n$. It is easy to check that any element X of the Cantor Set can be expressed as

$$X = \sum_{n=1}^{\infty} \frac{X_n}{3^n}$$

with $\text{supp}\{X_n\} = \{0, 2\}$[9] and that $Cs(0, 1)$ is a closed set, uncountable, nowhere dense in $[0, 1]$ and with zero measure. The Cantor Distribution, whose support is the Cantor Set, is defined assigning a probability $P(X_n = 0) = P(X_n = 2) = 1/2$. Thus, X is a continuous random quantity with support on a non-denumerable set of measure zero and can not be described by a probability density function. The Distribution Function $F(x) = P(X \leq x)$ (Cantor Function; Fig. 1.1) is an example of singular Distribution.

1.3.2 Distributions in More Dimensions

The previous considerations can be extended to random quantities in more dimensions but with some care. Let's consider the the two-dimensional case: $\boldsymbol{X} = (X_1, X_2)$. The Distribution Function will be defined as:

$$F(x_1, x_2) = P(X_1 \leq x_1, X_2 \leq x_2); \quad \forall (x_1, x_2) \in \mathcal{R}^2$$

[9] Note that the representation of a real number $r \in [0, 1]$ as $(a_1, a_2, \ldots) : \sum_{n=1}^{\infty} a_n 3^{-n}$ with $a_i = \{0, 1, 2\}$ is not unique. In fact $x = 1/3 \in Cs(0, 1)$ and can be represented by $(1, 0, 0, 0, \ldots)$ or $(0, 2, 2, 2, \ldots)$.

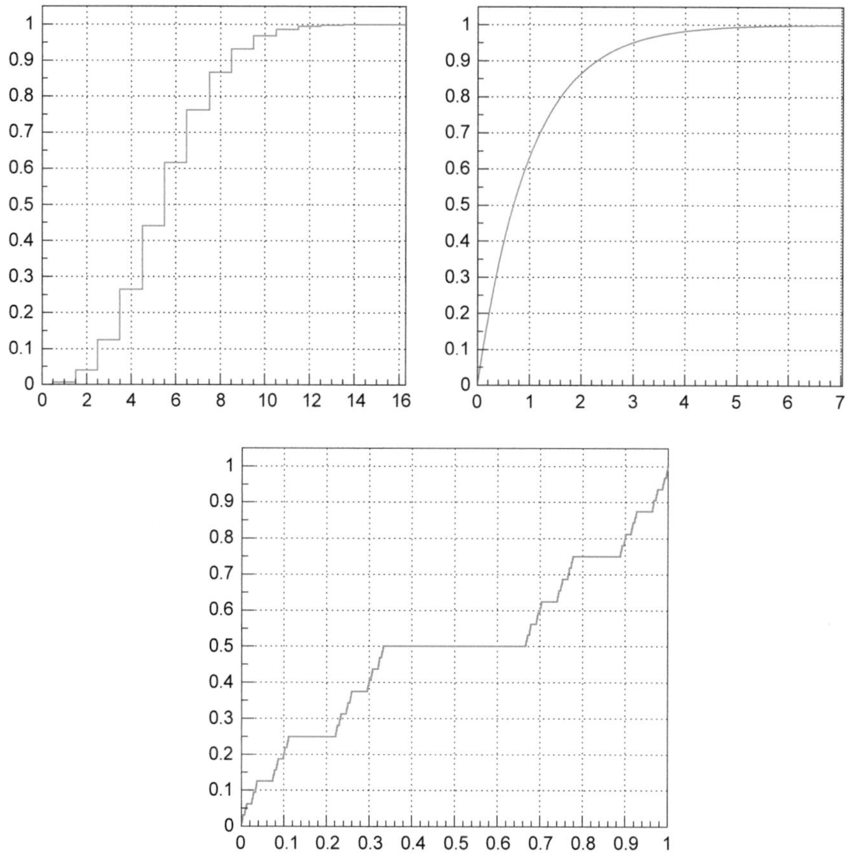

Fig. 1.1 Empiric distribution functions (ordinate) form a Monte Carlo sampling (10^6 events) of the Poisson $Po(x|5)$ (discrete; *upper left*), Exponential $Ex(x|1)$ (absolute continuous; *upper right*) and Cantor (singular; *bottom*) Distributions

and satisfies:

(i) monotonous no-decreasing in both variables; that is, if $(x_1, x_2), (x'_1, x'_2) \in \mathcal{R}^2$:

$$x_1 \leq x'_1 \longrightarrow F(x_1, x_2) \leq F(x'_1, x_2) \quad \text{and} \quad x_2 \leq x'_2 \longrightarrow F(x_1, x_2) \leq F(x_1, x'_2)$$

(ii) continuous on the right at $(x_1, x_2) \in \mathcal{R}^2$:

$$F(x_1 + \epsilon, x_2) = F(x_1, x_2 + \epsilon) = F(x_1, x_2)$$

(iii) $F(-\infty, x_2) = F(x_1, -\infty) = 0$ and $F(+\infty, +\infty) = 1$.

Now, if $(x_1, x_2), (x_1', x_2') \in \mathcal{R}^2$ with $x_1 < x_1'$ and $x_2 < x_2'$ we have that:

$$P(x_1 < X_1 \le x_1', x_2 < X_2 \le x_2') = F(x_1', x_2') - F(x_1, x_2') - F(x_1', x_2) + F(x_1, x_2) \ge 0$$

and

$$P(x_1 \le X_1 \le x_1', x_2 \le X_2 \le x_2') = F(x_1', x_2') - F(x_1 - \epsilon, x_2') - F(x_1', x_2 - \epsilon) + F(x_1 - \epsilon, x_2 - \epsilon) \ge 0$$

so, for discrete random quantities, if $x_1 = x_1'$ and $x_2 = x_2'$:

$$P(X_1 = x_1, X_2 = x_2) = F(x_1, x_2) - F(x_1 - \epsilon_1, x_2) - F(x_1, x_2 - \epsilon) + F(x_1 - \epsilon, x_2 - \epsilon) \ge 0$$

will give the amplitude of the jump of the Distribution Function at the points of discontinuity.

As for the one-dimensional case, for absolutely continuous random quantities we can introduce a two-dimensional probability density function $p(\boldsymbol{x}) : \mathcal{R}^2 \longrightarrow \mathcal{R}$:

(i) $p(\boldsymbol{x}) \ge 0; \quad \forall \boldsymbol{x} \in \mathcal{R}^2$;
(ii) At every bounded interval of $\mathcal{R}^2$, $p(\boldsymbol{x})$ is bounded and Riemann integrable;
(iii) $\int_{\mathcal{R}^2} p(\boldsymbol{x}) d\boldsymbol{x} = 1$

such that:

$$F(x_1, x_2) = \int_{-\infty}^{x_1} du_1 \int_{-\infty}^{x_2} du_2 \, p(u_1, u_2) \quad \longleftrightarrow \quad p(x_1, x_2) = \frac{\partial^2}{\partial x_1 \, \partial x_2} F(x_1, x_2).$$

1.3.2.1 Marginal and Conditional Distributions

It may happen that we are interested only in one of the two random quantities say, for instance, X_1. Then we ignore all aspects concerning X_2 and obtain the one-dimensional Distribution Function

$$F_1(x_1) = P(X_1 \le x_1) = P(X_1 \le x_1, X_2 \le +\infty) = F(x_1, +\infty)$$

that is called the **Marginal Distribution Function** of the random quantity X_1. In the same manner, we have $F_2(x_2) = F(+\infty, x_2)$ for the random quantity X_2. For absolutely continuous random quantities,

$$F_1(x_1) = F(x_1, +\infty) = \int_{-\infty}^{x_1} du_1 \int_{-\infty}^{+\infty} p(u_1, u_2) \, du_2 = \int_{-\infty}^{x_1} p(u_1) \, du_1$$

with $p(x_1)$ the **marginal probability density function**[10] of the random quantity X_1:

$$p_1(x_1) = \frac{\partial}{\partial x_1} F_1(x_1) = \int_{-\infty}^{+\infty} p(x_1, u_2)\, du_2$$

In the same manner, we have for X_2

$$p_2(x_2) = \frac{\partial}{\partial x_2} F_2(x_2) = \int_{-\infty}^{+\infty} p(u_1, x_2)\, du_1$$

As we have seen, given a probability space (Ω, B_Ω, P), for any two sets $A, B \in B_\Omega$ the conditional probability for A given B was *defined* as

$$P(A|B) \stackrel{def.}{=} \frac{P(A \cap B)}{P(B)} \equiv \frac{P(A, B)}{P(B)}$$

provided $P(B) \neq 0$. Intimately related to this definition is the Bayes' rule:

$$P(A, B) = P(A|B)\, P(B) = P(B|A)\, P(A)$$

Consider now the discrete random quantity $\boldsymbol{X} = (X_1, X_2)$ with values on $\Omega \subset \mathcal{R}^2$. It is then natural to define

$$P(X_1 = x_1 | X_2 = x_2) \stackrel{def.}{=} \frac{P(X_1 = x_1, X_2 = x_2)}{P(X_2 = x_2)}$$

and therefore

$$F(x_1|x_2) = \frac{P(X_1 \leq x_1, X_2 = x_2)}{P(X_2 = x_2)}$$

whenever $P(X_2 = x_2) \neq 0$. For absolutely continuous random quantities we can express the probability density as

$$p(x_1, x_2) = p(x_1|x_2)\, p(x_2) = p(x_2|x_1)\, p(x_1)$$

and define the **conditional probability density function** as

$$p(x_1|x_2) \stackrel{def.}{=} \frac{p(x_1, x_2)}{p(x_2)} \qquad = \frac{\partial}{\partial x_1} F(x_1|x_2)$$

[10] It is habitual to avoid the indices and write $p(x)$ meaning *"the probability density function of the variable x"* since the distinctive features are clear within the context.

provided again that $p_2(x_2) \neq 0$. This is certainly is an admissible density.[11] since $p(x_1|x_2) \geq 0 \; \forall (x_1, x_2) \in \mathcal{R}^2$ and $\int_{\mathcal{R}} p(x_1|x_2) dx_1 = 1$.

As stated already, two events $A, B \in B_\Omega$ are statistically independent iff:

$$P(A, B) \equiv P(A \cap B) = P(A) \cdot P(B)$$

Then, we shall say that two discrete random quantities X_1 and X_2 are **statistically independent** if $F(x_1, x_2) = F_1(x_1) F_2(x_2)$; that is, if

$$P(X_1 = x_1, X_2 = x_2) = P(X_1 = x_1)\, P(X_2 = x_2)$$

for discrete random quantities and

$$p(x_1, x_2) = \frac{\partial^2}{\partial x_1 \, \partial x_2} F(x_1) F(x_2) = p(x_1)\, p(x_2) \quad \longleftrightarrow \quad p(x_1|x_2) = p(x_1)$$

and for absolutely continuous random quantities.

Example 1.10 Consider the probability space $(\Omega, \mathcal{B}_\Omega, \lambda)$ with $\Omega = [0, 1]$ and λ the Lebesgue measure. If F is an arbitrary Distribution Function, $X : w \in [0, 1] \longrightarrow F^{-1}(w) \in \mathcal{R}$ is a random quantity and is distributed as $F(w)$. Take the Borel set $I = (-\infty, r]$ with $r \in \mathcal{R}$. Since F is a Distribution Function is monotonous and non-decreasing we have that:

$$\begin{aligned} X^{-1}(I) &= \{w \in \Omega \,|\, X(w) \leq r\} = \{w \in [0, 1] \,|\, F^{-1}(w) \leq r\} \\ &= \{w \in \Omega \,|\, w \leq F(r)\} == [0, F(r)] \in \mathcal{B}_\Omega \end{aligned}$$

and therefore $X(w) = F^{-1}(w)$ is measurable over $\mathcal{B}_\mathcal{R}$ and is distributed as

$$P(X(w) \leq x\} = P(F^{-1}(w) \leq x\} = P(w \leq F(x)\} = \int_0^{F(x)} d\lambda = F(x)$$

Example 1.11 Consider the probability space $(\mathcal{R}, \mathcal{B}, \mu)$ with μ the probability measure

$$\mu(A) = \int_{A \in \mathcal{B}} dF$$

[11] Recall that for continuous random quantities $P(X_2 = x_2) = P(X_1 = x_1) = 0)$. One can *justify* this expression with kind of heuristic arguments; essentially considering $X_1 \in \Lambda_1 = (-\infty, x_1]$, $X_2 \in \Delta_\epsilon(x_2) = [x_2, x_2 + \epsilon]$ and taking the limit $\epsilon \to 0^+$ of

$$P(X_1 \leq x_1 | X_2 \in \Delta_\epsilon(x_2)) = \frac{P(X_1 \leq x_1, X_2 \in \Delta_\epsilon(x_2))}{P(X_2 \in \Delta_\epsilon(x_2))} = \frac{F(x_1, x_2 + \epsilon) - F(x_1, x_2)}{F_2(x_2 + \epsilon) - F_2(x_2)}$$

See however [1]; Vol 2; Chap. 10, for the Radon–Nikodym density with conditional measures.

The function $X : w \in \mathcal{R} \longrightarrow F(w) \in [0, 1]$ is measurable on $\mathcal{B}$. Take $I = [a, b) \in \mathcal{B}_{[0,1]}$. Then

$$X^{-1}(I) = \{w \in \mathcal{R} \,|\, a \leq F(w) < b\} = \{w \in \mathcal{R} \,|\, F^{-1}(a) \leq w < F^{-1}(b)\} = [w_a, w_b) \in \mathcal{B}_{\mathcal{R}}$$

It is distributed as $X \sim Un(x|0, 1)$:

$$P(X(w) \leq x\} = P(F(w) \leq x\} = P(w \leq F^{-1}(x)\} = \int_{-\infty}^{F^{-1}(x)} dF = x$$

This is the basis of the Inverse Transform sampling method that we shall see in Chap. 3 on Monte Carlo techniques.

Example 1.12 Suppose that the number of eggs a particular insect may lay (X_1) follows a Poisson distribution $X_1 \sim Po(x_1|\mu)$:

$$P(X_1 = x_1|\mu) = e^{-\mu} \frac{\mu^{x_1}}{\Gamma(x_1 + 1)}; \quad x_1 = 0, 1, 2, \ldots$$

Now, if the probability for an egg to hatch is θ and X_2 represent the number of off springs, given x_1 eggs the probability to have x_2 descendants follows a Binomial law $X_2 \sim Bi(x_2|x_1, \theta)$:

$$P(X_2 = x_2|x_1, \theta) = \binom{x_1}{x_2} \theta^{x_2} (1 - \theta)^{x_1 - x_2}; \quad 0 \leq x_2 \leq x_1$$

In consequence

$$\begin{aligned} P(X_1 = x_1, X_2 = x_2|\mu, \theta) &= P(X_2 = x_2|X_1 = x_1, \theta)\, P(X_1 = x_1|\mu) = \\ &= \binom{x_1}{x_2} \theta^{x_2} (1 - \theta)^{x_1 - x_2} e^{-\mu} \frac{\mu^{x_1}}{\Gamma(x_1 + 1)}; \quad 0 \leq x_2 \leq x_1 \end{aligned}$$

Suppose that we have not observed the number of eggs that were laid. What is the distribution of the number of off springs? This is given by the marginal probability

$$P(X_2 = x_2|\theta, \mu) = \sum_{x_1 = x_2}^{\infty} P(X_1 = x_1, X_2 = x_2) = e^{-\mu\theta} \frac{(\mu\theta)^{x_2}}{\Gamma(x_2 + 1)} = Po(x_2|\mu\theta)$$

Now, suppose that we have found x_2 new insects. What is the distribution of the number of eggs laid? This will be the conditional probability $P(X_1 = x_1|X_2 = x_2, \theta, \mu)$ and, since $P(X_1 = x_1, X_2 = x_2) = P(X_1 = x_1|X_2 = x_2)P(X_2 = x_2)$ we have that:

$$P(X_1 = x_1 | X_2 = x_2, \mu, \theta) = \frac{P(X_1 = x_1, X_2 = x_2)}{P(X_2 = x_2)}$$
$$= \frac{1}{(x_1 - x_2)!} \; (\mu(1-\theta))^{x_1 - x_2} \; e^{-\mu(1-\theta)}$$

with $0 \leq x_2 \leq x_1$; that is, again a Poisson with parameter $\mu(1-\theta)$.

Example 1.13 Let X_1 and X_2 two independent Poisson distributed random quantities with parameters μ_1 and μ_2. How is $Y = X_1 + X_2$ distributed? Since they are independent:

$$P(X_1 = x_1, X_2 = x_2 | \mu_1, \mu_2) = e^{-(\mu_1 + \mu_2)} \frac{\mu_1^{x_1}}{\Gamma(x_1+1)} \frac{\mu_2^{x_2}}{\Gamma(x_2+1)}$$

Then, since $X_2 = Y - X_1$:

$$P(X_1 = x, Y = y) = P(X_1 = x, X_2 = y - x) = e^{-(\mu_1 + \mu_2)} \frac{\mu_1^{x}}{\Gamma(x+1)} \frac{\mu_2^{(y-x)}}{\Gamma(y-x+1)}$$

Being $X_2 = y - x \geq 0$ we have the condition $y \geq x$ so the marginal probability for Y will be

$$P(Y = y) = e^{-(\mu_1 + \mu_2)} \sum_{x=0}^{y} \frac{\mu_1^{x}}{\Gamma(x+1)} \frac{\mu_2^{(y-x)}}{\Gamma(y-x+1)} = e^{-(\mu_1 + \mu_2)} \frac{(\mu_1 + \mu_2)^y}{\Gamma(y+1)}$$

that is, $Po(y|\mu_1 + \mu_2)$.

Example 1.14 Consider a two-dimensional random quantity $X = (X_1, X_2)$ that takes values in $\mathcal{R}^2$ with the probability density function $N(x_1, x_2 | \boldsymbol{\mu} = \mathbf{0}, \boldsymbol{\sigma} = \mathbf{1}, \rho)$:

$$p(x_1, x_2 | \rho) = \frac{1}{2\pi} \frac{1}{\sqrt{1-\rho^2}} \; e^{-\frac{1}{2(1-\rho^2)} (x_1^2 - 2\rho x_1 x_2 + x_2^2)}$$

being $\rho \in (-1, 1)$. The marginal densities are:

$$X_1 \sim p(x_1) = \int_{-\infty}^{+\infty} p(x_1, u_2) \, du_2 = \frac{1}{\sqrt{2\pi}} \, e^{-\frac{1}{2} x_1^2}$$
$$X_2 \sim p(x_2) = \int_{-\infty}^{+\infty} p(u_1, x_2) \, du_1 = \frac{1}{\sqrt{2\pi}} \, e^{-\frac{1}{2} x_2^2}$$

and since

$$p(x_1, x_2 | \rho) = p(x_1) \, p(x_2) \frac{1}{\sqrt{1-\rho^2}} \; e^{-\frac{\rho}{2(1-\rho^2)} (x_1^2 \rho - 2 x_1 x_2 + x_2^2 \rho)}$$

both quantities will be independent iff $\rho = 0$. The conditional densities are

$$p(x_1|x_2,\rho) = \frac{p(x_1,x_2)}{p(x_2)} = \frac{1}{\sqrt{2\pi}}\frac{1}{\sqrt{1-\rho^2}}\, e^{-\frac{1}{2(1-\rho^2)}(x_1 - x_2\rho)^2}$$

$$p(x_2|x_1,\rho) = \frac{f(x_1,x_2)}{f_1(x_1)} = \frac{1}{\sqrt{2\pi}}\frac{1}{\sqrt{1-\rho^2}}\, e^{-\frac{1}{2(1-\rho^2)}(x_2 - x_1\rho)^2}$$

and when $\rho = 0$ (thus independent) $p(x_1|x_2) = p(x_1)$ and $p(x_2|x_1) = p(x_2)$. Last, it is clear that

$$p(x_1, x_2|\rho) = p(x_2|x_1,\rho)\, p(x_1) = p(x_1|x_2,\rho)\, p(x_2).$$

1.4 Stochastic Characteristics

1.4.1 Mathematical Expectation

Consider a random quantity $X(w) : \Omega \to \mathcal{R}$ that can be either discrete

$$X(w) = \begin{cases} X(w) = \sum_{k=1}^{n} x_k \mathbf{1}_{A_k}(w) \\ \\ X(w) = \sum_{k=1}^{\infty} x_k \mathbf{1}_{A_k}(w) \end{cases} \longrightarrow P(X = x_k) = P(A_k) = \int_{\mathcal{R}} \mathbf{1}_{A_k}(w)\, dP(w)$$

or absolutely continuous for which

$$P(X(w) \in A) = \int_{\mathcal{R}} \mathbf{1}_A(w)\, dP(w) = \int_A dP(w) = \int_A p(w) dw$$

The **mathematical expectation** of a n-dimensional random quantity $\boldsymbol{Y} = g(\boldsymbol{X})$ is defined as[12]:

$$E[\boldsymbol{Y}] = E[g(\boldsymbol{X})] \stackrel{def.}{=} \int_{\mathcal{R}^n} g(\boldsymbol{x})\, dP(\boldsymbol{x}) = \int_{\mathcal{R}^n} g(\boldsymbol{x})\, p(\boldsymbol{x})\, d\boldsymbol{x}$$

[12]In what follows we consider the Stieltjes-Lebesgue integral so $\int \to \sum$ for discrete random quantities and in consequence:

$$\int_{-\infty}^{\infty} g(x)\, dP(x) = \int_{-\infty}^{\infty} g(x)\, p(x)\, dx \longrightarrow \sum_{\forall x_k} g(x_k)\, P(X = x_k).$$

In general, the function $g(x)$ will be unbounded on supp$\{X\}$ so both the sum and the integral have to be **absolutely convergent** for the *mathematical expectation* to exist.

In a similar way, we define the *conditional expectation.* If $\boldsymbol{X} = (X_1, \ldots, X_m \ldots, X_n)$, $\boldsymbol{W} = (X_1 \ldots, X_m)$ and $\boldsymbol{Z} = (X_{m+1} \ldots, X_n)$ we have for $\boldsymbol{Y} = g(\boldsymbol{W})$ that

$$E[\boldsymbol{Y}|\boldsymbol{Z}_0] = \int_{\mathcal{R}^m} g(\boldsymbol{w})\, p(\boldsymbol{w}|\boldsymbol{z}_0)\, d\boldsymbol{w} = \int_{\mathcal{R}^m} g(\boldsymbol{w})\, \frac{p(\boldsymbol{w}, \boldsymbol{z}_0)}{p(\boldsymbol{z}_0)}\, d\boldsymbol{w}.$$

1.4.2 Moments of a Distribution

Given a random quantity $X \sim p(x)$, we define the *moment or order n* (α_n) as:

$$\alpha_n = E[X^n] \stackrel{def.}{=} \int_{-\infty}^{\infty} x^n\, p(x)\, dx$$

Obviously, they exist if $x^n\, p(x) \in L_1(\mathcal{R})$ so it may happen that a particular probability distribution has only a finite number of moments. It is also clear that if the moment of order n exists, so do the moments of lower order and, if it does not, neither those of higher order. In particular, the moment of order 0 always exists (that, due to the normalization condition, is $\alpha_0 = 1$) and those of even order, if exist, are non-negative. A specially important moment is that order 1: the **mean** (*mean value*) $\mu = E[X]$ that has two important properties:

- It is a **linear operator** since $X = c_0 + \sum_{i=1}^{n} c_i X_i \longrightarrow E[X] = c_0 + \sum_{i=1}^{n} c_i\, E[X_i]$
- If $X = \prod_{i=1}^{n} c_i X_i$ with $\{X_i\}_{i=1}^{n}$ independent random quantities, then $E[X] = \prod_{i=1}^{n} c_i\, E[X_i]$.

We can define as well the moments (β_n) with respect to any point $c \in \mathcal{R}$ as:

$$\beta_n = E[(X - c)^n] \stackrel{def.}{=} \int_{-\infty}^{\infty} (x - c)^n\, p(x)\, dx$$

so α_n are also called *central moments or moments with respect to the origin*. It is easy to see that the non-central moment of second order, $\beta_2 = E[(X - c)^2]$, is minimal for $c = \mu = E[X]$. Thus, of special relevance are the *moments or order n with respect to the mean*

$$\mu_n \equiv E[(X - \mu)^n] = \int_{-\infty}^{\infty} (x - \mu)^n\, p(x)\, dx$$

and, among them, the moment of order 2: the **variance** $\mu_2 = V[X] = \sigma^2$. It is clear that $\mu_0 = 1$ and, if exists, $\mu_1 = 0$. Note that:

- $V[X] = \sigma^2 = E[(X-\mu)^2] > 0$
- It is **not a linear operator** since $X = c_0 + c_1 X_1 \longrightarrow V[X] = \sigma_X^2 = c_1^2\, V[X_1] = c_1^2 \sigma_{X_1}^2$
- If $X = \sum_{i=1}^n c_i X_i$ and $\{X_i\}_{i=1}^n$ are independent random quantities, $V[X] = \sum_{i=1}^n c_i^2\, V[X_i]$.

Usually, is less tedious to calculate the moments with respect to the origin and evidently they are related so, from the binomial expansion

$$(X-\mu)^n = \sum_{k=0}^{n} \binom{n}{k} X^k\,(-\mu)^{n-k} \longrightarrow \mu_n = \sum_{k=0}^{n} \binom{n}{k} \alpha_k\,(-\mu)^{n-k}$$

The previous definitions are trivially extended to n-dimensional random quantities. In particular, for 2 dimensions, $\boldsymbol{X} = (X_1, X_2)$, we have the moments of order (n, m) with respect to the origin:

$$\alpha_{nm} = E[X_1^n\, X_2^m] = \int_{\mathcal{R}^2} x_1^n\, x_2^m\; p(x_1, x_2)\, dx_1\, dx_2$$

so that $\alpha_{01} = \mu_1$ and $\alpha_{02} = \mu_2$, and the moments order (n, m) with respect to the mean:

$$\mu_{nm} = E[(X_1-\mu_1)^n\,(X_2-\mu_2)^m] = \int_{\mathcal{R}^2} (x_1-\mu_1)^n\,(x_2-\mu_2)^m\; p(x_1, x_2)\, dx_1\, dx_2$$

for which

$$\mu_{20} = E[(X_1-\mu_1)^2] = V[X_1] = \sigma_1^2 \quad \text{and} \quad \mu_{02} = E[(X_2-\mu_2)^2] = V[X_2] = \sigma_2^2$$

The moment

$$\mu_{11} = E[(X_1-\mu_1)\,(X_2-\mu)] = \alpha_{11} - \alpha_{10}\,\alpha_{01} = V[X_1, X_2] = V[X_2, X_1]$$

is called **covariance** between the random quantities X_1 and X_2 and, if they are independent, $\mu_{11} = 0$. The second order moments with respect to the mean can be condensed in matrix form, the **covariance matrix** defined as:

$$\boldsymbol{V}[\boldsymbol{X}] = \begin{pmatrix} \mu_{20} & \mu_{11} \\ \mu_{11} & \mu_{02} \end{pmatrix} = \begin{pmatrix} V[X_1, X_1] & V[X_1, X_2] \\ V[X_1, X_2] & V[X_2, X_2] \end{pmatrix}$$

Similarly, for $\boldsymbol{X} = (X_1, X_2, \ldots, X_n)$ we have the moments with respect to the origin

$$\alpha_{k_1,k_2,\ldots,k_n} = E[X_1^{k_1}\, X_2^{k_2} \cdots X_n^{k_n}];$$

the moments with respect to the mean

$$\mu_{k_1,k_2,\ldots,k_n} = E[(X_1-\mu_1)^{k_1}\,(X_2-\mu_2)^{k_2}\cdots(X_n-\mu_n)^{k_n}]$$

and the covariance matrix:

$$V[X] = \begin{pmatrix} \mu_{20\ldots0} & \mu_{11\ldots0} & \cdots & \mu_{10\ldots1} \\ \mu_{11\ldots0} & \mu_{02\ldots0} & \cdots & \mu_{01\ldots1} \\ \vdots & \vdots & \cdots & \vdots \\ \mu_{10\ldots1} & \mu_{01\ldots1} & \cdots & \mu_{00\ldots2} \end{pmatrix} = \begin{pmatrix} V[X_1,X_1] & V[X_1,X_2] & \cdots & V[X_1,X_n] \\ V[X_1,X_2] & V[X_2,X_2] & \cdots & V[X_2,X_n] \\ \vdots & \vdots & \cdots & \vdots \\ V[X_1,X_n] & V[X_2,X_n] & \cdots & V[X_n,X_n] \end{pmatrix}$$

The covariance matrix $\boldsymbol{V}[\boldsymbol{X}] = E[(\boldsymbol{X}-\boldsymbol{\mu})(\boldsymbol{X}-\boldsymbol{\mu})^T]$ has the following properties that are easy to prove from basic matrix algebra relations:

(1) It is a **symmetric** matrix ($\boldsymbol{V} = \boldsymbol{V}^T$) with **positive diagonal elements** ($\boldsymbol{V}_{ii} \geq 0$);
(2) It is **positive defined** ($\boldsymbol{x}^T\boldsymbol{V}\boldsymbol{x} \geq 0;\ \forall \boldsymbol{x} \in \mathcal{R}^n$ with the equality when $\forall i\ x_i = 0$);
(3) Being $\boldsymbol{V}$ symmetric, all the eigenvalues are real and the corresponding eigenvectors orthogonal. Furthermore, since it is positive defined all eigenvalues are positive;
(4) If $\mathbf{J}$ is a diagonal matrix whose elements are the eigenvalues of $\boldsymbol{V}$ and $\mathbf{H}$ a matrix whose columns are the corresponding eigenvectors, then $\mathbf{V} = \mathbf{HJH}^{-1}$ (Jordan dixit);
(5) Since $\boldsymbol{V}$ is symmetric, there is an orthogonal matrix $\mathbf{C}$ ($\mathbf{C}^T = \mathbf{C}^{-1}$) such that $\mathbf{CVC}^T = \mathbf{D}$ with $\mathbf{D}$ a diagonal matrix whose elements are the eigenvalues of $\boldsymbol{V}$;
(6) Since $\mathbf{V}$ is symmetric and positive defined, there is a non-singular matrix $\mathbf{C}$ such that $\mathbf{V} = \mathbf{CC}^T$;
(7) Since $\mathbf{V}$ is symmetric and positive defined, the inverse $\mathbf{V}^{-1}$ is also symmetric and positive defined;
(8) (Cholesky Factorization) Since $\mathbf{V}$ is symmetric and positive defined, there exists a unique lower triangular matrix $\mathbf{C}$ ($\mathbf{C}_{ij} = 0;\ \forall i < j$) with positive diagonal elements such that $\mathbf{V} = \mathbf{CC}^T$ (more about this in Chap. 3).

Among other things to be discussed later, the moments of the distribution are interesting because they give an idea of the shape and location of the probability distribution and, in many cases, the distribution parameters are expressed in terms of the moments.

1.4.2.1 Position Parameters

Let $X \sim p(x)$ with support in $\Omega \subset \mathcal{R}$. The position parameters *choose* a *characteristic* value of X and indicate more or less where the distribution is located. Among them we have the **mean value**

$$\mu = \alpha_1 = E[X] = \int_{-\infty}^{\infty} x\,p(x)\,dx$$

The mean is bounded by the minimum and maximum values the random quantity can take but, clearly, if $\Omega \subset \mathcal{R}$ it may happen that $\mu \notin \Omega$. If, for instance, $\Omega = \Omega_1 \cup \Omega_2$ is the union of two disconnected regions, μ may lay in between and therefore $\mu \notin \Omega$. On the other hand, as has been mentioned the integral has to be absolute convergent and there are some probability distributions for which there is no mean value. There are however other interesting location quantities. The **mode** is the value x_0 of X for which the distribution is maximum; that is,

$$x_0 = \sup_{x \in \Omega} p(x)$$

Nevertheless, it may happen that there are several relative maximums so we talk about uni-modal, bi-modal,... distributions. The **median** is the value x_m such that

$$F(x_m) = P(X \leq x_m) = 1/2 \longrightarrow \int_{-\infty}^{x_m} p(x)dx = \int_{x_m}^{\infty} p(x)dx = P(X > x_m) = 1/2$$

For discrete random quantities, the distribution function is either a finite or countable combination of indicator functions $\mathbf{1}_{A_k}(x)$ with $\{A_k\}_{k=1}^{n,\infty}$ a partition of Ω so it may happen that $F(x) = 1/2 \ \forall x \in A_k$. Then, any value of the interval A_k can be considered the median. Last, we may consider the **quantiles** α defined as the value q_α of the random quantity such that $F(q_\alpha) = P(X \leq q_\alpha) = \alpha$ so the *median* is the *quantile* $q_{1/2}$.

1.4.2.2 Dispersion Parameters

There are many ways to give an idea of how *dispersed* are the values the random quantity may take. Usually they are based on the mathematical expectation of a function that depends on the difference between X and some characteristic value it may take; for instance $E[|X - \mu|]$. By far, the most usual and important one is the already defined **variance**

$$V[X] = \sigma^2 = E[(X - E[X])^2] = \int_{\mathcal{R}} (x - \mu)^2 p(x)dx$$

provided it exists. Note that if the random quantity X has dimension $D[X] = d_X$, the variance has dimension $D[\sigma^2] = d_X^2$ so to have a quantity that gives an idea of the *dispersion* and has the same dimension one defines the **standard deviation** $\sigma = +\sqrt{V[X]} = +\sqrt{\sigma^2}$ and, if both the mean value (μ) and the variance exist, the **standardized** random quantity

$$Y = \frac{X - \mu}{\sigma}$$

for which $E[Y] = 0$ and $V[Y] = \sigma_Y^2 = 1$.

1.4.2.3 Asymmetry and Peakiness Parameters

Related to higher order non-central moments, there are two dimensionless quantities of interest: the skewness and the kurtosis. The first non-trivial odd moment with respect to the mean is that of order 3: μ_3. Since it has dimension $D[\mu_3] = d_X^3$ we define the **skewness** (γ_1) as the dimensionless quantity

$$\gamma_1 \stackrel{def}{=} \frac{\mu_3}{\mu_2^{3/2}} = \frac{\mu_3}{\sigma^3} = \frac{E[(X-\mu)^3]}{\sigma^3}$$

The skewness is $\gamma_1 = 0$ for distributions that are symmetric with respect to the mean, $\gamma_1 > 0$ if the probability content is more concentrated on the right of the mean and $\gamma_1 < 0$ if it is to the left of the mean. Note however that there are many asymmetric distributions for which $\mu_3 = 0$ and therefore $\gamma_1 = 0$. For unimodal distributions, it is easy to see that

$$\gamma_1 = 0 \qquad mode = median = mean$$

$$\gamma_1 > 0 \qquad mode < median < mean$$

$$\gamma_1 < 0 \qquad mode > median > mean$$

The **kurtosis** is defined, again for dimensional considerations, as

$$\gamma_2 = \frac{\mu_4}{\mu_2^2} = \frac{\mu_4}{\sigma^4} = \frac{E[(X-\mu)^4]}{\sigma^4}$$

and gives an idea of how *peaked* is the distribution. For the Normal distribution $\gamma_2 = 3$ so in order to have a reference one defines the *extended kurtosis* as $\gamma_2^{ext} = \gamma_2 - 3$. Thus, $\gamma_2^{ext} > 0$ (<0) indicates that the distribution is *more (less) peaked* than the Normal. Again, $\gamma_2^{ext} = 0$ for the Normal density and for any other distribution for which $\mu_4 = 3\,\sigma^4$. Last you can check that $\forall a, b \in \mathcal{R}\; E[(X-\mu-a)^2(X-\mu-b)^2] > 0$ so, for instance, defining $u = a + b$, $w = ab$ and taking derivatives, $\gamma_2 \geq 1 + \gamma_1^2$.

Example 1.15 Consider the discrete random quantity $X \sim Pn(k|\lambda)$ with

$$P(X = k) \equiv Pn(k|\lambda) = e^{-\lambda}\,\frac{\lambda^k}{\Gamma(k+1)}; \qquad \lambda \in \mathcal{R}^+;\; k = 0, 1, 2, \ldots$$

The moments with respect to the origin are

$$\alpha_n(\lambda) \equiv E[X^n] = e^{-\lambda} \sum_{k=0}^{\infty} k^n\,\frac{\lambda^k}{k!}$$

If a_k denotes the kth term of the sum, then

$$a_{k+1} = \frac{\lambda}{k+1}\left(1+\frac{1}{k}\right)^n a_k \longrightarrow \lim_{k\to\infty}\left|\frac{a_{k+1}}{a_k}\right| \to 0$$

so being the series absolute convergent all order moments exist. Taking the derivative of $\alpha_n(\lambda)$ with respect to λ one gets the recurrence relation

$$\alpha_{n+1}(\lambda) = \lambda\left(\alpha_n(\lambda) + \frac{d\alpha_n(\lambda)}{d\lambda}\right); \qquad \alpha_0(\lambda) = 1$$

so we can easily get

$$\alpha_0 = 1; \quad \alpha_1 = \lambda; \quad \alpha_2 = \lambda(\lambda+1); \quad \alpha_3 = \lambda(\lambda^2+3\lambda+1); \quad \alpha_4 = \lambda(\lambda^3+6\lambda^2+7\lambda+1)$$

and from them

$$\mu_0 = 1; \qquad \mu_1 = 0; \qquad \mu_2 = \lambda; \qquad \mu_3 = \lambda; \qquad \mu_4 = \lambda(3\lambda+1)$$

Thus, for the Poisson distribution $Po(n|\lambda)$ we have that:

$$E[X] = \lambda; \qquad V[X] = \lambda; \qquad \gamma_1 = \lambda^{-1/2}; \qquad \gamma_2 = 3 + \lambda^{-1}$$

Example 1.16 Consider $X \sim Ga(x|a,b)$ with:

$$p(x) = \frac{a^b}{\Gamma(b)}\, e^{-ax}\, x^{b-1}\mathbf{1}_{(0,\infty)}(x)\lambda; \qquad a, b \in \mathcal{R}^+$$

The moments with respect to the origin are

$$\alpha_n = E[X^n] = \frac{a^b}{\Gamma(b)}\int_0^\infty e^{-ax}\, x^{b+n-1}\, dx = \frac{\Gamma(b+n)}{\Gamma(b)}\, a^{-n}$$

being the integral absolute convergent. Thus we have:

$$\mu_n = \frac{1}{a^n\Gamma(b)}\sum_{k=0}^{n}\binom{n}{k}(-b)^{n-k}\,\Gamma(b+k)$$

and in consequence

$$E[X] = \frac{b}{a}; \qquad V[X] = \frac{b}{a^2}; \qquad \gamma_1 = \frac{2}{\sqrt{b}}; \qquad \gamma_2^{ext.} = \frac{6}{b}$$

Example 1.17 For the Cauchy distribution $X \sim Ca(x|1, 1)$,

$$p(x) == \frac{1}{\pi} \frac{1}{1+x^2} \mathbf{1}_{(-\infty,\infty)}(x)$$

we have that

$$\alpha_n = E[X^n] = \frac{1}{\pi} \int_{-\infty}^{\infty} \frac{x^n}{1+x^2}\, dx$$

and clearly the integral diverges for $n > 1$ so there are no moments but the trivial one α_0. Even for $n = 1$, the integral

$$\int_{-\infty}^{\infty} \frac{|x|}{(1+x^2)}\, dx = 2 \int_{0}^{\infty} \frac{x}{(1+x^2)}\, dx = \lim_{a\to\infty} \ln\,(1+a^2)$$

is not absolute convergent so, in strict sense, there is no mean value. However, the mode and the median are $x_0 = x_m = 0$, the distribution is symmetric about $x = 0$ and for $n = 1$ there exists the Cauchy's Principal Value and is equal to 0. Had we introduced the Probability Distributions as a subset of Generalized Distributions, the Principal Value is an admissible distribution. It is left as an exercise to show that for:

- **Pareto**: $X \sim Pa(x|\nu, x_m)$ with $p(x|x_m, \nu) \propto x^{-(\nu+1)} \mathbf{1}_{[x_m,\infty)}(x)$; $x_m, \nu \in \mathcal{R}^+$
- **Student**: $X \sim St(x|\nu)$ with $p(x|\nu) \propto \left(1 + x^2/\nu\right)^{-(\nu+1)/2} \mathbf{1}_{(-\infty,\infty)}(x)$; $\nu \in \mathcal{R}^+$

the moments $\alpha_n = E[X^n]$ exist iff $n < \nu$.
Another distribution of interest in physics is the Landau Distribution that describes the energy lost by a particle when traversing a material under certain conditions. The probability density, given as the inverse Laplace Transform, is:

$$p(x) = \frac{1}{2\pi i} \int_{c-i\infty}^{c+i\infty} e^{s \log s + xs}\, ds$$

with $c \in \mathcal{R}^+$ and closing the contour on the left along a counterclockwise semicircle with a branch-cut along the negative real axis it has a real representation

$$p(x) = \frac{1}{\pi} \int_{0}^{\infty} e^{-(r \log r + xr)}\, \sin(\pi r)\, dr$$

The actual expression of the distribution of the energy loss is quite involved and some simplifying assumptions have been made; among other things, that the energy transfer in the collisions is unbounded (no kinematic constraint). But nothing is for free and the price to pay is that the Landau Distribution has no moments other than the trivial of order zero. This is why instead of mean and variance one talks about the *most probable energy loss* and the *full-width-half-maximum*.

1.4.2.4 Correlation Coefficient

The **covariance** between the random quantities X_i and X_j was defined as:

$$V[X_i, X_j] = V[X_j, X_i] = E[(X_i - \mu_i)(X_i - \mu_j)] = E[X_i X_j] - E[X_i]E[X_j]$$

If X_i and X_j are independent, then $E[X_i X_j] = E[X_i]E[X_j]$ and $V[X_i, X_j] = 0$. Conversely, if $V[X_i, X_j] \neq 0$ then $E[X_i X_j] \neq E[X_i]E[X_j]$ and in consequence X_i and X_j are not statistically independent. Thus, the *covariance* $V[X_i, X_j]$ serves to quantify, to some extent, the degree of *statistical dependence* between the random quantities X_i and X_j. Again, for dimensional considerations one defines the *correlation coefficient*

$$\rho_{ij} = \frac{V[X_i, X_j]}{\sqrt{V[X_i]\,V[X_j]}} = \frac{E[X_i X_j] - E[X_i]E[X_j]}{\sigma_i\,\sigma_j}$$

Since $p(x_i, x_j)$ is a non-negative function we can write

$$V[X_i, X_j] = \int_{\mathcal{R}^2} \left\{(x_i - \mu_i)\sqrt{p(x_i, x_j)}\right\} \left\{(x_j - \mu_j)\sqrt{p(x_i, x_j)}\right\} dx_i dx_j$$

and from the Cauchy–Schwarz inequality:

$$-1 \leq \rho_{ij} \leq 1$$

The extreme values $(+1, -1)$ will be taken when $E[X_i X_j] = E[X_i]E[X_j] \pm \sigma_i \sigma_j$ and $\rho_{ij} = 0$ when $E[X_i X_j] = E[X_i]E[X_j]$. In particular, it is immediate to see that if here is a linear relation between both random quantities; that is, $X_i = aX_j + b$, then $\rho_{ij} = \pm 1$. Therefore, it is a **linear correlation coefficient**. Note however that:

- If X_i and X_j are linearly related, $\rho_{ij} = \pm 1$, but $\rho_{ij} = \pm 1$ **does not** imply necessarily a linear relation;
- If X_i and X_j are statistically independent, then $\rho_{ij} = 0$ but $\rho_{ij} = 0$ **does not** imply necessarily statistical independence as the following example shows.

Example 1.18 Let $X_1 \sim p(x_1)$ and define a random quantity X_2 as

$$X_2 = g(X_1) = a + bX_1 + c{X_1}^2$$

Obviously, X_1 and X_2 are not statistically independent for there is a clear parabolic relation. However

$$V[X_1, X_2] = E[X_1 X_2] - E[X_1]\,E[X_2] = b\sigma^2 + c(\alpha_3 - \mu^3 - \mu\sigma^2)$$

with μ, σ^2 and α_3 respectively the mean, variance and moment of order 3 with respect to the origin of X_1 and, if we take $b = c\sigma^{-2}(\mu^3 + \mu\sigma^2 - \alpha_3)$ then $V[Y, X] = 0$ and so is the (linear) correlation coefficient.

NOTE 2: Information as a measure of independence.
The *Mutual Information* (see Sect. 4.4) serves also to quantify the degree of statistical dependence between random quantities. Consider for instance the two-dimensional random quantity $\boldsymbol{X} = (X_1, X_2) \sim p(x_1, x_2)$. Then:

$$I(X_1 : X_2) = \int_X dx_1\, dx_2 p(x_1, x_2) \ln\left(\frac{p(x_1, x_2)}{p(x_1)\, p(x_2)}\right)$$

and $I(X_1 : X_2) \geq 0$ with equality iff $p(x_1, x_2) = p(x_1)p(x_2)$. Let's look as an example to the bi-variate normal distribution: $N(\boldsymbol{x}|\boldsymbol{\mu}, \boldsymbol{\Sigma})$:

$$p(\boldsymbol{x}|\phi) = (2\pi)^{-1}\, |\det[\boldsymbol{\Sigma}]|^{-1/2} \exp\left\{-\frac{1}{2}\left((\boldsymbol{x}-\boldsymbol{\mu})^T\, \boldsymbol{\Sigma}^{-1}\, (\boldsymbol{x}-\boldsymbol{\mu})\right)\right\}$$

with covariance matrix

$$\boldsymbol{\Sigma} = \begin{pmatrix} \sigma_1^2 & \rho\sigma_1\sigma_2 \\ \rho\sigma_1\sigma_2 & \sigma_2^2 \end{pmatrix} \quad \text{and} \quad \det[\boldsymbol{\Sigma}] = \sigma_1^2\sigma_2^2(1-\rho)^2$$

Since $X_i \sim N(x_i|\mu_i, \sigma_i)$; $i = 1, 2$ we have that:

$$I(X_1 : X_2) = \int_X dx_1\, dx_2\; p(\boldsymbol{x}|\boldsymbol{\mu}, \boldsymbol{\Sigma}) \ln\left(\frac{p(\boldsymbol{x}|\boldsymbol{\mu}, \boldsymbol{\Sigma})}{p(x_1|\mu_1, \sigma_1)\, p(x_2|\mu 2, \sigma_2)}\right) = -\frac{1}{2}\ln(1-\rho^2)$$

Thus, if X_1 and X_2 are independent ($\rho = 0$), $I(X_1 : X_2) = 0$ and when $\rho \rightarrow \pm 1$, $I(X_1 : X_2) \rightarrow \infty$.

1.4.3 The "Error Propagation Expression"

Consider a n-dimensional random quantity $\boldsymbol{X} = (X_1, \ldots, X_n)$ with $E[X_i] = \mu_i$ and the random quantity $Y = g(\boldsymbol{X})$ with $g(x)$ an infinitely differentiable function. If we make a Taylor expansion of $g(\boldsymbol{X})$ around $E[\boldsymbol{X}] = \boldsymbol{\mu}$ we have

$$Y = g(\boldsymbol{X}) = g(\boldsymbol{\mu}) + \sum_{i=1}^{n}\left(\frac{\partial g(\boldsymbol{x})}{\partial x_i}\right)_{\mu} Z_i + \frac{1}{2!}\sum_{i=1}^{n}\sum_{j=1}^{n}\left(\frac{\partial^2 g(\boldsymbol{x})}{\partial x_i \partial x_j}\right)_{\mu} Z_i Z_j + R$$

where $Z_i = X_i - \mu_i$. Now, $E[Z_i] = 0$ and $E[Z_i Z_j] = V[X_i, X_j] = V_{ij}$ so

$$E[Y] = E[g(\boldsymbol{X})] = g(\boldsymbol{\mu}) + + \frac{1}{2!}\sum_{i=1}^{n}\sum_{j=1}^{n}\left(\frac{\partial^2 g(\boldsymbol{x})}{\partial x_i \partial x_j}\right)_{\boldsymbol{\mu}} V_{ij} + \cdots$$

and therefore

$$Y - E[Y] = \sum_{i=1}^{n}\left(\frac{\partial g(\boldsymbol{x})}{\partial x_i}\right)_{\boldsymbol{\mu}} Z_i + \frac{1}{2!}\sum_{i=1}^{n}\sum_{j=1}^{n}\left(\frac{\partial^2 g(\boldsymbol{x})}{\partial x_i \partial x_j}\right)_{\boldsymbol{\mu}} (Z_i Z_j - V_{ij}) + \cdots$$

Neglecting all but the first term

$$V[Y] = E[(Y - E[Y])^2] = \sum_{i=1}^{n}\sum_{j=1}^{n}\left(\frac{\partial g(x)}{\partial x_i}\right)_{\boldsymbol{\mu}}\left(\frac{\partial g(x)}{\partial x_j}\right)_{\boldsymbol{\mu}} V[X_i, X_j] + \cdots$$

This is the first order approximation to $V[Y]$ and usually is reasonable but has to be used with care. On the one hand, we have assumed that higher order terms are negligible and this is not always the case so further terms in the expansion may have to be considered. Take for instance the simple case $Y = X_1 X_2$ with X_1 and X_2 independent random quantities. The first order expansion gives $V[Y] \simeq \mu_1^2\sigma_2^2 + \mu_2^2\sigma_1^2$ and including second order terms (there are no more) $V[Y] = \mu_1^2\sigma_2^2 + \mu_2^2\sigma_1^2 + \sigma_1^2\sigma_2^2$; the correct result. On the other hand, all this is obviously meaningless if the random quantity Y has no variance. This is for instance the case for $Y = X_1 X_2^{-1}$ when $X_{1,2}$ are Normal distributed.

1.5 Integral Transforms

The Integral Transforms of Fourier, Laplace and Mellin are a very useful tool to study the properties of the random quantities and their distribution functions. In particular, they will allow us to obtain the distribution of the sum, product and ratio of random quantities, the moments of the distributions and to study the convergence of a sequence $\{F_k(x)\}_{k=1}^{\infty}$ of distribution functions to $F(x)$.

1.5.1 The Fourier Transform

Let $f : \mathcal{R} \rightarrow \mathcal{C}$ be a complex and integrable function ($f \in L_1(\mathcal{R})$). The *Fourier Transform* $\mathcal{F}(t)$ with $t \in \mathcal{R}$ of $f(x)$ is defined as:

$$\mathcal{F}(t) = \int_{-\infty}^{\infty} f(x)\, e^{ixt}\, dx$$

The class of functions for which the Fourier Transform exists is certainly much wider than the probability density functions $p(x) \in L_1(\mathcal{R})$ (normalized real functions of real argument) we are interested in for which the transform always exists. If $X \sim p(x)$, the Fourier Transform is nothing else but the mathematical expectation

$$\mathcal{F}(t) = E[e^{itX}]; \quad t \in \mathcal{R}$$

and it is called *Characteristic Function* $\Phi(t)$. Thus, depending on the character of the random quantity X, we shall have:

- if X is **discrete**: $\Phi(t) = \sum_{x_k} e^{itx_k}\, P(X = x_k)$
- if X is **continuous**: $\Phi(t) = \int_{-\infty}^{+\infty} e^{itx}\, dP(x) = \int_{-\infty}^{+\infty} e^{itx}\, p(x)\, dx$

Attending to its definition, the Characteristic Function $\Phi(t)$, with $t \in \mathcal{R}$, is a complex function and has the following properties:

(1) $\Phi(0) = 1$;
(2) $\Phi(t)$ is bounded: $|\Phi(t)| \leq 1$;
(3) $\Phi(t)$ has schwarzian symmetry: $\Phi(-t) = \overline{\Phi}(t)$;
(4) $\Phi(t)$ is uniformly continuous in $\mathcal{R}$.

The first three properties are obvious. For the fourth one, observe that for any $\epsilon > 0$ there exists a $\delta > 0$ such that $|\Phi(t_1) - \Phi(t_2)| < \epsilon$ when $|t_1 - t_2| < \delta$ with t_1 and t_2 arbitrary in $\mathcal{R}$ because

$$|\Phi(t+\delta) - \Phi(t)| \leq \int_{-\infty}^{+\infty} |1 - e^{-i\delta x}| dP(x) = 2 \int_{-\infty}^{+\infty} |\sin\, \delta x/2|\, dP(x)$$

and this integral can be made arbitrarily small taking a sufficiently small δ.

These properties, that obviously hold also for a discrete random quantity, are **necessary** but **not sufficient** for a function $\Phi(t)$ to be the Characteristic Function of a distribution $P(x)$ (see Example 1.9). Generalizing for a n-dimensional random quantity $\mathbf{X} = (X_1, \ldots, X_n)$:

$$\Phi(t_1, \ldots, t_n) = E[e^{i\mathbf{tX}}] = E[e^{i(t_1X_1 + \cdots + t_nX_n)}]$$

so, for the discrete case:

$$\Phi(t_1, \ldots, t_n) = \sum_{x_1} \cdots \sum_{x_n} e^{i(t_1x_1 + \cdots + t_nx_n)}\, P(X_1 = x_1, \ldots, X_n = x_n)$$

and for the continuous case:

$$\Phi(t_1, \ldots, t_n) = \int_{-\infty}^{+\infty} dx_1 \ldots \int_{-\infty}^{+\infty} dx_n \, e^{i(t_1 x_1 + \cdots + t_n x_n)} \, p(x_1, \ldots, x_n)$$

The n-dimensional Characteristic Function is such that:

(1) $\Phi(0, \ldots, 0) = 1$
(2) $|\Phi(t_1, \ldots, t_n)| \leq 1$
(3) $\Phi(-t_1, \ldots, -t_n) = \overline{\Phi}(t_1, \ldots, t_n)$

Laplace Transform: For a function $f(x) : \mathcal{R}^+ \rightarrow \mathcal{C}$ defined as $f(x) = 0$ for $x < 0$, we may consider also the *Laplace Transform* defined as

$$L(s) = \int_0^{\infty} e^{-sx} f(x) \, dx$$

with $s \in \mathcal{C}$ provided it exists. For a non-negative random quantity $X \sim p(x)$ this is just the mathematical expectation $E[e^{-sx}]$ and is named *Moment Generating Function* since the derivatives give the moments of the distribution (see Sect. 1.5.1.4). While the Fourier Transform exists for $f(x) \in L_1(\mathcal{R})$, the Laplace Transform exists if $e^{-sx} f(x) \in L_1(\mathcal{R}^+)$ and thus, for a wider class of functions and although it is formally defined for functions with non-negative support, it may be possible to extend the limits of integration to the whole real line (*Bilateral Laplace Transform*). However, for the functions we shall be interested in (probability density functions), both Fourier and Laplace Transforms exist and usually there is no major advantage in using one or the other.

Example 1.19 There are several criteria (Bochner, Kintchine, Cramèr,...) specifying sufficient and necessary conditions for a function $\Phi(t)$, that satisfies the four aforementioned conditions, to be the Characteristic Function of a random quantity $X \sim F(x)$. However, it is easy to find simple functions like

$$g_1(t) = e^{-t^4} \qquad \text{and} \qquad g_2(t) = \frac{1}{1+t^4};$$

that satisfy four stated conditions and that can not be Characteristic Functions associated to any distribution. Let's calculate the moments of order one with respect to the origin and the central one of order two. In both cases (see Sect. 1.5.1.4) we have that:

$$\alpha_1 = \mu = E[X] = 0 \qquad \text{and} \qquad \mu_2 = \sigma^2 = E[(X-\mu)^2] = 0$$

that is, the mean value and the variance are zero so the distribution function is zero almost everywhere but for $X = 0$ where $P(X = 0) = 1$... but this is the *Singular Distribution* $\text{Sn}(x|0)$ that takes the value 1 if $X = 0$ and 0 otherwise whose Characteristic Function is $\Phi(t) = 1$. In general, any function $\Phi(t)$ that in a boundary of $t = 0$ behaves as $\Phi(t) = 1 + O(t^{2+\epsilon})$ with $\epsilon > 0$ can not be the Characteristic Function associated to a distribution $F(x)$ unless $\Phi(t) = 1$ for all $t \in \mathcal{R}$.

Example 1.20 The elements of the Cantor Set $C_S(0, 1)$ can be represented in base 3 as:

$$X = \sum_{n=1}^{\infty} \frac{X_n}{3^n}$$

with $X_n \in \{0, 2\}$. This set is non-denumerable and has zero Lebesgue measure so any distribution with support on it is *singular* and, in consequence, has no pdf. The Uniform Distribution on $C_S(0, 1)$ is defined assigning a probability $P(X_n = 0) = P(X_n = 2) = 1/2$ (Geometric Distribution). Then, for the random quantity X_n we have that

$$\Phi_{X_n}(t) = E[e^{itX}] = \frac{1}{2}\left(1 + e^{2it}\right)$$

and for $Y_n = X_n/3^n$:

$$\Phi_{Y_n}(t) = \Phi_{X_n}(t/3^n) = \frac{1}{2}\left(1 + e^{2it/3^n}\right)$$

Being all X_n statistically independent, we have that

$$\Phi_X(t) = \prod_{n=1}^{\infty} \frac{1}{2}\left(1 + e^{2it/3^n}\right) = \prod_{n=1}^{\infty} \frac{1}{2} e^{it/3^n} \cos(t/3^n) = e^{it/2} \prod_{n=1}^{\infty} \cos(t/3^n)$$

and, from the derivatives (Sect. 1.5.1.4) it is straight forward to calculate the moments of the distribution. In particular:

$$\Phi_X^{(1)}(0) = \frac{i}{2} \quad \longrightarrow \quad E[X] = \frac{1}{2} \qquad \text{and} \qquad \Phi_X^{(2)}(0) = -\frac{3}{8} \quad \longrightarrow \quad E[X^2] = \frac{3}{8}$$

so $V[X] = 1/8$.

1.5.1.1 Inversion Theorem (Lévy 1925)

The Inverse Fourier Transform allows us to obtain the distribution function of a random quantity from the Characteristic Function. If X is a continuous random quantity and $\Phi(t)$ its Characteristic Function, then the pdf $p(x)$ will be given by

$$p(x) = \frac{1}{2\pi} \int_{-\infty}^{+\infty} e^{-itx} \, \Phi(t) \, dt$$

provided that $p(x)$ is continuous at x and, if X is discrete:

$$P(X = x_k) = \frac{1}{2\pi} \int_{-\infty}^{+\infty} e^{-itx_k} \, \Phi(t) \, dt$$

In particular, if the discrete distribution is *reticular* (that is, all the possible values that the random quantity X may take can be expressed as $a + bn$ with $a, b \in \mathcal{R}$; $b \neq 0$ and n integer) we have that:

$$P(X = x_k) = \frac{b}{2\pi} \int_{-\pi/b}^{\pi/b} e^{-itx_k} \, \Phi(t) \, dt$$

From this expressions, we can obtain also the relation between the Characteristic Function and the Distribution Function. For discrete random quantities we shall have:

$$F(x_k) = \sum_{x \leq x_k} P(X = x_k) = \frac{1}{2\pi} \int_{-\infty}^{+\infty} \sum_{x \leq x_k} e^{-itx} \, \Phi(t) \, dt$$

and, in the continuous case, for $x_1 < x_2 \in \mathcal{R}$ we have that:

$$F(x_2) - F(x_1) = \int_{x_1}^{x_2} p(x) \, dx = \frac{1}{2\pi i} \int_{-\infty}^{+\infty} \Phi(t) \, \frac{1}{t} \, (e^{-itx_1} - e^{-itx_2}) \, dt$$

so taking $x_1 = 0$ we have that (Lévy, 1925):

$$F(x) = F(0) + \frac{1}{2\pi i} \int_{-\infty}^{+\infty} \Phi(t) \, \frac{1}{t} \, (1 - e^{-itx}) \, dt$$

The **Inversion Theorem** states that there is a one-to-one correspondence between a distribution function and its Characteristic Function so to each Characteristic Function corresponds **one** and **only one** distribution function that can be either discrete or continuous but not a combination of both. Therefore, two distribution functions with the same Characteristic Function may differ, at most, on their points of discontinuity that, as we have seen, are a set of zero measure. In consequence, if we have two random quantities X_1 and X_2 with distribution functions $P_1(x)$ and $P_2(x)$, a **necessary** and **sufficient** condition for $P_1(x) = P_2(x)$ a.e. is that $\Phi_1(t) = \Phi_2(t)$ for all $t \in \mathcal{R}$.

1.5.1.2 Changes of Variable

Let $X \sim P(x)$ be a random quantity with Characteristic Function $\Phi_X(t)$ and $g(X)$ a one-to-one finite real function defined for all real values of X. The Characteristic Function of the random quantity $Y = g(X)$ will be given by:

$$\Phi_Y(t) = E_Y[e^{itY}] = E_X[e^{itg(X)}]$$

that is:

$$\Phi_Y(t) = \int_{-\infty}^{+\infty} e^{itg(x)}\, dP(x) \qquad \text{or} \qquad \Phi_Y(t) = \sum_{x_k} e^{itg(x_k)}\, P(X = x_k)$$

depending on whether X is continuous or discrete. In the particular case of a linear transformation $Y = aX + b$ with a and b real constants, we have that:

$$\Phi_Y(t) = E_X[e^{it(aX+b)}] = e^{itb}\, \Phi_X(at).$$

1.5.1.3 Sum of Random Quantities

The Characteristic Function is particularly useful to obtain the Distribution Function of a random quantity defined as the sum of independent random quantities. If $X_1, \ldots, X_n$ are n independent random quantities with Characteristic Functions $\Phi_1(t_1), \ldots, \Phi_n(t_n)$, then the Characteristic Function of $X = X_1 + \cdots + X_n$ will be:

$$\Phi_X(t) = E[e^{itX}] = E[e^{it(X_1 + \cdots + X_n)}] = \Phi_1(t) \cdots \Phi_n(t)$$

that is, the product of the Characteristic Functions of each one, a **necessary** but **not sufficient** condition for the random quantities $X_1, \ldots, X_n$ to be independent. In a similar way, we have that if $X = X_1 - X_2$ with X_1 and X_2 independent random quantities, then

$$\Phi_X(t) = E[e^{it(X_1 - X_2)}] = \Phi_1(t)\, \Phi_2(-t) = \Phi_1(t)\, \overline{\Phi_2}(t)$$

Form these considerations, it is left as an exercise to show that:

- **Poisson**: The sum of n independent random quantities, each distributed as $Po(x_k|\mu_k)$ with $k = 1, \ldots, n$ is Poisson distributed with parameter $\mu_s = \mu_1 + \cdots + \mu_n$.
- **Normal**: The sum of n independent random quantities, each distributed as $N(x_k|\mu_k, \sigma_k)$ with $k = 1, \ldots, n$ is Normal distributed with mean $\mu_s = \mu_1 + \cdots + \mu_n$ and variance $\sigma_s^2 = \sigma_1^2 + \cdots + \sigma_n^2$.

- **Cauchy**: The sum of n independent random quantities, each Cauchy distributed $Ca(x_k|\alpha_k, \beta_k)$ with $k = 1, \ldots, n$ is is Cauchy distributed with parameters $\alpha_s = \alpha_1 + \cdots + \alpha_n$ and $\beta_s = \beta_1 + \cdots + \beta_n$.
- **Gamma**: The sum of n independent random quantities, each distributed as $Ga(x_k|\alpha, \beta_k)$ with $k = 1, \ldots, n$ is Gamma distributed with parameters $(\alpha, \beta_1 + \cdots + \beta_n)$.

Example 1.21 (*Difference of Poisson distributed random quantities*) Consider two independent random quantities $X_1 \sim Po(X_1|\mu_1)$ and $X_2 \sim Po(X_2|\mu_2)$ and let us find the distribution of $X = X_1 - X_2$. Since for the Poisson distribution:

$$X_i \sim Po(\mu_i) \quad \longrightarrow \quad \Phi_i(t) = e^{-\mu_i(1 - e^{it})}$$

we have that

$$\Phi_X(t) = \Phi_1(t)\,\overline{\Phi}_2(t) = e^{-(\mu_1 + \mu_2)} e^{(\mu_1 e^{it} + \mu_2 e^{-it})}$$

Obviously, X is a discrete random quantity with integer support $\Omega_X = \{\ldots, -2, -1, 0, 1, 2, \ldots\}$; that is, a *reticular* random quantity with $a = 0$ and $b = 1$. Then

$$P(X = n) = \frac{1}{2\pi} e^{-\mu_S} \int_{-\pi}^{\pi} e^{-itn} e^{(\mu_1 e^{it} + \mu_2 e^{-it})} dt$$

being $\mu_S = \mu_1 + \mu_2$. If we take:

$$z = \sqrt{\frac{\mu_1}{\mu_2}}\, e^{it}$$

we have

$$P(X = n) = \left(\frac{\mu_1}{\mu_2}\right)^{n/2} e^{-\mu_S} \frac{1}{2\pi i} \oint_C z^{-n-1} e^{\frac{w}{2}(z + 1/z)} dz$$

with $w = 2\sqrt{\mu_1 \mu_2}$ and C the circle $|z| = \sqrt{\mu_1/\mu_2}$ around the origin. From the definition of the Modified Bessel Function of first kind

$$I_n(z) = \frac{1}{2\pi i} \oint_C t^{-n-1} e^{\frac{z}{2}(t + 1/t)} dt$$

with C a circle enclosing the origin anticlockwise and considering that $I_{-n}(z) = I_n(z)$ we have finally:

$$P(X = n) = \left(\frac{\mu_1}{\mu_2}\right)^{n/2} e^{-(\mu_1 + \mu_2)} I_{|n|}(2\sqrt{\mu_1 \mu_2}).$$

1.5.1.4 Moments of a Distribution

Consider a continuous random quantity $X\sim P(x)$ and Characteristic Function

$$\Phi(t) = E[e^{itX}] = \int_{-\infty}^{+\infty} e^{itx}\, dP(x)$$

and let us assume that there exists the moment of order k. Then, upon derivation of the Characteristic Function k times with respect to t we have:

$$\frac{\partial^k}{\partial^k t}\Phi(t) = E[i^k\, X^k\, e^{itX}] = \int_{-\infty}^{+\infty} (ix)^k\, e^{itx}\, dP(x)$$

and taking $t = 0$ we get the *moments with respect to the origin*:

$$E[X^k] = \frac{1}{i^k}\left(\frac{\partial^k}{\partial^k t}\Phi(t)\right)_{t=0}$$

Consider now the Characteristic Function referred to an arbitrary point $a \in \mathcal{R}$; that is:

$$\Phi(t,a) = E[e^{it(X-a)}] = \int_{-\infty}^{+\infty} e^{it(x-a)}\, dP(x) = e^{-ita}\,\Phi(t)$$

In a similar way, upon k times derivation with respect to t we get the moments with respect to an arbitrary point a:

$$E[(X-a)^k] = \frac{1}{i^k}\left(\frac{\partial^k}{\partial^k t}\Phi(t,a)\right)_{t=0}$$

and the *central moments* if $a = E(X) = \mu$. The extension to n dimensions immediate: for a n dimensional random quantity $\mathbf{X}$ we shall have the for the moment $\alpha_{k_1\ldots k_n}$ with respect to the origin that

$$\alpha_{k_1\ldots k_n} = E[X_1^{k_1}\cdots X_n^{k_n}] = \frac{1}{i^{k_1+\cdots+k_n}}\left(\frac{\partial^{k_1+\cdots+k_n}}{\partial^{k_1}t_1\cdots\partial^{k_n}t_n}\Phi(t_1,\ldots,t_n)\right)_{t_1=\cdots=t_n=0}$$

Example 1.22 For the difference of Poisson distributed random quantities analyzed in the previous example, one can easily derive the moments from the derivatives of the Characteristic Function. Since

$$\log\Phi_X(t) = -(\mu_1+\mu_2) + (\mu_1 e^{it} + \mu_2 e^{-it})$$

we have that

$$\begin{aligned} \Phi'_X(0) &= i\,(\mu_1-\mu_2) & \longrightarrow \quad & E[X]=\mu_1-\mu_2 \\ \Phi''_X(0) &= (\Phi'_X(0))^2-(\mu_1+\mu_2) & \longrightarrow \quad & V[X]=\mu_1+\mu_2 \end{aligned}$$

and so on.

Problem 1.3 The Moyal Distribution, with density

$$p(x)=\frac{1}{\sqrt{2\pi}}\exp\left\{-\frac{1}{2}\left(x+e^{-x}\right)\right\}\,\mathbf{1}_{(-\infty,\infty)}(x)$$

is sometimes used as an approximation to the Landau Distribution. Obtain the Characteristic Function $\Phi(t)=\pi^{-1/2}2^{-it}\Gamma(1/2-it)$ and show that $E[x]=\gamma_E+\ln 2$ and $V[X]=\pi^2/2$.

1.5.2 The Mellin Transform

Let $f:\mathcal{R}^+\rightarrow\mathcal{C}$ be a complex and integrable function with support on the real positive axis. The *Mellin Transform* is defined as:

$$M(f;s)=M_f(s)=\int_0^\infty f(x)\,x^{s-1}\,dx$$

with $s\in\mathcal{C}$, provided that the integral exists. In general, we shall be interested in continuous probability density functions $f(x)$ such that

$$\lim_{x\rightarrow 0+}f(x)=O(x^{\alpha}) \qquad \text{and} \qquad \lim_{x\rightarrow\infty}f(x)=O(x^{\beta})$$

and therefore

$$\begin{aligned} |M(f;s)| &\leq \int_0^\infty |f(x)|x^{Re(s)-1}dx=\int_0^1 |f(x)|x^{Re(s)-1}dx+\int_1^\infty |f(x)|x^{Re(s)-1}dx\leq \\ &\leq C_1\int_0^1 x^{Re(s)-1+\alpha}\,dx+C_2\int_1^\infty x^{Re(s)-1+\beta}\,dx \end{aligned}$$

The first integral converges for $-\alpha<Re(s)$ and the second for $Re(s)<-\beta$ so the Mellin Transform exists and is holomorphic on the band $-\alpha<Re(s)<-\beta$, parallel to the imaginary axis $\Im(s)$ and determined by the conditions of convergence of the integral. We shall denote the holomorphy band (that can be a half of the complex plane or the whole complex plane) by $S_f=\langle-\alpha,-\beta\rangle$. Last, to simplify the notation when dealing with several random quantities, we shall write for $X_n\sim p_n(x)$ $M_n(s)$ of $M_X(s)$ instead of $M(p_n;s)$.

1.5.2.1 Inversion

For a given function $f(t)$ we have that

$$M(f;s) = \int_0^\infty f(t)\, t^{s-1}\, dt = \int_0^\infty f(t)\, e^{(s-1)\ln t}\, dt = \int_{-\infty}^\infty f(e^u)\, e^{su}\, du$$

assuming that the integral exists. Since $s \in \mathcal{C}$, we can write $s = x + iy$ so: the Mellin Transform of $f(t)$ is the Fourier Transform of $g(u) = f(e^u)e^{xu}$. Setting now $t = e^u$ we have that

$$f(t) = \frac{1}{2\pi}\int_{-\infty}^\infty M(f; s = x + iy)\, t^{-(x+iy)}\, dy = \frac{1}{2\pi i}\int_{\sigma - i\infty}^{\sigma + i\infty} M(f;s)\, t^{-s}\, ds$$

where, due to Chauchy's Theorem, σ lies anywhere within the holomorphy band. The uniqueness of the result holds with respect to this strip so, in fact, the Mellin Transform consists on the pair $M(s)$ together with the band $\langle a, b \rangle$.

Example 1.23 It is clear that to determine the function $f(x)$ from the transform $F(s)$ we have to specify the strip of analyticity for, otherwise, we do not know which poles should be included. Let's see as an example $f_1(x) = e^{-x}$. We have that

$$M_1(z) = \int_0^\infty e^{-x}\, x^{z-1}\, dx = \Gamma(z)$$

holomorphic in the band $\langle 0, \infty \rangle$ so, for the inverse transform, we shall include the poles $z = 0, -1, -2, \ldots$. For $f_2(x) = e^{-x} - 1$ we get $M_2(s) = \Gamma(s)$, the same function, but

$$\lim_{x\to 0+} f(x) \simeq O(x^1) \longrightarrow \alpha = 1 \qquad \text{and} \qquad \lim_{x\to\infty} f(x) \simeq O(x^0) \longrightarrow \beta = 0$$

Thus, the holomorphy strip is $\langle -1, 0 \rangle$ and for the inverse transform we shall include the poles $z = -1, -2, \ldots$. For $f_3(x) = e^{-x} - 1 + x$ we get $M_3(s) = \Gamma(s)$, again the same function, but

$$\lim_{x\to 0+} f(x) \simeq O(x^2) \longrightarrow \alpha = 2 \qquad \text{and} \qquad \lim_{x\to\infty} f(x) \simeq O(x^1) \longrightarrow \beta = 1$$

Thus, the holomorphy strip is $\langle -2, -1 \rangle$ and for the inverse transform we include the poles $z = -2, -3, \ldots$.

1.5.2.2 Useful Properties

Consider a positive random quantity X with continuous density $p(x)$ and $x \in [0, \infty)$, the Mellin Transform $M_X(s)$ (defined only for $x \geq 0$)

$$M(p;s) = \int_0^\infty x^{s-1}\, p(x)\, dx = E[X^{s-1}]$$

and the Inverse Transform

$$p(x) = \frac{1}{2\pi i}\int_{c-i\infty}^{c+i\infty} x^{-s}\, M(p;s)\, ds$$

defined for all x where $p(x)$ is continuous with the line of integration contained in the strip of analyticity of $M(p;s)$. Then:

• **Moments**: $E[X^n] = M_X(n+1)$;
• For the positive random quantity $Z = aX^b$ ($a, b \in \mathcal{R}$ and $a > 0$) we have that

$$M_Z(s) = \int_0^\infty z^{s-1}\, f(z)\, dz = \int_0^\infty a^{s-1}\, x^{b(s-1)}\, p(x)\, dx = a^{s-1}\, M_X(bs-b+1)$$
$$2\pi i\; p(z) = \int_{c-i\infty}^{c+i\infty} z^{-s}\, M_X(bs-b+1)\, ds$$

In particular, for $Z = 1/X$ ($a = 1$ and $b = -1$) we have that

$$M_{Z=1/X}(s) = M_X(2-s)$$

• If $Z = X_1 X_2 \cdots X_n$ with $\{X_i\}_{i=1}^n$ n independent positive defined random quantities, each distributed as $p_i(x_i)$, we have that

$$M_Z(s) = \int_0^\infty z^{s-1}\, p(z)\, dz = \prod_{i=1}^n \int_0^\infty x_i^{s-1}\, p_i(x_i)\, dx_i = \prod_{i=1}^n E[X_i^{s-1}] = \prod_{i=1}^n M_i(s)$$
$$2\pi i\; p(z) = \int_{c-i\infty}^{c+i\infty} z^{-s}\, M_1(s)\cdots M_n(s)\, ds$$

In particular, for $n = 2$, $X = X_1 X_2$ it is easy to check that

$$p(x) = \int_0^\infty p_1(w)\, p_2(x/w)\, dw/w = \frac{1}{2\pi i}\int_{c-i\infty}^{c+i\infty} x^{-s}\, M_1(s) M_2(s)\, ds$$

Obviously, the strip of holomorphy is $S_1 \cap S_2$.
• For $X = X_1/X_2$, with both X_1 and X_2 positive defined and independent, we have that

$$M_X(s) = M_1(s)\, M_2(2-s)$$

and therefore

$$p(x) = \int_0^\infty p_1(wx)\, p_2(w)\, w\, dw = \frac{1}{2\pi i}\int_{c-i\infty}^{c+i\infty} x^{-s}\, M_1(s)\, M_2(2-s)\, ds$$

• Consider the distribution function $F(x) = \int_0^x p(u)du$ of the random quantity X. Since $dF(x) = p(x)dx$ we have that

$$M(p(x); s) = \int_0^{\infty} x^{s-1}\, dF(x) = \left[x^{s-1}F(x)\right]_0^{\infty} - (s-1)\int_0^{\infty} x^{s-2}\, F(x)\, dx$$

and therefore, if $\lim_{x\to 0+}[x^{s-1}F(x)] = 0$ and $\lim_{x\to\infty}[x^{s-1}F(x)] = 0$ we have, shifting $s\to s-1$, that

$$M(F(x); s) = M(\int_0^{x} p(u)\, du; s) = -\frac{1}{s}\, M(p(x); s+1).$$

1.5.2.3 Some Useful Examples

• Ratio and product of two independent Exponential distributed random quantities

Consider $X_1 \sim Ex(x_1|a_1)$ and $X_2 \sim Ex(x_2|a_2)$. The Mellin transform of $X \sim Ex(x|a)$ is

$$M_X(s) = \int_0^{\infty} x^{s-1} p(x|a)\, dx = a\int_0^{\infty} x^{s-1} e^{-ax}\, dx = \frac{\Gamma(s)}{a^{s-1}}$$

and therefore, for $Z = 1/X$:

$$M_Z(s) = M_X(2-s) = \frac{\Gamma(2-s)}{a^{1-s}}$$

In consequence, we have that

• $X = X_1 X_2 \longrightarrow M_X(z) = M_1(z)M_2(z) = \dfrac{\Gamma(z)^2}{(a_1\, a_2)^{z-1}}$

$$p(x) = \frac{a_1\, a_2}{2\pi i}\int_{c-i\infty}^{c+i\infty} (a_1 a_2 x)^{-z}\, \Gamma(z)^2\, dz$$

The poles of the integrand are at $z_n = -n$ and the residuals[13] are

$$Res(f(z), z_n) = \frac{(a_1 a_2 x)^n}{(n!)^2}\, (2\psi(n+1) - \ln(a_1 a_2 x))$$

and therefore

$$p(x) = a_1\, a_2 \sum_{n=0}^{\infty} \frac{(a_1 a_2 x)^n}{(n!)^2}\, (2\psi(n+1) - \ln(a_1 a_2 x))$$

[13] In the following examples, $-\pi \le arg(z) < \pi$.

If we define $w = 2\sqrt{a_1 a_2 x}$

$$p(x) = 2\, a_1\, a_2\, K_0(2\sqrt{a_1 a_2 x})\mathbf{1}_{(0,\infty)}(x)$$

from the Neumann Series expansion the Modified Bessel Function $K_0(w)$.

• $Y = X_1 X_2^{-1} \longrightarrow M_Y(z) = M_1(z) M_2(2-z) = \left(\frac{a_2}{a_1}\right)^{z-1} \frac{\pi(1-z)}{\sin(z\pi)}$

$$p(x) = \frac{a_1\, a_2^{-1}}{2i} \int_{c-i\infty}^{c+i\infty} (a_1 a_2^{-1} x)^{-z}\, \frac{1-z}{\sin(z\pi)} dz$$

Considering again the poles of $M_Y(z)$ at $z_n = -n$ we get the residuals

$$Res(f(z), z_n) = (1+n)\,(-1)^n \left(\frac{a_1}{a_2}\right)^{n+1} x^n$$

and therefore:

$$p(x) = \frac{a_1}{a_2} \sum_{n=0}^{\infty} (1+n)\,(-1)^n \left(\frac{a_1\, x}{a_2}\right)^n = \frac{a_1\, a_2}{(a_2 + a_1 x)^2} \mathbf{1}_{(1,\infty)}(x)$$

To summarize, if $X_1 \sim Ex(x_1|a_1)$ and $X_2 \sim Ex(x_2|a_2)$ are independent random quantities:

$$X = X_1 X_2 \sim 2\, a_1\, a_2\, K_0(2\sqrt{a_1 a_2 x})\mathbf{1}_{(1,\infty)}(x)$$
$$Y = X_1/X_2 \sim \frac{a_1\, a_2}{(a_2 + a_1 x)^2} \mathbf{1}_{(0,\infty)}(x)$$

• **Ratio and product of two independent Gamma distributed random quantities**
Consider $Y \sim Ga(x|a, b)$. Then $X = aY \sim Ga(x|1, b)$ with Mellin Transform

$$M_X(s) = \frac{\Gamma(b+s-1)}{\Gamma(b)}$$

The, if $X_1 \sim Ga(x_1|1, b_1)$ and $X_2 \sim Ga(x_2|1, b_2)$; $b_1 \neq b_2$:

• $X = X_1 X_2^{-1} \longrightarrow M_X(z) = M_1(z) M_2(z) = \frac{\Gamma(b_1 - 1 + z)}{\Gamma(b_1)} \frac{\Gamma(b_2 + 1 - z)}{\Gamma(b_2)}$

$$2\pi i\, \Gamma(b_1)\, \Gamma(b_2) p(x) = \int_{c-i\infty}^{c+i\infty} x^{-z}\, \Gamma(b_1 - 1 + z)\, \Gamma(b_2 + 1 - z)\, dz$$

Closing the contour on the left of the line $Re(z) = c$ contained in the strip of holomorphy $\langle 0, \infty \rangle$ we have poles of order one at $b_i - 1 + z_n = -n$ with $n = 0, 1, 2, \ldots$, that is, at $z_n = 1 - b_i - n$. Expansion around $z = z_n + \epsilon$ gives the residuals

$$Res(f(z), z_n) = \frac{(-1)^n}{n!}\,\Gamma(b_1 + b_2 + n)\, x^{n+b_1-1}$$

and therefore the quantity $X = X_1/X_2$ is distributed as

$$p(x) = \frac{x^{b_1-1}}{\Gamma(b_1)\,\Gamma(b_2)} \sum_{n=0}^{\infty} \frac{(-1)^n}{n!}\,\Gamma(b_1 + b_2 + n)\, x^n = \frac{\Gamma(b_1 + b_2)}{\Gamma(b_1)\Gamma(b_2)}\,\frac{x^{b_1-1}}{(1+x)^{b_1+b_2}}\mathbf{1}_{(0,\infty)}(x)$$

• $X = X_1 X_2 \longrightarrow M_X(z) = M_1(z) M_2(z) = \frac{\Gamma(b_1 - 1 + z)}{\Gamma(b_1)}\frac{\Gamma(b_2 - 1 + z)}{\Gamma(b_2)}$
Without loss of generality, we may assume that $b_2 > b_1$ so the strip of holomorphy is $\langle 1 - b_1, \infty\rangle$. Then, with $c > 1 - b_1$ real

$$\Gamma(b_1)\,\Gamma(b_2) p(x) = \frac{1}{2\pi i}\int_{c-i\infty}^{c+i\infty} x^{-z}\,\Gamma(b_1 - 1 + z)\,\Gamma(b_2 - 1 + z)\, dz$$

Considering the definition of the Modified Bessel Functions

$$I_\nu(x) = \sum_{n=0}^{\infty} \frac{1}{n!\,\Gamma(1 + n + \nu)}\left(\frac{x}{2}\right)^{2n+\nu} \quad \text{and} \quad K_\nu(x) = \frac{\pi}{2}\,\frac{I_{-\nu}(x) - I_\nu(x)}{\sin(\nu\pi)}$$

we get that

$$p(x) = \frac{2}{\Gamma(b_1)\,\Gamma(b_2)}\, x^{(b_1+b_2)/2-1}\, K_\nu(2\sqrt{x})\mathbf{1}_{(0,\infty)}(x)$$

with $\nu = b_2 - b_1 > 0$.

To summarize, if $X_1 \sim Ga(x_1|a_1, b_1)$ and $X_2 \sim Ga(x_2|a_2, b_2)$ are two independent random quantities and $\nu = b_2 - b_1 > 0$ we have that

$$X = X_1 X_2 \sim \frac{2a_1^{b_1} a_2^{b_2}}{\Gamma(b_1)\,\Gamma(b_2)}\left(\frac{a_2}{a_1}\right)^{\nu/2} x^{(b_1+b_2)/2-1}\, K_\nu(2\sqrt{a_1 a_2 x})\mathbf{1}_{(0,\infty)}(x)$$

$$X = X_1/X_2 \sim \frac{\Gamma(b_1 + b_2)}{\Gamma(b_1)\Gamma(b_2)}\,\frac{a_1^{b_1} a_2^{b_2} x^{b_1-1}}{(a_2 + a_1 x)^{b_1+b_2}}\mathbf{1}_{(0,\infty)}(x)$$

• **Ratio and product of two independent Uniform distributed random quantities**

Consider $X \sim Un(x|0, 1)$. Then $M_X(z) = 1/z$ with with $S = \langle 0, \infty\rangle$. For $X = X_1 \cdots X_n$ we have

$$p(x) = \frac{1}{2\pi i}\int_{c-i\infty}^{c+i\infty} e^{-z\ln x}\, z^{-n}\, dz = \frac{(-\ln x)^{n-1}}{\Gamma(n)}\mathbf{1}_{(0,1]}(x)$$

being $z = 0$ the only pole or order n.

For $X = X_1/X_2$ one has to be careful when defining the contours. In principle,

$$M_X(s) = M_1(s)M_2(2-s) = \frac{1}{s}\frac{1}{2-s}$$

so the strip of holomorphy is $S = \langle 0, 2\rangle$ and there are two poles, at $s = 0$ and $s = 2$. If $\ln x < 0 \rightarrow x < 1$ we shall close the Bromwich the contour on the left enclosing the pole at $s = 0$ and if $\ln x > 0 \rightarrow x > 1$ we shall close the contour on the right enclosing the pole at $s = 2$ so the integrals converge. Then it is easy to get that

$$p(x) = \frac{1}{2}\left[\mathbf{1}_{(0,1]}(x) + x^{-2}\,\mathbf{1}_{(1,\infty)}(x)\right] = Un(x|0,1) + Pa(x|1,1)$$

Note that

$$E[X^n] = M_X(n+1) = \frac{1}{n+1}\frac{1}{1-n}$$

and therefore there are no moments for $n \geq 1$.

Example 1.24 Show that if $X_i \sim Be(x_i|a_i, b_i)$ with $a_i, b_i > 0$, then

$$M_i(s) = \frac{\Gamma(a_i+b_i)}{\Gamma(a_i)}\frac{\Gamma(s+a_i-1)}{\Gamma(s+a_i+b_i-1)}$$

with $S = \langle 1-a_i, \infty\rangle$ and therefore:

- $X = X_1X_2$

$$p(x) = N_p\,x^{a_1-1}\,(1-x)^{b_1+b_2-1}\,F(a_1-a_2+b_1, b_2, b_1+b_2, 1-x)\mathbf{1}_{(0,1)}(x)$$

with

$$N_p = \frac{\Gamma(a_1+b_1)\Gamma(a_2+b_2)}{\Gamma(a_1)\Gamma(a_2)\Gamma(b_1+b_2)}$$

- $X = X_1/X_2$

$$p(x) = N_1x^{-(a_2+1)}\,F(1-b_2, a_1+a_2, b_1+a_1+a_2, x^{-1})\mathbf{1}_{(1,\infty)}(x) + \\ +\; N_2\,x^{a_1-1}\,F(1-b_1, a_1+a_2, b_2+a_1+a_2, x)\mathbf{1}_{(0,1)}(x)$$

with

$$N_k = \frac{B(a_1+a_2, b_k)}{B(a_1+b_1)B(a_2+b_2)}$$

Example 1.25 Consider a random quantity

$$X \sim p(x|a,b) = \frac{2\,a^{(b+1)/2}}{\Gamma(b/2+1/2)}\,e^{-ax^2}\,x^b$$

with $a, b > 0$ and $x \in [0, \infty)$. Show that

$$M(s) \sim p(x|a,b) = \frac{\Gamma(b/2+s/2)}{\Gamma(b/2+1/2)}\,a^{-(s-1)/2}$$

with $S = \langle -b, \infty \rangle$ and, from this, derive that the probability density function of $X = X_1 X_2$, with $X_1 \sim p(x_1|a_1, b_1)$ and $X_2 \sim p(x_2|a_2, b_2)$ independent, is given by:

$$p(x) = \frac{4\sqrt{a_1 a_2}}{\Gamma(b_1/2+1/2)\,\Gamma(b_2/2+1/2)}\,(\sqrt{a_1 a_2}x)^{(b_1+b_2)/2}\,K_{|\nu|}\left(2\sqrt{a_1 a_2}x\right)$$

with $\nu = (b_2 - b_1)/2$ and for $X = X_1/X_2$ by

$$p(x) = \frac{2\,\Gamma(b+1)}{\Gamma(b_1/2+1/2)\,\Gamma(b_2/2+1/2)}\,a^{1/2}\,\frac{(a\,x^2)^{b_1/2}}{(1+a\,x^2)^{b+1}}$$

with $a = a_1/a_2$ and $b = (b_1 + b_2)/2$.

Problem 1.4 Show that if $X_{1,2} \sim Un(x|0,1)$, then for $X = X_1^{X_2}$ we have that $p(x) = -x^{-1}\text{Ei}(\ln x)$, with $\text{Ei}(z)$ the exponential integral, and $E[X^m] = m^{-1}\ln(1+m)$. Hint: Consider $Z = \log X = X_1 \log X_2 = -X_1 W_2$ and the Mellin Transform for the Uniform and Exponential densities.

1.5.2.4 Distributions with Support in $\mathcal{R}$

The Mellin Transform is defined for integrable functions with non-negative support. To deal with the more general case $X \sim p(x)$ with $\text{supp}\{X\} = \Omega_{x\geq 0} + \Omega_{x<0} \subseteq \mathcal{R}$ we have to

(1) Express the density as $p(x) = \underbrace{p(x)\,\mathbf{1}_{x\geq 0}(x)}_{p^+(x)} + \underbrace{p(x)\,\mathbf{1}_{x<0}(x)}_{p^-(x)}$;

(2) Define $Y_1 = X$ when $x \geq 0$ and $Y_2 = -X$ when $x < 0$ so $\text{supp}\{Y_2\}$ is positive and find $M_{Y_1}(s)$ and $M_{Y_2}(s)$;

(3) Get from the inverse transform the corresponding densities $p_1(z)$ for the quantity of interest $Z_1 = Z(Y_1, X_2, \ldots)$ with $M_{Y_1}(s)$ and $p_2(z)$ for $Z_2 = Z(Y_2, X_2, \ldots)$ with $M_{Y_2}(s)$ and at the end for $p_2(z)$ make the corresponding change for $X \to -X$.

This is usually quite messy and for most cases of interest it is far easier to find the distribution for the product and ratio of random quantities with a simple change of variables.

• **Ratio of Normal and χ^2 distributed random quantities** Let's study the random quantity $X = X_1(X_2/n)^{-1/2}$ where $X_1 \sim N(x_1|0, 1)$ with $\sup\{X_1\} = \mathcal{R}$ and $X_2 \sim \chi^2(x_2|n)$ with $\sup\{X_2\} = \mathcal{R}^+$. Then, for X_1 we have

$$p(x_1) = \underbrace{p(x_1)\,\mathbf{1}_{[0,\infty)}(x_1)}_{p^+(x_1)} + \underbrace{p(x_1)\,\mathbf{1}_{(-\infty,0)}(x_1)}_{p^-(x_1)}$$

and therefore for X

$$X \sim p(x) = p(x)\,\mathbf{1}_{[0,\infty)}(x) + p(x)\,\mathbf{1}_{(-\infty,0)}(x) = p^+(x) + p^-(x)$$

Since

$$M_2(s) = \frac{2^{s-1}\,\Gamma(n/2 + s - 1)}{\Gamma(n/2)}$$

we have for $Z = (X_2/n)^{-1/2}$ that

$$M_Z(s) = n^{(s-1)/2}\,M_2((3-s)/2) = \left(\frac{n}{2}\right)^{(s-1)/2}\frac{\Gamma((n+1-s)/2)}{\Gamma(n/2)}$$

for $0 < \Re(s) < n + 1$. For $X_1 \in [0, \infty)$ we have that

$$M_1^+(s) = \frac{2^{s/2}\Gamma(s/2)}{2\sqrt{2\pi}}; \qquad 0 < \Re(s)$$

and therefore

$$M_X^+(s) = M_1^+(s)\,M_Z(s) = \frac{n^{s/2}\,\Gamma(s/2)\,\Gamma((n+1-s)/2)}{2\sqrt{n\pi}}$$

with holomorphy stripe $0 < \Re(s) < n + 1$. There are poles at $s_m = -2m$ with $m = 0, 1, 2, \ldots$ on the negative real axis and $s_k = n + 1 + 2k$ with $k = 0, 1, 2, \ldots$ on the positive real axis. Closing the contour on the left we include only s_m so

$$\begin{aligned} p^+(x) &= \frac{1}{\sqrt{n\pi}\Gamma(n/2)}\sum_{m=0}^{\infty}\frac{(-1)^m}{\Gamma(m+1)}\left(\frac{x^2}{n}\right)^m\Gamma\left(m + \frac{n+1}{2}\right) = \\ &= \frac{\Gamma((n+1)/2)}{\sqrt{n\pi}\Gamma(n/2)}\left(1 + \frac{x^2}{n}\right)^{-(n+1)/2}\mathbf{1}_{[0,\infty)}(x) \end{aligned}$$

For $X_1 \in (-\infty, 0)$ we should in principle define $Y = -X_1$ with support in $(0, \infty)$, find $M_Y(s)$, obtain the density for $X' = Y/Z$ and then obtain the corresponding one for $X = -X'$. However, in this case it is clear by symmetry that $p^+(x) = p^-(x)$ and therefore

$$X \sim p(x) = \frac{\Gamma((n+1)/2)}{\sqrt{n\pi}\Gamma(n/2)} \left(1 + \frac{x^2}{n}\right)^{-(n+1)/2} \mathbf{1}_{(-\infty,\infty)}(x) = St(x|n)$$

• **Ratio and product of Normal distributed random quantities** Consider $X_1 \sim N(x_1|\mu_1, \sigma_1)$ and $X_2 \sim N(x_2|\mu_2, \sigma_2)$. The Mellin Transform is

$$M_Y(s) = \frac{e^{-\mu^2/4\sigma^2}}{\sqrt{2\pi}} \sigma^{s-1} \Gamma(s) D_{-s}(\mp\mu/\sigma)$$

with $D_a(x)$ the Whittaker Parabolic Cylinder Functions. The upper sign $(-)$ of the argument corresponds to $X \in [0, \infty)$ and the lower one $(+)$ to the quantity $Y = -X \in (0, \infty)$. Again, the problem is considerably simplified if $\mu_1 = \mu_2 = 0$ because

$$M_Y(z) = \frac{2^{z/2}}{2\sqrt{2\pi}} \sigma^{z-1} \Gamma(z/2)$$

with $S = \langle 0, \infty \rangle$ and, due to symmetry, all contributions are the same. Thus, summing over the poles at $z_n = -2n$ for $n = 0, 1, 2, \ldots$ we have that for $X = X_1 X_2$ and $a^{-1} = 4\sigma_1^2\sigma_2^2$:

$$p(x) = \frac{2\sqrt{a}}{\pi} \sum_{n=0}^{\infty} \frac{(\sqrt{a}|x|)^{2n}}{\Gamma(n+1)^2} \left(2\Psi(1+n) - \ln(\sqrt{a}|x|)\right) = \frac{2\sqrt{a}}{\pi} K_0(2\sqrt{a}|x|)$$

Dealing with the general case of $\mu_i \neq 0$ it is much more messy to get compact expressions and life is easier with a simple change of variables. Thus, for instance for $X = X_1/X_2$ we have that

$$p(x) = \frac{\sqrt{a_1 a_2}}{\pi} \int_{-\infty}^{\infty} e^{-\{a_1(xw - \mu_1)^2 + a_2(w - \mu_2)^2\}} |w| \, dw$$

where $a_i = 1/(2\sigma_i^2)$ and if we define:

$$w_0 = a_2 + a_1 x^2; \qquad w_1 = a_1 a_2 \, (x\mu_2 - \mu_1)^2 \qquad \text{and} \qquad w_2 = (a_1\mu_1 x + a_2\mu_2)/\sqrt{w_0}$$

one has:

$$p(x) = \frac{\sqrt{a_1 a_2}}{\pi} \frac{1}{w_0} e^{-w_1/w_0} \left(e^{-w_2^2} + \sqrt{\pi}\, w_2 \,\mathrm{erf}(w_2)\right) \mathbf{1}_{(-\infty,\infty)}(x).$$

1.6 Ordered Samples

Let $X \sim p(x|\boldsymbol{\theta})$ be a one-dimensional random quantity and the experiment $e(n)$ that consists on n independent observations and results in the exchangeable sequence $\{x_1, x_2, \ldots, x_n\}$ are equivalent to an observation of the n-dimensional random quantity $\boldsymbol{X} \sim p(\boldsymbol{x}|\boldsymbol{\theta})$ where

$$p(\boldsymbol{x}|\boldsymbol{\theta}) = p(x_1, x_2, \ldots, x_n|\boldsymbol{\theta}) = \prod_{i=1}^{n} p(x_i|\boldsymbol{\theta})$$

Consider now a monotonic non-decreasing ordering of the observations

$$\underbrace{x_1 \leq x_2 \leq \cdots x_{k-1}}_{k-1} \leq x_k \leq \underbrace{x_{k+1} \leq \cdots \leq x_{n-1} \leq x_n}_{n-k}$$

and the *Statistic of Order* k; that is, the random quantity $X_{(k)}$ associated with the kth observation $(1 \leq k \leq n)$ of the *ordered sample* such that there are $k-1$ observations smaller than x_k and $n-k$ above x_k. Since

$$P(X \leq x_k|\boldsymbol{\theta}) = \int_{-\infty}^{x_k} p(x|\boldsymbol{\theta})dx = F(x_k|\boldsymbol{\theta}) \quad \text{and} \quad P(X > x_k|\boldsymbol{\theta}) = 1 - F(x_k|\boldsymbol{\theta})$$

we have that

$$\begin{aligned} X_{(k)} \sim p(x_k|\boldsymbol{\theta}, n, k) &= C_{n,k}\, p(x_k|\boldsymbol{\theta}) \left[F(x_k|\boldsymbol{\theta})\right]^{k-1} \left[1 - F(x_k|\boldsymbol{\theta})\right]^{n-k} \\ &= C_{n,k}\, p(x_k|\boldsymbol{\theta}) \underbrace{\left[\int_{-\infty}^{x_k} p(x|\boldsymbol{\theta})\, dx\right]^{k-1}}_{[P(X \leq x_k)]^{k-1}} \underbrace{\left[\int_{x_k}^{\infty} p(x|\boldsymbol{\theta})\, dx\right]^{n-k}}_{[P(X > x_k)]^{n-k}} \end{aligned}$$

The normalization factor

$$C_{n,k} = k \binom{n}{k}$$

is given by combinatorial analysis although in general it is easier to get by normalization of the final density. With a similar reasoning we have that the density function of the two dimensional random quantity $X_{(ij)} = (X_i, X_j)$; $j > i$, associated to the observations x_i and x_j (*Statistic of Order* i, j; $i < j$)[14] will be:

[14]If the random quantities X_i are not identically distributed the idea is the same but one hast to deal with permutations and the expressions are more involved.

$$X_{(ij)} \sim p(x_i, x_j|\boldsymbol{\theta}, i, j, n) = C_{n,i,j} \underbrace{\left[\int_{-\infty}^{x_i} p(x|\boldsymbol{\theta})dx\right]^{i-1}}_{[P(X<x_i)]^{i-1}} p(x_i|\boldsymbol{\theta}) \underbrace{\left[\int_{x_i}^{x_j} p(x|\boldsymbol{\theta})dx\right]^{j-i-1}}_{[P(x_i<X\leq<x_j)]^{j-i-1}}$$

$$p(x_j|\boldsymbol{\theta}) \underbrace{\left[\int_{x_j}^{\infty} p(x|\boldsymbol{\theta})dx\right]^{n-j}}_{[P(x_j<X)]^{n-j}}$$

where $(x_i, x_j) \in (-\infty, x_j] \times (-\infty, \infty)$ or $(x_i, x_j) \in (-\infty, \infty) \times [x_i, \infty)$. Again by combinatorial analysis or integration we have that

$$C_{n,i,j} = \frac{n!}{(i-1)!\,(j-i-1)!\,(n-j)!}$$

The main *Order Statistics* we are usually interested in are

- **Maximum** $X_{(n)} = \max\{X_1, X_2, \ldots, X_n\}$:

$$p(x_n|\cdot) = n\, p(x_n|\boldsymbol{\theta}) \left[\int_{-\infty}^{x_n} p(x|\boldsymbol{\theta})\, dx\right]^{n-1}$$

- **Minimum** $X_{(1)} = \min\{X_1, X_2, \ldots, X_n\}$:

$$p(x_1|\cdot) = n\, p(x_1|\boldsymbol{\theta}) \left[\int_{x_1}^{\infty} p(x|\boldsymbol{\theta})\, dx\right]^{n-1}$$

- **Range** $R = X_{(n)} - X_{(1)}$

$$p(x_1, x_n|\cdot) = n(n-1)\, p(x_1|\boldsymbol{\theta})\, p(x_n|\boldsymbol{\theta}) \left[\int_{x_1}^{x_n} p(x|\boldsymbol{\theta})\, dx\right]^{n-2}$$

If $\text{supp}(X) = [a, b]$, then $R \in (0, b-a)$ and

$$p(r) = n(n-1) \left\{\int_a^{b-r} p(w+r)\, p(w)\, [F(w+r) - F(w)]^{n-2}\, dw\right\}$$

There is no explicit form unless we specify the Distribution Function $F(x|\theta)$.

- **Difference** $S = X_{(i+1)} - X_{(i)}$. If $\text{supp}(X) = [a, b]$, then $S \in (0, b-a)$ and

$$p(s) = \frac{\Gamma(n+1)}{\Gamma(i)\Gamma(n-i)} \left\{\int_a^{b-s} p(w+s)\, p(w)\, [F(w)]^{i-1}\, [1 - F(w+s)]^{n-i-1}\, dw\right\}$$

In the case of discrete random quantities, the idea is the same but a bit more messy because one has to watch for the discontinuities of the Distribution Function. Thus, for instance:

- **Maximum** $X_{(n)} = \max\{X_1, X_2, \ldots, X_n\}$:

  $X_{(n)} \le x$ iff **all** x_i are **less or equal** x and this happens with probability

$$P(x_n \le x) = [F(x)]^n$$

  $X_{(n)} < x$ iff **all** x_i are **less than** x and this happens with probability

$$P(x_n < x) = [F(x-1)]^n$$

  Therefore

$$P(x_n = x) = P(x_n \le x) - P(x_n < x) = [F(x)]^n - [F(x-1)]^n$$

- **Minimum** $X_{(1)} = \min\{X_1, X_2, \ldots, X_n\}$:

  $X_{(1)} \ge x$ iff **all** x_i are **grater or equal** x and this happens with probability

$$P(x_1 \ge x) = 1 - P(x_1 < x) = [1 - F(x-1)]^n$$

  $X_{(1)} > x$ iff **all** x_i are **greater than** x and this happens with probability

$$P(x_1 > x) = 1 - P(x_1 \le x) = [1 - F(x)]^n$$

  Therefore

$$P(x_1 = x) = P(x_1 \le x) - P(x_1 < x) = [1 - P(x_1 > x)] - [1 - P(x_1 \ge x)] = \\ = [1 - F(x-1)]^n - [1 - F(x)]^n$$

Example 1.26 Let $X \sim Un(x|a, b)$ and an iid sample of size n. Then, if $L = b - a$:

- Maximum: $p(x_n) = n\,\dfrac{(x_n - a)^{n-1}}{(b-a)^n}\mathbf{1}_{(a,b)}(x_n)$
- Minimum: $p(x_1) = n\,\dfrac{(b - x_1)^{n-1}}{(b-a)^n}\mathbf{1}_{(a,b)}(x_1)$
- Range: $R = X_{(n)} - X_{(1)}:\ p(r) = \frac{n(n-1)}{L}\left(\frac{r}{L}\right)^{n-2}\left(1 - \frac{r}{L}\right)\mathbf{1}_{(0,L)}(r)$
- Difference: $S = X_{(k+1)} - X_{(k)}:\ p(s) = \frac{n}{L}\left(1 - \frac{s}{L}\right)^{n-1}\mathbf{1}_{(0,L)}(s)$

Example 1.27 Let's look at the Uniform distribution in more detail. Consider a random quantity $X \sim Un(x|a, b)$, the experiment $e(n)$ that provides a sample of n independent events and the ordered sample

$$\mathbf{X}_n = \{x_1 \le x_2 \le \cdots \le x_k \le \cdots \le x_{k+p} \le \cdots \le x_{n-1} \le x_n\}$$

Then, for the ordered statistics X_k, X_{k+p} and X_{k+p+1} with $k, p \in \mathcal{N}$, $1 \le k \le n-1$ and $p \le n-k-1$ we have that

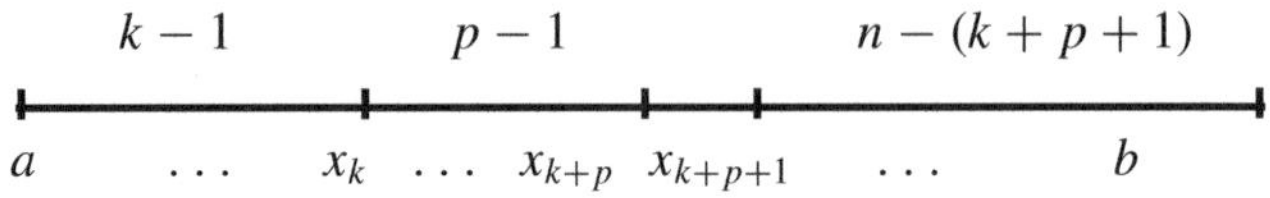

$$p(x_k, x_{k+p}|a, b, n, p) \propto \left[\int_a^{x_k} ds_1\right]^{k-1} \left[\int_{x_k}^{x_{k+p}} ds_2\right]^{p-1} \left[\int_{x_{k+p+1}}^{b} ds_3\right]^{n-(k+p+1)}$$

Let's think for instance that those are the arrival times of n events collected with a detector in a time window $[a = 0, b = T]$. If we define $w_1 = x_{k+p} - x_k$ and $w_2 = x_{k+p+1} - x_k$ we have that

$$p(x_k, w_1, w_2|T, n, p) = x_k^{k-1} w_1^{p-1} (T - x_k - w_2)^{n-k-p-1} \mathbf{1}_{[0,T-w_2]}(x_k) \mathbf{1}_{[0,w_2]}(w_1) \mathbf{1}_{[0,T]}(w_2)$$

and, after integration of x_k:

$$p(w_1, w_2|T, n, p) = \binom{n}{p} \frac{p(n-p)}{T^n}\, w_1^{p-1}\, (T - w_2)^{n-p-1} \mathbf{1}_{[0,w_2]}(w_1) \mathbf{1}_{[0,T]}(w_2)$$

Observe that the support can be expressed also as $\mathbf{1}_{[0,T]}(w_1)\mathbf{1}_{[w_1,T]}(w_2)$ and that the distribution of (W_1, W_2) does not depend on k. The marginal densities are given by:

$$p(w_1|T, n, p) = \binom{n}{p} \frac{p}{T^n}\, w_1^{p-1}\, (T - w_1)^{n-p}\; \mathbf{1}_{[0,T]}(w_1)$$

$$p(w_2|T, n, p) = \binom{n}{p} \frac{n-p}{T^n}\, w_2^{p}\, (T - w_2)^{n-p-1}\; \mathbf{1}_{[0,T]}(w_2)$$

and if we take the limit $T \to \infty$ and $n \to \infty$ keeping the rate $\lambda = n/T$ constant we have

$$\lim_{T,n\to\infty} p(w_1, w_2|T, n, p) = p(w_1, w_2|\lambda, p) = \frac{\lambda^{p+1}}{\Gamma(p)}\, e^{-\lambda w_2}\, w_1^{p-1}\; \mathbf{1}_{[0,w_2)}(w_1)\, \mathbf{1}_{[0,\infty)}(w_2)$$

and

$$p(w_1|\lambda, p) = \frac{\lambda^p}{\Gamma(p)}\, e^{-\lambda w_1}\, w_1^{p-1}\; \mathbf{1}_{[0,\infty)}(w_1)$$

In consequence, under the stated conditions the time difference between two consecutive events ($p = 1$) tends to an exponential distribution. Let's consider for simplicity this limiting behaviour in what follows and leave as an exercise the more involved case of finite time window T.

Suppose now that after having observed one event, say x_k, we have a dead-time of size a in the detector during which we can not process any data. All the events that fall in $(x_k, x_k + a)$ are lost (unless we play with buffers). If the next observed event is at time x_{k+p+1}, we have lost p events and the probability for this to happen is

$$\mathcal{P}(w_1 \leq a, w_2 \geq a|\lambda, p) = e^{-\lambda a} \frac{(\lambda a)^p}{\Gamma(p+1)}$$

that is, $N_{\text{lost}} \sim Po(p|\lambda a)$ regardless the position of the last recorded time (x_k) in the ordered sequence. As one could easily have intuited, the expected number of events lost for each observed one is $E[N_{\text{lost}}] = \lambda a$. Last, it is clear that the density for the time difference between two consecutive observed events when p are lost due to the dead-time is

$$p(w_2|w_1 \leq a, \lambda, p) = \lambda e^{-\lambda(w_2 - a)} \, \mathbf{1}_{[a,\infty)}(w_2)$$

Note that it depends on the dead-time window a and not on the number of events lost.

Example 1.28 Let $X \sim Ex(x|\lambda)$ and an iid sample of size n. Then:

- *Maximum*: $p(x_n) = n\,\lambda\,e^{-\lambda x_n}\,(1 - e^{-\lambda x_n})^{n-1}\mathbf{1}_{(0,\infty)}(x_n)$
- *Minimum*: $p(x_1) = n\,\lambda\,e^{-\lambda n x_1}\mathbf{1}_{(0,\infty)}(x_1)$
- *Range*: $R = X_{(n)} - X_{(1)} : p(r) = (n-1)\lambda^{n-1}e^{-\lambda r}\left[1 - e^{-\lambda r}\right]^{n-2}\mathbf{1}_{(0,\infty)}(r)$
- *Difference*: $S = X_{(k+1)} - X_{(k)} : p(s) = (n-k)\,\lambda e^{-\lambda x(n-k)}\mathbf{1}_{(0,\infty)}(s)$.

1.7 Limit Theorems and Convergence

In Probability, the Limit Theorems are statements that, under the conditions of applicability, describe the behavior of a sequence of random quantities or of Distribution Functions. In principle, whenever we can define a distance (or at least a positive defined set function) we can establish a convergence criteria and, obviously, some will be stronger than others so, for instance, a sequence of random quantities $\{X_i\}_{i=1}^{\infty}$ may converge according to one criteria and not to other. The most usual types of convergence, their relation and the Theorems derived from them are:

Distribution	$\Longrightarrow$ Central Limit Theorem
$\Uparrow$	$\Longrightarrow$ Glivenko–Cantelly Theorem (weak form)
$\Uparrow$	
Probability	$\Longrightarrow$ Weak Law of Large Numbers
$\Uparrow$ $\Uparrow$	
$\Uparrow$ Almost Sure	$\Longrightarrow$ Strong Law of Large Numbers
$\Uparrow$	
$L_p(\mathcal{R})$ Norm	$\Longrightarrow$ Convergence in Quadratic Mean
Uniform	$\Longrightarrow$ Glivenko–Cantelly Theorem

so *Convergence in Distribution* is the weakest of all since does not imply any of the others. In principle, there will be no explicit mention to statistical independence of the random quantities of the sequence nor to an specific Distribution Function. In most cases we shall just state the different criteria for convergence and refer to the literature, for instance [2], for further details and demonstrations. Let's start with the very useful Chebyshev's Theorem.

1.7.1 Chebyshev's Theorem

Let X be a random quantity that takes values in $\Omega\subset\mathcal{R}$ with Distribution Function $F(x)$ and consider the random quantity $Y = g(X)$ with $g(X)$ a non-negative single valued function for all $X \in \Omega$. Then, for $\alpha \in \mathcal{R}^+$

$$P\left(g(X) \geq \alpha\right) \;\leq\; \frac{E[g(X)]}{\alpha}$$

In fact, given a measure space $(\Omega, \mathcal{B}_\Omega, \mu)$, for any μ-integrable function $f(x)$ and $c > 0$ we have for $A = \{x : |f(x)| \geq c\}$ that $c\mathbf{1}_A(x) \leq |f(x)|$ for all x and therefore

$$c\mu(A) = \int c\mathbf{1}_A(x)d\mu \leq \int |f(x)|d\mu$$

Let's see two particular cases. First, consider $g(X) = (X-\mu)^{2n}$ where $\mu = E[X]$ and n a positive integer such that $g(X) \geq 0\,\forall X \in \Omega$. Applying Chebishev's Theorem:

$$P((X-\mu)^{2n} \geq \alpha) = P(|X-\mu| \geq \alpha^{1/2n}) \;\leq\; \frac{E[(X-\mu)^{2n}]}{\alpha} = \frac{\mu_{2n}}{\alpha}$$

For $n = 1$, if we take $\alpha = k^2\sigma^2$ we get the *Bienaymé–Chebishev*'s inequality

$$P(|X-\mu| \geq k\sigma) \;\leq\; 1/k^2$$

that is, whatever the Distribution Function of the random quantity X is, the probability that X differs from its expected value μ more than k times its standard deviation is less or equal than $1/k^2$. As a second case, assume X takes only positive real values and has a first order moment $E[X] = \mu$. Then (Markov's inequality):

$$P(X \geq \alpha) \leq \frac{\mu}{\alpha} \qquad \overset{\alpha=k\mu}{\longrightarrow} \qquad P(X \geq k\mu) \leq 1/k$$

The Markov and Bienaymé–Chebishev's inequalities provide upper bounds for the probability knowing just mean value and the variance although they are usually very conservative. They can be considerably improved if we have more information about the Distribution Function but, as we shall see, the main interest of Chebishev's inequality lies on its importance to prove Limit Theorems.

1.7.2 Convergence in Probability

The sequence of random quantities $\{X_n(w)\}_{n=1}^{\infty}$ *converges in probability* to $X(w)$ iff:

$$\lim_{n\to\infty} P(|X_n(w) - X(w)| \geq \epsilon) = 0; \quad \forall\epsilon > 0;$$

or, equivalently, iff:

$$\lim_{n\to\infty} P(|X_n(w) - X(w)| < \epsilon) = 1 \quad \forall\epsilon > 0;$$

Note that $P(|X_n(w) - X(w)| \geq \epsilon)$ is a a real number so this is is the usual limit for a sequence of real numbers and, in consequence, for all $\epsilon > 0$ and $\delta > 0$ $\exists\, n_0(\epsilon, \delta)$ such that for all $n > n_0(\epsilon, \delta)$ it holds that $P(|X_n(w) - X(w)| \geq \epsilon) < \delta$. For a sequence of n-dimensional random quantities, this can be generalized to $\lim_{n\to\infty} P(\|X_n(w), X(w)\|)$ and, as said earlier, Convergence in Probability implies Convergence in Distribution but the converse is not true. An important consequence of the Convergence in Probability is the

• **Weak Law of Large Numbers**: Consider a sequence of independent random quantities $\{X_i(w)\}_{i=1}^{\infty}$, all with the same Distribution Function and first order moment $E[X_i(w)] = \mu$, and define a new random quantity

$$Z_n(w) = \frac{1}{n}\sum_{i=i}^{n} X_i(w)$$

The, the sequence $\{Z_n(w)\}_{n=1}^{\infty}$ converges in probability to μ; that is:

$$\lim_{n\to\infty} P(|Z_n(w) - \mu| \geq \epsilon) = 0; \quad \forall\epsilon > 0;$$

The Law of Large Numbers was stated first by J. Bernoulli in 1713 for the Binomial Distribution, generalized (and named *Law of Large Numbers*) by S.D. Poisson and shown in the general case by A. Khinchin in 1929. In the case $X_i(w)$ have variance $V(X_i) = \sigma^2$ it is straight forward from Chebishev's inequality:

$$P\left(|Z_n - \mu| \geq \epsilon\right) = P\left((Z_n - \mu)^2 \geq \epsilon^2\right) \leq \frac{E[(Z_n - \mu)^2]}{\epsilon^2} = \frac{\sigma^2}{n\epsilon^2}$$

Intuitively, Convergence in Probability means that when n is very large, the probability that $Z_n(w)$ differs from μ by a small amount is very small; that is, $Z_n(w)$ gets more concentrated around μ. But "*very small*" is not zero and it may happen that for some $k > n$ Z_k differs from μ by more than ϵ. An stronger criteria of convergence is the *Almost Sure Convergence*.

1.7.3 Almost Sure Convergence

A sequence $\{X_n(w)\}_{n=1}^{\infty}$ of random quantities converges *almost sure* to $X(w)$ if, and only if:

$$\lim_{n\to\infty} X_n(w) = X(w)$$

for all $w \in \Omega$ except at most on a set $W \subset \Omega$ of zero measure ($P(W) = 0$ so it is also referred to as *convergence almost everywhere*). This means that for all $\epsilon > 0$ and all $w \in W^c = \Omega - W$, $\exists n_0(\epsilon, w) > 0$ such that $|X_n(w) - X(w)| < \epsilon$ for all $n > n_0(\epsilon, w)$. Thus, we have the equivalent forms:

$$P\left[\lim_{n\to\infty} |X_n(w) - X(w)| \geq \epsilon\right] = 0 \quad \text{or} \quad P\left[\lim_{n\to\infty} |X_n(w) - X(w)| < \epsilon\right] = 1$$

for all $\epsilon > 0$. Needed less to say that the random quantities $X_1, X_2 \ldots$ and X are defined on the same probability space. Again, Almost Sure Convergence implies Convergence in Probability but the converse is not true. An important consequence of the Almost Sure Convergence is the:

- **Strong Law of Large Numbers** (E. Borel 1909, A.N. Kolmogorov,...): Let $\{X_i(w)\}_{i=1}^{\infty}$ be a sequence of independent random quantities all with the same Distribution Function and first order moment $E[X_i(w)] = \mu$. Then the sequence $\{Z_n(w)\}_{n=1}^{\infty}$ with

$$Z_n(w) = \frac{1}{n}\sum_{i=i}^{n} X_i(w)$$

converges almost sure to μ; that is:

$$P\left[\lim_{n\to\infty}|Z_n(w)-\mu|\geq\epsilon\right]=0 \qquad \forall\epsilon>0$$

Intuitively, Almost Sure Convergence means that the probability that for some $k>n$, Z_k differs from μ by more than ϵ becomes smaller as n grows.

1.7.4 Convergence in Distribution

Consider the sequence of random quantities $\{X_n(\omega)\}_{n=1}^{\infty}$ and of their corresponding Distribution Functions $\{F_n(x)\}_{n=1}^{\infty}$. In the limit $n\to\infty$, the random quantity $X_n(w)$ tends to be distributed as $X(w)\sim F(x)$ iff

$$\lim_{n\to\infty}F_n(x)=F(x) \quad \Leftrightarrow \quad \lim_{n\to\infty}P(X_n\leq x)=P(X\leq x); \qquad \forall x\in C(F)$$

with $C(F)$ the set of points of continuity of $F(x)$. Expressed in a different manner, the sequence $\{X_n(w)\}_{n=1}^{\infty}$ Converges in Distribution to $X(w)$ if, and only if, for all $\epsilon>0$ and $x\in C(F)$, $\exists\, n_0(\epsilon,x)$ such that $|F_n(x)-F(x)|<\epsilon$, $\forall n>n_0(\epsilon,x)$. Note that, in general, n_0 depends on x so it is possible that, given an $\epsilon>0$, the value of n_0 for which the condition $|F_n(x)-F(x)|<\epsilon$ is satisfied for certain values of x may not be valid for others. It is important to note also that we have not made any statement about the statistical independence of the random quantities and that the Convergence in Distribution is determined only by the Distribution Functions so the corresponding random quantities do not have to be defined on the same probability space. To study the Convergence in Distribution, the following theorem it is very useful:

• **Theorem** (Lévy 1937; Cramèr 1937): Consider a sequence of Distribution Functions $\{F_n(x)\}_{n=1}^{\infty}$ and of the corresponding Characteristic Functions $\{\Phi_n(t)\}_{n=1}^{\infty}$. Then

▷ if $\lim_{n\to\infty}F_n(x)=F(x)$, then $\lim_{n\to\infty}\Phi_n(t)=\Phi(t)$ for all $t\in\mathcal{R}$ with $\Phi(t)$ the Characteristic Function of $F(x)$.
▷ Conversely, if $\Phi_n(t)\stackrel{n\to\infty}{\longrightarrow}\Phi(t)$ $\forall t\in\mathcal{R}$ and $\Phi(t)$ is continuous at $t=0$, then $F_n(x)\stackrel{n\to\infty}{\longrightarrow}F(x)$

This criteria of convergence is weak in the sense that if there is convergence if probability or almost sure or in quadratic mean then there is convergence in distribution but the converse is not necessarily true. However, there is a very important consequence of the Convergence in Distribution:

• **Central Limit Theorem** (Lindberg-Levy): Let $\{X_i(w)\}_{i=1}^{\infty}$ be a sequence of independent random quantities all with the same Distribution Function and with second order moments so $E[X_i(w)]=\mu$ and $V[X_i(w)]=\sigma^2$. Then the sequence $\{Z_n(w)\}_{n=1}^{\infty}$ of random quantities

$$Z_n(w) = \frac{1}{n} \sum_{i=i}^{n} X_i(w)$$

with

$$E[Z_n] = \frac{1}{n} \sum_{i=i}^{n} E[X_i] = \mu \quad \text{and} \quad V[Z_n] = \frac{1}{n^2} \sum_{i=i}^{n} V[X_i] = \frac{\sigma^2}{n}$$

tends, in the limit $n \rightarrow \infty$, to be distributed as $N(z|\mu, \sigma/\sqrt{n})$ or, what is the same, the *standardized* random quantity

$$\tilde{Z}_n = \frac{Z_n - \mu}{\sqrt{V[Z_n]}} = \frac{\frac{1}{n} \sum_{i=1}^{n} X_i - \mu}{\sigma/\sqrt{n}}$$

tends to be distributed as $N(x|0, 1)$.
Consider, without loss of generality, the random quantity $W_i = X_i - \mu$ so that $E[W_i] = E[X_i] - \mu = 0$ and $V[W_i] = V[X_i] = \sigma^2$. Then,

$$\Phi_W(t) = 1 - \frac{1}{2} t^2 \sigma^2 + \mathcal{O}(t^k)$$

Since we require that the random quantities X_i have at least moments of order two, the remaining terms $\mathcal{O}(t^k)$ are either zero or powers of t larger than 2. Then,

$$Z_n = \frac{1}{n} \sum_{i=i}^{n} X_i = \frac{1}{n} \sum_{i=i}^{n} W_i + \mu; \qquad E[Z_n] = \mu; \qquad V[Z_n] = \sigma^2_{Z_n} = \frac{\sigma^2}{n}$$

so

$$\Phi_{Z_n}(t) = e^{it\mu} \, [\Phi_W(t/n)]^n \longrightarrow \lim_{n \rightarrow \infty} \Phi_{Z_n}(t) = e^{it\mu} \lim_{n \rightarrow \infty} [\Phi_W(t/n)]^n$$

Now, since:

$$\Phi_W(t/n) = 1 - \frac{1}{2} \left(\frac{t}{n}\right)^2 \sigma^2 + \mathcal{O}(t^k/n^k) = 1 - \frac{1}{2} \frac{t^2}{n} \sigma^2_{Z_n} + \mathcal{O}(t^k/n^k)$$

we have that:

$$\lim_{n \rightarrow \infty} [\Phi_W(t/n)]^n = \lim_{n \rightarrow \infty} \left[1 - \frac{1}{2} \frac{t^2}{n} \sigma^2_{Z_n} + \mathcal{O}(t^k/n^k)\right]^n = \exp\left\{-\frac{1}{2} t^2 \sigma^2_{Z_n}\right\}$$

and therefore:

$$\lim_{n\to\infty} \Phi_{Z_n}(t) = e^{it\mu}\, e^{-\frac{1}{2}t^2\sigma^2/n}$$

so, $\lim_{n\to\infty} Z_n \sim N(x|\mu, \sigma/\sqrt{n})$.

The first indications about the Central Limit Theorem are due to A. De Moivre (1733). Later, C.F. Gauss and P.S. Laplace enunciated the behavior in a general way and, in 1901, A. Lyapunov gave the first rigorous demonstration under more restrictive conditions. The theorem in the form we have presented here is due to Lindeberg and Lévy and requires that the random quantities X_i are:

(i) Statistically Independent;
(ii) have the same Distribution Function;
(iii) First and Second order moments exist (i.e. they have mean value and variance).

In general, there is a set of Central Limit Theorems depending on which of the previous conditions are satisfied and justify the empirical fact that many natural phenomena are adequately described by the Normal Distribution. To quote E.T. Whittaker and G. Robinson (*Calculus of Observations*):

> "Everybody believes in the exponential law of errors;
> The experimenters because they think that it can be proved by mathematics;
> and the mathematicians because they believe it has been established by
> observation"

Example 1.29 From the limiting behavior of the Characteristic Function, show that:

- If $X \sim Bi(r|n, p)$, in the limit $p \to 0$ with np constant tends to a Poisson Distribution $Po(r|\mu = np)$;
- If $X \sim Bi(r|n, p)$, in the limit $n \to \infty$ the standardized random quantity

$$Z = \frac{X - \mu_X}{\sigma_X} = \frac{X - np}{\sqrt{npq}} \quad \overset{n\to\infty}{\sim} \quad N(x|0, 1)$$

- If $X \sim Po(r|\mu)$, then

$$Z = \frac{X - \mu_X}{\sigma_X} = \frac{X - \mu}{\sqrt{\mu}} \quad \overset{\mu\to\infty}{\sim} \quad N(x|0, 1)$$

- $X \sim \chi^2(x|n)$, then $n\to\infty$ the standardized random quantity

$$Z = \frac{X - \mu_X}{\sigma_X} = \frac{X - \nu}{\sqrt{2\nu}} \quad \overset{n\to\infty}{\sim} \quad N(x|0, 1)$$

- The Student's Distribution $St(x|0, 1, \nu)$ converges to $N(x|0, 1)$ in the limit $\nu\to\infty$;
- The Snedecor's Distribution $Sn(x|\nu_1, \nu_2)$ converges to $\chi^2(x|\nu_1)$ in the limit $\nu_2\to\infty$, to $St(x|0, 1, \nu_2)$ in the limit $\nu_1\to\infty$ and to $N(x|0, 1)$ in the limit $\nu_1, \nu_2\to\infty$.

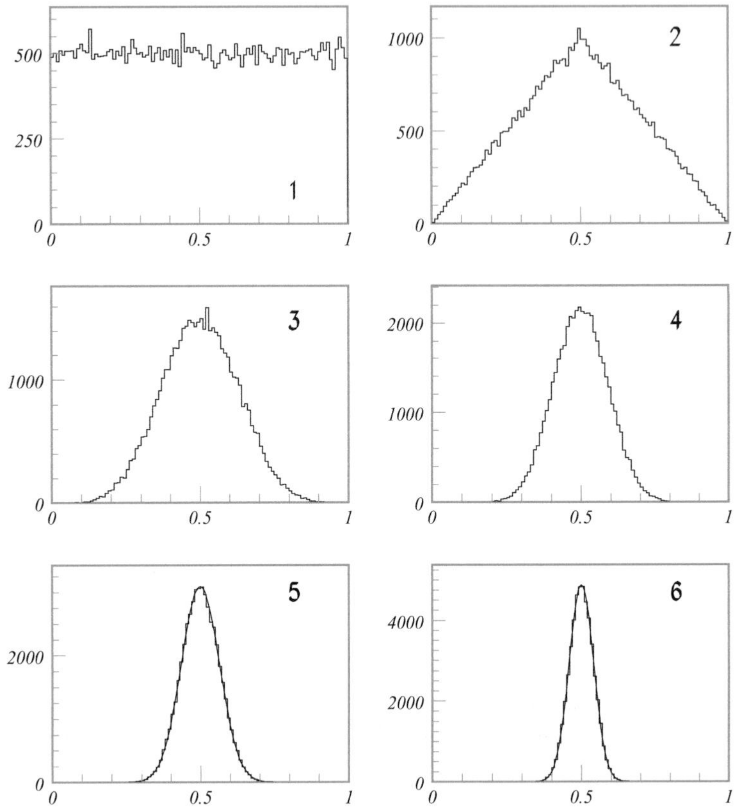

Fig. 1.2 Generated sample from $Un(x|0, 1)$ (**1**) and sampling distribution of the mean of 2 (**2**), 5 (**3**), 10 (**4**), 20 (**5**) y 50 (**6**) generated values

Example 1.30 It is interesting to see the Central Limit Theorem at work. For this, we have done a Monte Carlo sampling of the random quantity $X \sim Un(x|0, 1)$. The sampling distribution is shown in the Fig. 1.2(1) and the following ones show the sample mean of $n = 2$ (Fig. 1.2(2)), 5 (Fig. 1.2(3)), 10 (Fig. 1.2(4)), 20 (Fig. 1.2(5)) y 50 (Fig. 1.2(6)) consecutive values. Each histogram has 500000 events and, as you can see, as n grows the distribution "looks" more Normal. For $n = 20$ and $n = 50$ the Normal distribution is superimposed.

The same behavior is observed in Fig. 1.3 where we have generated a sequence of values from a parabolic distribution with minimum at $x = 1$ and support on $\Omega = [0, 2]$.

Last, Fig. 1.4 shows the results for a sampling from the Cauchy Distribution $X \sim Ca(x|0, 1)$. As you can see, the sampling averages follow a Cauchy Distribution regardless the value of n. For $n = 20$ and $n = 50$ a Cauchy and a Normal distributions have been superimposed. In this case, since the Cauchy Distribution has no moments the Central Limit Theorem does not apply.

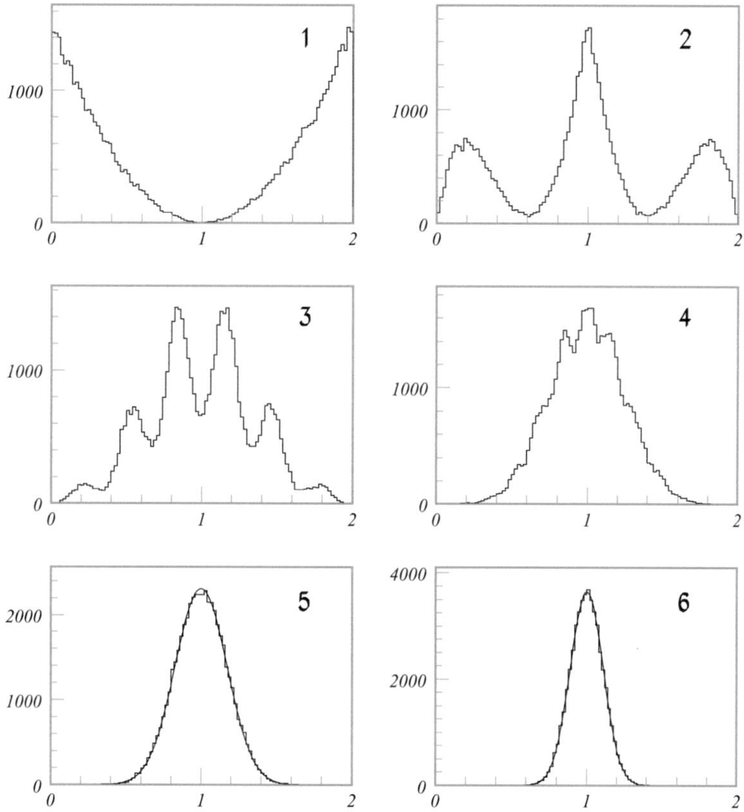

Fig. 1.3 Generated sample from a parabolic distribution with minimum at $x = 1$ and support on $\Omega = [0, 2]$ (**1**) and sampling distribution of the mean of 2 (**2**), 5 (**3**), 10 (**4**), 20 (**5**) y 50 (**6**) generated values

Example 1.31 Let $\{X_i(w)\}_{i=1}^{\infty}$ be a sequence of independent random quantities all with the same Distribution Function, mean value μ and variance σ^2 and consider the random quantity

$$Z(w) = \frac{1}{n} \sum_{i=i}^{n} X_i(w)$$

What is the value of n such that the probability that Z differs from μ more than ϵ is less than $\delta = 0.01$?

From the Central Limit Theorem we know that in the limit $n \rightarrow \infty$, $Z \sim N(x|\mu, \sigma/\sqrt{n})$ so we may consider that, for large n:

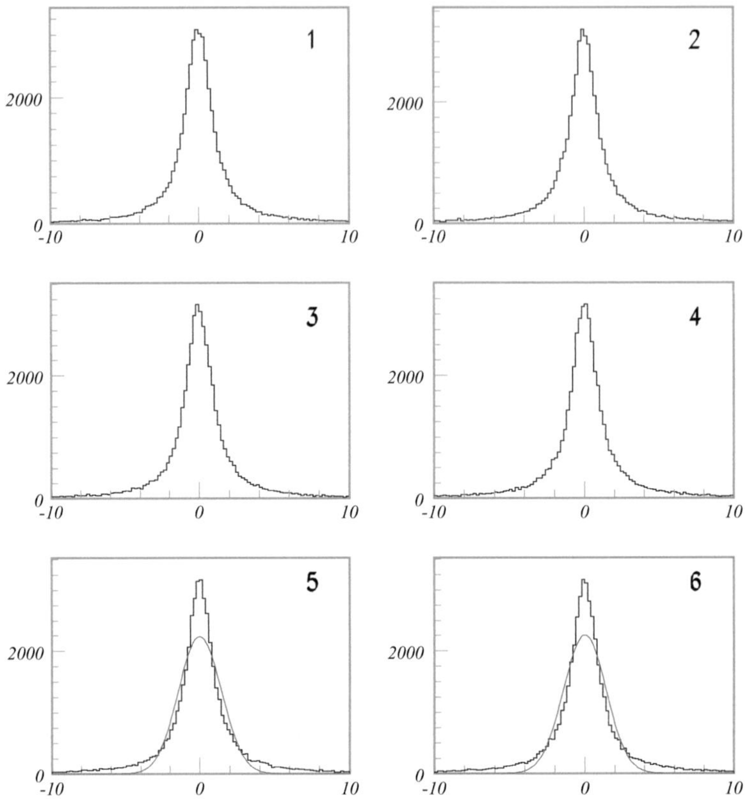

Fig. 1.4 Generated sample from a Cauchy distribution $Ca(x|0, 1)$ (**1**) and sampling distribution of the mean of 2 (**2**), 5 (**3**), 10 (**4**), 20 (**5**) y 50 (**6**) generated values

$$P(|Z-\mu| \geq \epsilon) = P(\mu-\epsilon \geq Z \geq \mu+\epsilon) \simeq$$
$$\simeq \int_{-\infty}^{\mu-\epsilon} N(x|\mu,\sigma)\,dx + \int_{\mu+\epsilon}^{+\infty} N(x|\mu,\sigma)\,dx = 1 - \text{erf}\left[\frac{\sqrt{n}\epsilon}{\sigma\sqrt{2}}\right] < \delta$$

For $\delta = 0.01$ we have that

$$\frac{\sqrt{n}\epsilon}{\sigma} \geq 2.575 \longrightarrow n \geq \frac{6.63\,\sigma^2}{\epsilon^2}.$$

1.7.5 *Convergence in L_p Norm*

A sequence of random quantities $\{X_n(w)\}_{n=1}^{\infty}$ converges to $X(w)$ in $L_p(\mathcal{R})$ $(p \geq 1)$ norm iff,

$$X(w) \in L_p(\mathcal{R}), \quad X_n(w) \in L_p(\mathcal{R}) \; \forall n \quad \text{and} \quad \lim_{n\to\infty} E[|X_n(w) - X(w)|^p] = 0$$

that is, iff for any real $\epsilon > 0$ there exists a natural $n_0(\epsilon) > 0$ such that for all $n \geq n_0(\epsilon)$ it holds that $E[|X_n(w) - X(w)|^p] < \epsilon$. In the particular case that $p = 2$ it is called *Convergence in Quadratic Mean.*

From Chebyshev's Theorem

$$P(|X_n(w) - X(w)| \geq \alpha^{1/p}) \leq \frac{E[(X_n(w) - X(w))^p]}{\alpha}$$

so, taking $\alpha = \epsilon^p$, if there is convergence in $L_p(\mathcal{R})$ norm:

$$\lim_{n\to\infty} P(|X_n(w) - X(w)| \geq \epsilon) \leq \lim_{n\to\infty} \frac{E[(X_n(w) - X(w))^p]}{\epsilon^p} = 0 \quad \forall \epsilon > 0$$

and, in consequence, we have convergence in probability.

1.7.6 Uniform Convergence

In some cases, point-wise convergence of Distribution Functions is not strong enough to guarantee the desired behavior and we require a stronger type of convergence. To some extent one may think that, more than a criteria of convergence, Uniform Convergence refers to the way in which it is achieved. Point-wise convergence requires the existence of an n_0 that may depend on ϵ and on x so that the condition $|f_n(x) - f(x)| < \epsilon$ for $n \geq n_0$ may be satisfied for some values of x and not for others, for which a different value of n_0 is needed. The idea behind uniform convergence is that we can find a value of n_0 for which the condition is satisfied regardless the value of x. Thus, we say that a sequence $\{f_n(x)\}_{n=1}^{\infty}$ converges uniformly to $f(x)$ iff:

$$\forall \epsilon > 0 \,, \; \exists n_0 \in \mathcal{N} \;\; \text{such that} \;\; |f_n(x) - f(x)| < \epsilon \qquad \forall n > n_0 \;\; \text{and} \;\; \forall x$$

or, in other words, iff:

$$\sup_x \; |f_n(x) - f(x)| \;\; \stackrel{n\to\infty}{\longrightarrow} \;\; 0$$

Thus, it is a stronger type of convergence that implies point-wise convergence. Intuitively, one may visualize the uniform convergence of $f_n(x)$ to $f(x)$ if one can draw a band $f(x) \pm \epsilon$ that contains all $f_n(x)$ for any n sufficiently large. Look for instance at the sequence of functions $f_n(x) = x(1 + 1/n)$ with $n = 1, 2, \ldots$ and $x \in \mathcal{R}$. It is clear that converges point-wise to $f(x) = x$ because $\lim_{n\to\infty} f_n(x) = f(x)$ for all $x \in \mathcal{R}$; that is, if we take $n_0(x, \epsilon) = x/\epsilon$, for all $n > n_0(x, \epsilon)$ it is true that

$|f_n(x) - f(x)| < \epsilon$ but for larger values of x we need larger values of n. Thus, the the convergence is not uniform because

$$\sup_x |f_n(x) - f(x)| = \sup_x |x/n| = \infty \quad \forall n \in \mathcal{N}$$

Intuitively, for whatever small a given ϵ is, the band $f(x) \pm \epsilon = x \pm \epsilon$ does not contain $f_n(x)$ for all n sufficiently large. As a second example, take $f_n(x) = x^n$ with $x \in (0, 1)$. We have that $\lim_{n\to\infty} f_n(x) = 0$ but $\sup_x |g_n(x)| = 1$ so the convergence is not uniform. For the cases we shall be interested in, if a Distribution Function $F(x)$ is continuous and the sequence of $\{F_n(x)\}_{n=1}^{\infty}$ *converges in distribution* to $F(x)$ (i.e. point-wise) then it does *uniformly* too. An important case of uniform convergence is the (sometimes called *Fundamental Theorem of Statistics*):

• **Glivenko–Cantelli Theorem** (V. Glivenko-F.P. Cantelli; 1933): Consider the random quantity $X \sim F(x)$ and a statistically independent (essential point) sampling of size n $\{x_1, x_2, \ldots, x_n\}$. The *empirical Distribution Function*

$$F_n(x) = \frac{1}{n} \sum_{i=1}^{n} \mathbf{1}_{(-\infty,x]}(x_i)$$

converges uniformly to $F(x)$; that is (Kolmogorov–Smirnov Statistic):

$$\lim_{n\to\infty} \sup_x |F_n(x) - F(x)| = 0$$

Let's see the convergence in probability, in quadratic mean and, in consequence, in distribution. For a fixed value $x = x_0$, $Y = \mathbf{1}_{(-\infty,x_0]}(X)$ is a random quantity that follows a Bernoulli distribution with probability

$$\begin{aligned} p = P(Y = 1) = P(\mathbf{1}_{(-\infty,x_0]}(x) = 1) &= P(X \leq x_0) = F(x_0) \\ P(Y = 0) = P(\mathbf{1}_{(-\infty,x_0]}(x) = 0) &= P(X > x_0) = 1 - F(x_0) \end{aligned}$$

and Characteristic Function

$$\Phi_Y(t) = E[e^{itY}] = e^{it}\, p + (1 - p) = e^{it}\, F(x_0) + (1 - F(x_0))$$

Then, for a fixed value of x we have for the random quantity

$$Z_n(x) = \sum_{i=1}^{n} \mathbf{1}_{(-\infty,x]}(x_i) = n\, F_n(x) \longrightarrow \Phi_{Z_n}(t) = \left(e^{it}\, F(x) + (1 - F(x))\right)^n$$

and therefore $Z_n(x) \sim Bi(k|n, F(x))$ so, if $W = n\, F_n(x)$, then

$$P(W = k|n, F(x)) = \binom{n}{k} F(x)^k\, (1 - F(x))^{n-k}$$

with

$$E[W] = n\,F(x) \quad \longrightarrow \quad E[F_n(x)] = F(x)$$
$$V[W] = n\,F(x)\,(1 - F(x)) \quad \longrightarrow \quad V[F_n(x)] = \frac{1}{n}\,F(x)\,(1 - F(x))$$

From Chebishev's Theorem

$$P\left(|F_n(x) - F(x)| \geq \epsilon\right) \leq \frac{1}{n\,\epsilon^2}\,F(x)\,(1 - F(x))$$

and therefore

$$\lim_{n\to\infty} P[|F_n(x) - F(x)| \geq \epsilon] = 0; \quad \forall \epsilon > 0$$

so the empirical Distribution Function $F_n(x)$ converges in probability to $F(x)$. In fact, since

$$\lim_{n\to\infty} E[|F_n(x) - F(x)|^2] = \lim_{n\to\infty} \frac{F(x)\,(1 - F(x))}{n} = 0$$

converges also in quadratic mean and therefore in distribution.

Example 1.32 Let $X = X_1/X_2$ with $X_i \sim Un(x|0, 1)$; $i = 1, 2$ and Distribution Function

$$F(x) = \frac{x}{2}\mathbf{1}_{(0,1]}(x) + \left(1 - \frac{x}{2}\right)\mathbf{1}_{(1,\infty)}(x)$$

that you can get (exercise) from the Mellin Transform. This is depicted in black in Fig. 1.5. There are no moments for this distribution; that is $E[X^n]$ does not exist for $n \geq 1$. We have done Monte Carlo samplings of size $n = 10$, 50 and 100 and the corresponding empirical Distribution Functions

$$F_n(x) = \frac{1}{n}\sum_{i=1}^{n} \mathbf{1}_{(-\infty,x]}(x_i)$$

are shown in blue, read and green respectively.

NOTE 3: Divergence of measures.
Consider the measurable space $(\Omega, \mathcal{B}_\Omega)$ and the probability measures λ, $\mu << \lambda$ and $\nu << \lambda$. The Kullback's divergence between μ and ν is defined as (see Chap. 4)

$$K(\mu, \nu) = \int_\Omega \frac{d\mu}{d\lambda} \log\left(\frac{d\mu/d\lambda}{d\nu/d\lambda}\right) d\lambda$$

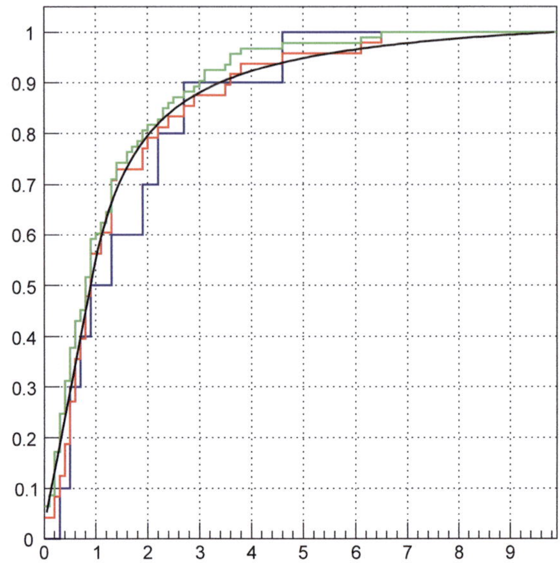

Fig. 1.5 Empirical distribution function of Example 1.32 for sample sizes 10 (*blue*), 50 (*green*) and 100 (*red*) together with the distribution function (*black*)

and the Hellinger distance as

$$d_H^2(\mu,\nu) = \frac{1}{2}\int_\Omega |\sqrt{d\mu/d\lambda} - \sqrt{d\nu/d\lambda}|^2 d\lambda$$

For λ Lebesgue measure, we can write

$$K(p,q) = \int_\Omega p(x)\log\left(\frac{p(x)}{q(x)}\right)dx \qquad \text{and} \qquad d_H^2(p,q) = 1 - \int_\Omega \sqrt{p(x)q(x)}dx$$

The Kullback's divergence will be relevant for Chap. 2. It is left as an exercise to show that the Normal density that best approximates in Kullback's sense a given density $p(x)$ is that with the same mean and variance (assuming they exist) and, using the Calculus of Variations, that

$$p(x|\boldsymbol{\lambda}) = f(x)\exp\left\{\sum_{i=0}^{k}\lambda_i\, h_i(x)\right\}$$

is the form (exponential family) that satisfies $k+1$ constraints $\int h_i(x)p(x)dx = c_i < \infty$; $i = 0, \ldots, k$ with specified constants $\{c_j\}_{j=0}^{k}$ ($c_0 = 1$ for $h_0(x) = 1$) and best approximates a given density $q(x)$. The Hellinger distance is a metric on the set of all probability measures on $\mathcal{B}_\Omega$ and we shall make use of it, for instance, as a further check for convergence in Markov Chain Monte Carlo sampling.

Appendices

Appendix 1: Indicator Function

This is one of the most useful functions in maths. Given subset $A \subset \Omega$ we define the *Indicator Function* $\mathbf{1}_A(x)$ for all elements $x \in \Omega$ as:

$$\mathbf{1}_A(x) = \begin{cases} 1 & \text{if } x \in A \\ 0 & \text{if } x \notin A \end{cases}$$

Given two sets $A, B \subset \Omega$, the following relations are obvious:

$$\begin{aligned} \mathbf{1}_{A\cap B}(x) &= \min\{\mathbf{1}_A(x), \mathbf{1}_B(x)\} = \mathbf{1}_A(x)\,\mathbf{1}_B(x) \\ \mathbf{1}_{A\cup B}(x) &= \max\{\mathbf{1}_A(x), \mathbf{1}_B(x)\} = \mathbf{1}_A(x) + \mathbf{1}_B(x) - \mathbf{1}_A(x)\,\mathbf{1}_B(x) \\ \mathbf{1}_{A^c}(x) &= 1 - \mathbf{1}_A(x) \end{aligned}$$

It is also called *"Characteristic Function"* but in Probability Theory we reserve this name for the Fourier Transform.

Appendix 2: Lebesgue Integral and Lebesgue Measure

The Lebesgue integral extends the Riemann theory of integration and is the natural integral in the Theory of Probability. There are thousands of good references on the subject; Chap. 2, Vol. 1 of *"Measure Theory"* by Bogachev [1] is a recommended reading but let's have a short and rather informal introduction for those who did not yet look into this.

Even though we use the Lebesgue integral in Probability ... Can we survive with a rough idea of what it is without entering in the mathematical formalism? Yes. The reason is that for the problems we have to deal with in experimental Particle Physics we have either probability density functions that are Riemann integrable (and, when it exists, coincides with that of Lebesgue) or we actually do Lebesgue integrals "unconsciously". Suppose for instance that we have a probability space $(\Omega, \mathcal{B}, \mu)$ a partition $\Omega = A_1 \cup A_2$, the algebra $\mathcal{B} = \{\emptyset, \Omega, A_1, A_2\}$ and a probability measure such that $\mu(A_1) = 1/3$ and $\mu(A_2) = 2/3$. What is the expected value of the random quantity $X(A_k) = k$ (that it is measurable with respect to $\mathcal{B}$)? We have that

$$E[X] = \sum_{k=1}^{2} X(A_k)\,\mu(A_k) = \sum_{k=1}^{2} k\,(k/3) = 5/3$$

Essentially, this is a Lebesgue integral so like Moliere's Bourgeois Gentleman, we have been speaking prose and didn't even know it! Let's look at another example following the original idea of Lebesgue [3]. In fact, there are more axiomatic ways to define the Lebesgue integral but this the most intuitive of all. Suppose we want to evaluate

$$\int_1^2 \ln x \, dx$$

In principle, for a real valued function $f(x)$ defined on $[a, b]$ the basic approach to evaluate $\int_a^b f dx$ is that of Riemann and goes as follows. Consider a partition of the interval $[a, b] = \cup_{k=0}^{n-1} \Delta_k$ with

$$\Delta_k = [x_k, x_{k+1}) \text{ for } k = 0, \dots n-2; \qquad \Delta_{n-1} = [x_{n-1}, x_n]; \qquad x_0 = a, \qquad x_n = b$$

and the sequence $\{x'_k \in \Delta_k\}_{k=0}^{n-1}$ of interior points. Note that the length of each subinterval Δ_k is $(x_{k+1} - x_k)$. Now, define the Riemann sum

$$S_n = \sum_{k=0}^{n-1} f(x'_k)\,(x_{k+1} - x_k)$$

and the limit of S_n as the partition gets finer and finer in such a way that $\max(x_{k+1} - x_k) \to 0$. If the limit exists, we say that the function $f(x)$ is Riemann integrable and the limit is the (Riemann) integral. Therefore, for the posed problem:

(1) Take a partition of the domain $[1, 2]$ where $x_k = 1 + k\epsilon$ with $x_0 = 1$ and $x_n = 2 \to \epsilon = 1/n$;
(2) For each subinterval Δ_k, of length ϵ, take $x'_k = x_k$ so $f(x'_k) = \ln x_k = \ln(1+k\epsilon)$;
(3) Evaluate the sum

$$S_n = \sum_{k=0}^{n-1} [\ln x_k]\,\epsilon = \frac{1}{n} \sum_{k=0}^{n-1} \ln(1 + k/n) = \frac{1}{n} \ln \prod_{k=0}^{n-1} (1 + k/n) = \frac{1}{n} \ln \left\{ \frac{\Gamma(2n)}{\Gamma(n)\, n^n} \right\}$$

and take the limit $\epsilon \to 0^+$ $(n \to \infty)$. You can check that $\lim_{n\to\infty} S_n = 2 \ln 2 - 1$.

Consider now a measure space $(\mathcal{R}, \mathcal{B}, \mu)$ and a non-negative, bounded and Borel measurable function $f(x)$. Lebesgue's definition of the integral rests on partitioning the range of $f(x)$ instead of the domain. Thus, we start with a partition of $[0, \sup f] = \cup_{k=0}^{n-1} \Delta_k$ where

$$\Delta_k = [y_k, y_{k+1}) \text{ for } k = 0, \dots n-2; \quad \Delta_{n-1} = [y_{n-1}, y_n]; \quad y_0 = 0, \quad y_n = \sup f$$

Being f Borel measurable, $\mu\{f^{-1}(\Delta_k)\}$ exists so we can evaluate the sum

$$S_n = \sum_{k=0}^{n-2} y_k \, \mu[f^{-1}(\Delta_k)] + y_{n-1} \, \mu[f^{-1}(\Delta_{n-1})]$$

Again, as the partition gets finer in such a way that $\max(y_{k+1} - y_k) \rightarrow 0$, the limit will be the Lebesgue integral provided it exists. For the problem at hand:

(1) Take a partition of the range $[0, \ln 2]$ where $y_k = k\epsilon$, $y_0 = 0$ and $y_n = \ln 2 \rightarrow \epsilon = n^{-1} \ln 2$;
(2) For each subinterval Δ_k determine the length of the corresponding interval on the support; that is, $\mu(\Delta_k) = f^{-1}(y_{k+1}) - f^{-1}(y_k) = e^{k\epsilon}(e^{\epsilon} - 1)$
(3) Evaluate the sum

$$S_n = \sum_{k=0}^{n-1} y_k\, \mu(\Delta_k) = \epsilon(e^{\epsilon} - 1) \sum_{k=0}^{n-1} k\, e^{k\epsilon} = 2\ln 2 + \epsilon e^{\epsilon}(1 - e^{\epsilon})^{-1}$$

and take the limit $\epsilon \rightarrow 0^+$ $(n \rightarrow \infty)$. As expected, $\lim_{\epsilon \rightarrow 0^+} S_n = 2\ln 2 - 1$.

Partitioning the range of the function and determining the *"length"* (measure) of each corresponding set on the domain allows to integrate functions defined over sets for which the Riemann integral does not exist. The typical example that you have almost certainly seen is the integral over $[0, 1]$ of the function $f(x) = \mathbf{1}_{\mathcal{Q}\cap[0,1]}(x)$. It is nowhere continuous and therefore is not Riemann integrable. but $\mathcal{Q}$ is countable so $\mu(\mathcal{Q}\cap[0, 1]) = 0$ and therefore

$$\int_{[0,1]} \mathbf{1}_{\mathcal{Q}\cap[0,1]}\, d\mu = \mu(\mathcal{Q}\cap[0, 1]) = 0$$

Nevertheless, the crucial difference with respect Riemann's integral is not the partition of the range but the possibility to perform integrals over *"wilder"* sets and, for us, the chance to consider arbitrary probability measures over arbitrary sets. But, for this, we have to clarify how to define the *measure* of a set. In general, we shall be concerned only with $\mathcal{R}^n$ and it turns out that there is a unique measure λ on R^n that is invariant under translations and such that for the unit cube $\lambda([0, 1]^n) = 1$: the **Lebesgue measure** that assigns to an interval $[a, b] \in \mathcal{R}$ what we intuitively would guess: $\lambda([a, b]) = (b - a)$. However, as explained in Sect. 1.1.2.2, if we want to satisfy these conditions there is a price to pay: not all subsets of $\mathcal{R}$ are measurable.

Let's finish with a more axiomatic introduction and some properties. Consider the measure space $(\Omega, \mathcal{B}, \mu)$; eventually a probability space with μ a probability measure. Then, for $S \subset \mathcal{B}$ we define

$$\mu(S) \stackrel{def}{=} \int_S d\mu = \int_\Omega \mathbf{1}_S\, d\mu$$

where $\mu(S)$ may be $+\infty$ (unless it is a finite measure). Now, given a finite partition $\{S_k;\, k = 1, \ldots, n\}$ of Ω and a simple function

$$S = \sum_{k=1}^{n} a_k\, \mathbf{1}_{S_k} \qquad \text{where} \qquad a_k \geq 0\ \forall k \quad \text{and} \quad \mu(S_k) < +\infty \ \text{ if } \ a_k \neq 0$$

it is natural to define for a measurable set $A \subset \Omega$:

$$\int_A S\, d\mu \stackrel{def}{=} \int_\Omega S\, \mathbf{1}_A\, d\mu = \sum_{k=1}^{n} a_k\, \mu(S_k \cap A)$$

Then:

(1) Let f be a non-negative measurable function with respect to $\mathcal{B}$ (that may take the value $+\infty$). We define:

$$\int_\Omega f\, d\mu \stackrel{def}{=} \sup \left(\int_\Omega S\, d\mu;\ 0 \leq S \leq f;\ S \text{ simple} \right)$$

that, obviously, it may be $+\infty$ in some cases.

(2) Let f be a measurable function that may take negative values and denote by

$$f^+ = f\, \mathbf{1}_{(f>0)} \quad \text{and} \quad f^- = -f\, \mathbf{1}_{(f<0)}$$

so that $f = f^+ - f^-$ and $|f| = f^+ + f^-$. Then, if

$$\int_\Omega f^+\, d\mu < +\infty \qquad \text{and} \qquad \int_\Omega f^-\, d\mu < +\infty$$

we have that

$$\int_\Omega |f|\, d\mu = \int_\Omega f^+\, d\mu + \int_\Omega f^-\, d\mu \ < \ +\infty$$

and we say that the function f is **Lebesgue integrable** with integral

$$\int_\Omega f\, d\mu = \int_\Omega f^+\, d\mu - \int_\Omega f^-\, d\mu$$

Observe that f defined on $\mathcal{R}$ is Lebesgue integrable iff it belongs to the Banach space $L_1(\mathcal{R})$.

Some of the main properties of the Lebesgue integral are:

(1) If $f(x)$ and $g(x)$ are two non-negative measurable functions such that $f = g$ almost everywhere; that is, $\mu(\{x \in \Omega | f(x) \neq g(x)\}) = 0$ then

$$\int f\, d\mu = \int g\, d\mu$$

The function $f(x)$ is integrable iff $g(x)$ is integrable and both integrals are the same;

(2) If $f(x)$ and $g(x)$ are two integrable functions then

- if $a, b \in \mathcal{R}$, it holds that $\int (a\, f + b\, g)\, d\mu = a \int f\, d\mu + b \int g\, d\mu$

- if $f \leq g$ it holds that $\int f\, d\mu \leq \int g\, d\mu$

(3) If $\{f_k(x)\}_{k\in\mathcal{N}}$ is a sequence of non-negative measurable functions such that $f_k(x) \leq f_{k+1}(x)$ for all $k \in \mathcal{N}$ and $x \in \Omega$, then

$$\lim_k \int f_k\, d\mu = \int \lim_k f_k\, d\mu$$

(The integrals can be infinite)

(4) If $\{f_k(x)\}_{k\in\mathcal{N}}$ is a sequence of functions that converge pointwise to $f(x)$ (i.e. $\lim_{k\to\infty} f_k(x) = f(x)$ for all x) and there exists an integrable function g such that $|f_k| \leq g$ for all k, then f is integrable and

$$\lim_{k\to\infty} \int f_k\, d\mu = \int f\, d\mu$$

Appendix 3: Some properties of Radon–Nikodym derivatives

Consider the σ-additive measures μ_1, μ_2 and μ_3 on the measurable space $(\Omega, \mathcal{B}_\Omega)$. Then

(1) If $\mu_1 << \mu_2$ and $\mu_2 << \mu_3$, then $\mu_1 << \mu_3$ and $\frac{d\mu_1}{d\mu_3} = \frac{d\mu_1}{d\mu_2}\frac{d\mu_2}{d\mu_3}$ so:

$$\mu_1(A) = \int_A d\mu_1 = \int_A \frac{d\mu_1}{d\mu_2}\, d\mu_2 = \int_A \frac{d\mu_1}{d\mu_2}\frac{d\mu_2}{d\mu_3}\, d\mu_3 = \int_A \frac{d\mu_1}{d\mu_3}\, d\mu_3$$

(2) If $\mu_1 << \mu_3$ and $\mu_2 << \mu_3$, then $\frac{d(\mu_1 + \mu_2)}{d\mu_3} = \frac{d\mu_1}{d\mu_3} + \frac{d\mu_2}{d\mu_3}$

(3) If $\mu_1 << \mu_2$ and g es a μ_1 integrable function, then

$$\int_A g\, d\mu_1 = \int_A g\, \frac{d\mu_1}{d\mu_2}\, d\mu_2$$

(4) If $\mu_1 << \mu_2$ and $\mu_2 << \mu_1$ (*equivalent*: $\mu_1 \sim \mu_2$) then $\frac{d\mu_1}{d\mu_2} = \left(\frac{d\mu_2}{d\mu_1}\right)^{-1}$

We shall use some of these properties in different places; for instance, relating mathematical expectations under different probability measures or justifying some techniques used in Monte Carlo Sampling.

References

1. V.I. Bogachev, *Measure Theory* (Springer, Berlin, 2006)
2. A. Gut, *Probability: A Graduate Course*, Springer Texts in Statistics (Springer, Berlin, 2013)
3. H.L. Lebesgue, *Sur le développment de la notion d'intégrale* (1926)

Chapter 2
Bayesian Inference

... some rule could be found, according to which we ought to estimate the chance that the probability for the happening of an event perfectly unknown, should lie between any two named degrees of probability, antecedently to any experiments made about it; ...
An Essay towards solving a Problem in the Doctrine of Chances
By the late Rev. Mr. Bayes...

The goal of *statistical inference* is to get information from experimental observations about quantities (parameters, models,...) on which we want to learn something, be them directly observable or not. Bayesian inference[1] is based on the *Bayes rule* and considers probability as a measure of the degree of knowledge we have on the quantities of interest. Bayesian methods provide a framework with enough freedom to analyze different models, as complex as needed, using in a natural and conceptually simple way all the information available from the experimental data within a scheme that allows to understand the different steps of the learning process:

(1) state the knowledge we have before we do the experiment;
(2) how the knowledge is modified after the data is taken;
(3) how to incorporate new experimental results.
(4) predict what shall we expect in a future experiment from the knowledge acquired.

It was Sir R.A Fisher, one of the greatest statisticians ever, who said that "The Theory of Inverse Probability (that is how Bayesianism was called at the beginning of the XX century) is founded upon an error and must be wholly rejected" although, as time went by, he became a little more acquiescent with Bayesianism. You will see that Bayesianism is great, rational, coherent, conceptually simple,... "even useful",... and worth to, at least, take a look at it and at the more detailed references on the subject

[1]For a gentle reading on the subject see [1].

C. Maña, *Probability and Statistics for Particle Physics*,
UNITEXT for Physics, DOI 10.1007/978-3-319-55738-0_2

given along the section. At the end, to quote Lindley, "Inside every non-Bayesian there is a Bayesian struggling to get out". For a more classical approach to Statistical Inference see [2] where most of what you will need in Experimental Physics is covered in detail.

2.1 Elements of Parametric Inference

Consider an experiment designed to provide information about the set of parameters $\boldsymbol{\theta} = \{\theta_1, \ldots, \theta_k\} \in \Theta \subseteq R^k$ and whose realization results in the random sample $\boldsymbol{x} = \{x_1, x_2, \ldots, x_n\}$. The inferential process entails:

(1) Specification of the probabilistic model for the random quantities of interest; that is, state the joint density:

$$p(\boldsymbol{\theta}, \boldsymbol{x}) = p(\theta_1, \theta_2, \ldots, \theta_k, x_1, x_2, \ldots, x_n); \qquad \boldsymbol{\theta} = \in \Theta \subseteq R^k; \qquad \boldsymbol{x} \in X$$

(2) Conditioning the observed data $(\boldsymbol{x})$ to the parameters $(\boldsymbol{\theta})$ of the model:

$$p(\boldsymbol{\theta}, \boldsymbol{x}) = p(\boldsymbol{x}|\boldsymbol{\theta})\, p(\boldsymbol{\theta})$$

(3) Last, since $p(\boldsymbol{\theta}, \boldsymbol{x}) = p(\boldsymbol{x}|\boldsymbol{\theta})\, p(\boldsymbol{\theta}) = p(\boldsymbol{\theta}|\boldsymbol{x})\, p(\boldsymbol{x})$ and

$$p(\boldsymbol{x}) = \int_{\Theta} p(\boldsymbol{x}, \boldsymbol{\theta})\, d\boldsymbol{\theta} = \int_{\Theta} p(\boldsymbol{x}|\boldsymbol{\theta})\, p(\boldsymbol{\theta})\, d\boldsymbol{\theta}$$

we have (*Bayes Rule*) that:

$$p(\boldsymbol{\theta}|\boldsymbol{x}) = \frac{p(\boldsymbol{x}|\boldsymbol{\theta})\, p(\boldsymbol{\theta})}{\int_{\Theta} p(\boldsymbol{x}|\boldsymbol{\theta})\, p(\boldsymbol{\theta}) d\boldsymbol{\theta}}$$

This is the basic equation for parametric inference. The integral of the denominator does not depend on the parameters $(\boldsymbol{\theta})$ of interest; is just a normalization factor so we can write in a general way;

$$p(\boldsymbol{\theta}|\boldsymbol{x}) \propto p(\boldsymbol{x}|\boldsymbol{\theta})\, p(\boldsymbol{\theta})$$

Let's see these elements in detail:

$p(\boldsymbol{\theta}|\boldsymbol{x})$: This is the *Posterior Distribution* that quantifies the knowledge we have on the parameters of interest $\boldsymbol{\theta}$ conditioned to the observed data $\boldsymbol{x}$ (that is, after the experiment has been done) and will allow to perform inferences about the parameters;

$p(\boldsymbol{x}|\boldsymbol{\theta})$: The *Likelihood*; the sampling distribution considered as a function of the parameters $\boldsymbol{\theta}$ for the *fixed* values (already observed) $\boldsymbol{x}$. Usually, it is written as $\ell(\boldsymbol{\theta}; \boldsymbol{x})$ to stress the fact that it is a function of the parameters. The experimental results modify the prior knowledge we have on the parameters $\boldsymbol{\theta}$ only through the likelihood so, for the inferential process, we can consider the likelihood function defined up to multiplicative factors provided they do not depend on the parameters.

$p(\boldsymbol{\theta})$: This is *a reference function*, independent of the results of the experiment, that quantifies or expresses, in a sense to be discussed later, the knowledge we have on the parameters $\boldsymbol{\theta}$ *before* the experiment is done. It is termed *Prior Density* although, in many cases, it is an improper function and therefore not a probability density.

2.2 Exchangeable Sequences

The inferential process to obtain information about a set of parameters $\boldsymbol{\theta} \in \Theta$ of a model $X \sim p(x|\boldsymbol{\theta})$ with $X \in \Omega_X$ is based on the realization of an experiment $e(1)$ that provides an observation $\{x_1\}$. The n−fold repetition of the experiment under the same conditions, $e(n)$, will provide the random sample $\boldsymbol{x} = \{x_1, x_2, \ldots, x_n\}$ and this can be considered as a draw of the n-dimensional random quantity $\boldsymbol{X} = (X_1, X_2, \ldots, X_n)$ where each $X_i \sim p(x|\boldsymbol{\theta})$.

In Classical Statistics, the inferential process makes extensive use of the idea that the observed sample is originated from a sequence of *independent and identically distributed* (iid) random quantities while Bayesian Inference rests on the less restrictive idea of *exchangeability* [3]. An infinite sequence of random quantities $\{X_i\}_{i=1}^{\infty}$ is said to be *exchangeable* if *any* finite sub-sequence $\{X_1, X_2, \ldots, X_n\}$ is *exchangeable*; that is, if the joint density $p(x_1, x_2, \ldots, x_n)$ is invariant under *any* permutation of the indices.

The hypothesis of *exchangeability* assumes a symmetry of the experimental observations $\{x_1, x_2, \ldots, x_n\}$ such that the subscripts which identify a particular observation (for instance the order in which they appear) are irrelevant for the inferences. Clearly, if $\{X_1, X_2, \ldots X_n\}$ are iid then the conditional joint density can be expressed as:

$$p(x_1, x_2, \ldots, x_n) = \prod_{i=1}^{n} p(x_i)$$

and therefore, since the product is invariant to reordering, is an *exchangeable* sequence. The converse is not necessarily true[2] so the hypothesis of exchangeability is weaker than the hypothesis of independence. Now, if $\{X_i\}_{i=1}^{\infty}$ is an exchangeable

[2]It is easy to check for instance that if X_0 is a non-trivial random quantity independent of the X_i, the sequence $\{X_0 + X_1, X_0 + X_2, \ldots X_0 + X_n\}$ is exchangeable but not iid.

sequence of real-valued random quantities it can be shown that, for *any* finite subset, there exists a parameter $\boldsymbol{\theta} \in \Theta$, a parametric model $p(x|\boldsymbol{\theta})$ and measure $d\mu(\boldsymbol{\theta})$ such that[3]:

$$p(x_1, x_2, \ldots, x_n) = \int_{\Theta} \prod_{i=1}^{n} p(x_i|\boldsymbol{\theta})\, d\mu(\boldsymbol{\theta})$$

Thus, any finite sequence of exchangeable observations is described by a model $p(x|\boldsymbol{\theta})$ and, if $d\mu(\boldsymbol{\theta}) = p(\boldsymbol{\theta})d\boldsymbol{\theta}$, there is a prior density $p(\boldsymbol{\theta})$ that we may consider as describing the available information on the parameter $\boldsymbol{\theta}$ before the experiment is done. This justifies and, in fact, leads to the Bayesian approach in which, by formally applying Bayes Theorem

$$p(\boldsymbol{x}, \boldsymbol{\theta}) = p(\boldsymbol{x}|\boldsymbol{\theta})\, p(\boldsymbol{\theta}) = p(\boldsymbol{\theta}|\boldsymbol{x})\, p(\boldsymbol{x})$$

we obtain the *posterior density* $p(\boldsymbol{\theta}|\boldsymbol{x})$ that accounts for the degree of knowledge we have on the parameter after the experiment has been performed. Note that the random quantities of the exchangeable sequence $\{X_1, X_2, \ldots, X_n\}$ are *conditionally independent given* $\boldsymbol{\theta}$ *but not iid* because

$$p(x_j) = \int_{\Theta} p(x_j|\boldsymbol{\theta})\, d\mu(\boldsymbol{\theta}) \left(\prod_{i(\neq j)=1}^{n} \int_{\Omega_X} p(x_i|\boldsymbol{\theta})\, dx_i \right)$$

and

$$p(x_1, x_2, \ldots, x_n) \neq \prod_{i=1}^{n} p(x_i)$$

There are situations for which the hypothesis of exchangeability can not be assumed to hold. That is the case, for instance, when the data collected by an experiment depends on the running conditions that may be different for different periods of time, for data provided by two different experiments with different acceptances, selection criteria, efficiencies,... or the same medical treatment when applied to individuals from different environments, sex, ethnic groups,... In these cases, we shall have different *units of observation* and it may be more sound to assume *partial exchangeability* within each unit (data taking periods, detectors, hospitals,...) and design a *hierarchical structure* with parameters that account for the relevant information from each unit analyzing all the data in a more global framework.

Note 4: Suppose that we have a parametric model $p_1(x|\boldsymbol{\theta})$ and the exchangeable sample $\boldsymbol{x}_1 = \{x_1, x_2, \ldots, x_n\}$ provided by the experiment $e_1(n)$. The inferences on

---

[3]This is referred as *De Finetti's Theorem* after B. de Finetti (1930s) and was generalized by E. Hewitt and L.J. Savage in the 1950s. See [4].

the parameters $\boldsymbol{\theta}$ will be drawn from the posterior density $p(\boldsymbol{\theta}|\boldsymbol{x}_1) \propto p_1(\boldsymbol{x}_1|\boldsymbol{\theta})p(\boldsymbol{\theta})$. Now, we do a second experiment $e_2(m)$, statistically independent of the first, that provides the exchangeable sample $\boldsymbol{x}_2 = \{x_{n+1}, x_{n+2}, \ldots, x_{n+m}\}$ from the model $p_2(x|\boldsymbol{\theta})$. It is sound to take as prior density for this second experiment the posterior of the first including therefore the information that we already have about $\boldsymbol{\theta}$ so

$$p(\boldsymbol{\theta}|\boldsymbol{x}_2) \propto p_2(\boldsymbol{x}_2|\boldsymbol{\theta})p(\boldsymbol{\theta}|\boldsymbol{x}_1) \propto p_2(\boldsymbol{x}_2|\boldsymbol{\theta})p_1(\boldsymbol{x}_1|\boldsymbol{\theta})p(\boldsymbol{\theta}).$$

Being the two experiments statistically independent and their sequences exchangeable, if they have the same sampling distribution $p(x|\boldsymbol{\theta})$ we have that $p_1(\boldsymbol{x}_1|\boldsymbol{\theta})p_2(\boldsymbol{x}_2|\boldsymbol{\theta}) = p(\boldsymbol{x}|\boldsymbol{\theta})$ where $\boldsymbol{x} = \{\boldsymbol{x}_1, \boldsymbol{x}_2\} = \{x_1, \ldots, x_n, x_{n+1}, \ldots, x_{n+m}\}$ and therefore $p(\boldsymbol{\theta}|\boldsymbol{x}_2) \propto p(\boldsymbol{x}|\boldsymbol{\theta})\, p(\boldsymbol{\theta})$. Thus, the knowledge we have on $\boldsymbol{\theta}$ including the information provided by the experiments $e_1(n)$ and $e_2(m)$ is determined by the likelihood function $p(\boldsymbol{x}|\boldsymbol{\theta})$ and, in consequence, under the aforementioned conditions the realization of $e_1(n)$ first and $e_2(m)$ after is equivalent, from the inferential point of view, to the realization of the experiment $e(n+m)$.

2.3 Predictive Inference

Consider the realization of the experiment $e_1(n)$ that provides the sample $\boldsymbol{x} = \{x_1, x_2, \ldots, x_n\}$ drawn from the model $p(x|\boldsymbol{\theta})$. Inferences about $\boldsymbol{\theta} \in \Theta$ are determined by the posterior density

$$p(\boldsymbol{\theta}|\boldsymbol{x}) \;\propto\; p(\boldsymbol{x}|\boldsymbol{\theta})\, \pi(\boldsymbol{\theta})$$

Now suppose that, under the same model and the same experimental conditions, we think about doing a new independent experiment $e_2(m)$. What will be the distribution of the random sample $\boldsymbol{y} = \{y_1, y_2, \ldots, y_m\}$ not yet observed? Consider the experiment $e(n+m)$ and the sampling density

$$p(\boldsymbol{\theta}, \boldsymbol{x}, \boldsymbol{y}) = p(\boldsymbol{x}, \boldsymbol{y}|\boldsymbol{\theta})\, \pi(\boldsymbol{\theta})$$

Since both experiments are independent and iid, we have the joint density

$$p(\boldsymbol{x}, \boldsymbol{y}|\boldsymbol{\theta}) = p(\boldsymbol{x}|\boldsymbol{\theta})\, p(\boldsymbol{y}|\boldsymbol{\theta}) \qquad \longrightarrow \qquad p(\boldsymbol{\theta}, \boldsymbol{x}, \boldsymbol{y}) = p(\boldsymbol{x}|\boldsymbol{\theta})\, p(\boldsymbol{y}|\boldsymbol{\theta})\, \pi(\boldsymbol{\theta})$$

and integrating the parameter $\boldsymbol{\theta} \in \Theta$:

$$p(\boldsymbol{y}, \boldsymbol{x}) = p(\boldsymbol{y}|\boldsymbol{x})\, p(\boldsymbol{x}) = \int_{\Theta} p(\boldsymbol{y}|\boldsymbol{\theta})\, p(\boldsymbol{x}|\boldsymbol{\theta})\pi(\boldsymbol{\theta})d\boldsymbol{\theta} = p(\boldsymbol{x}) \int_{\Theta} p(\boldsymbol{y}|\boldsymbol{\theta})\, p(\boldsymbol{\theta}|\boldsymbol{x})\, d\boldsymbol{\theta}$$

Thus, we have that

$$p(\boldsymbol{y}|\boldsymbol{x}) = \int_{\Theta} p(\boldsymbol{y}|\boldsymbol{\theta})\, p(\boldsymbol{\theta}|\boldsymbol{x})\, d\boldsymbol{\theta}$$

This is the basic expression for the *predictive inference* and allows us to predict the results $\boldsymbol{y}$ of a future experiment from the results $\boldsymbol{x}$ observed in a previous experiment within the same parametric model. Note that $p(\boldsymbol{y}|\boldsymbol{x})$ is the density of the quantities not yet observed conditioned to the observed sample. Thus, even though the experiments $e(\boldsymbol{y})$ and $e(\boldsymbol{x})$ are statistically independent, the realization of the first one ($e(\boldsymbol{x})$) modifies the knowledge we have on the parameters $\boldsymbol{\theta}$ of the model and therefore affect the prediction on future experiments for, if we do not consider the results of the first experiment or just don't do it, the predictive distribution for $e(\boldsymbol{y})$ would be

$$p(\boldsymbol{y}) = \int_{\Theta} p(\boldsymbol{y}|\boldsymbol{\theta})\, \pi(\boldsymbol{\theta})\, d\boldsymbol{\theta}$$

It is then clear from the expression of *predictive inference* that in practice it is equivalent to consider as prior density for the second experiment the proper density $\pi(\boldsymbol{\theta}) = p(\boldsymbol{\theta}|\boldsymbol{x})$. If the first experiment provides very little information on the parameters, then $p(\boldsymbol{\theta}|\boldsymbol{x}) \simeq \pi(\boldsymbol{\theta})$ and

$$p(\boldsymbol{y}|\boldsymbol{x}) \simeq \int_{\Theta} p(\boldsymbol{y}|\boldsymbol{\theta})\, \pi(\boldsymbol{\theta})\, d\boldsymbol{\theta} \simeq p(\boldsymbol{y})$$

On the other hand, if after the first experiment we know the parameters with high accuracy then, in distributional sense, $\langle p(\boldsymbol{\theta}|\boldsymbol{x}), \cdot\rangle \simeq \langle \delta(\boldsymbol{\theta}_0), \cdot\rangle$ and

$$p(\boldsymbol{y}|\boldsymbol{x}) \simeq \langle \delta(\boldsymbol{\theta}_0), p(\boldsymbol{y}|\boldsymbol{\theta})\rangle = p(\boldsymbol{y}|\theta_0).$$

2.4 Sufficient Statistics

Consider m random quantities $\{X_1, X_2, \ldots, X_m\}$ that take values in $\Omega_1 \times \cdots \times \Omega_m$ and a random vector

$$\boldsymbol{T} : \Omega_1 \times \cdots \times \Omega_m \longrightarrow \mathcal{R}^{k(m)}$$

whose $k(m) \leq m$ components are functions of the random quantities $\{X_i\}_{i=1}^{m}$. Given the sample $\{x_1, x_2, \ldots, x_m\}$, the vector $\boldsymbol{t} = \boldsymbol{t}(x_1, \ldots, x_m)$ is a $k(m)$-*dimensional statistic*. The practical interest lies in the existence of statistics that contain all the relevant information about the parameters so we don't have to work with the whole sample and simplify considerably the expressions. Thus, of special relevance are the *sufficient statistics*. Given the model $p(x_1, x_2, \ldots, x_n|\boldsymbol{\theta})$, the set of statistics

$\boldsymbol{t} = \boldsymbol{t}(x_1, \ldots, x_m)$ is *sufficient* for $\boldsymbol{\theta}$ if, and only if, $\forall m \geq 1$ and any prior distribution $\pi(\boldsymbol{\theta})$ it holds that

$$p(\boldsymbol{\theta}|x_1, x_2, \ldots, x_m) = p(\boldsymbol{\theta}|\boldsymbol{t})$$

Since the data act in the Bayes formula only through the likelihood, it is clear that to specify the posterior density of $\boldsymbol{\theta}$ we can consider

$$p(\boldsymbol{\theta}|x_1, x_2, \ldots, x_m) = p(\boldsymbol{\theta}|\boldsymbol{t}) \propto p(\boldsymbol{t}|\boldsymbol{\theta})\, \pi(\boldsymbol{\theta})$$

and all other aspects of the data but $\boldsymbol{t}$ are irrelevant. It is obvious however that $\boldsymbol{t} = \{x_1, \ldots, x_m\}$ is sufficient and, in principle, gives no simplification in the modeling. For this we should have $k(m) = \dim(\boldsymbol{t}) < m$ (*minimal sufficient statistics*) and, in the ideal case, we would like that $k(m) = k$ does not depend on m. Except some irregular cases, the only distributions that admit a fixed number of sufficient statistics independently of the sample size (that is, $k(m) = k < m\ \forall m$) are those that belong to the exponential family.

Example 2.1 (1) Consider the exponential model $X \sim Ex(x|\theta)$: and the iid experiment $e(m)$ that provides the sample $\boldsymbol{x} = \{x_1, \ldots, x_m\}$. The likelihood function is:

$$p(\boldsymbol{x}|\theta) = \theta^m\, e^{-\theta\,(x_1+\cdots+x_m)} = \theta^{t_1}\, e^{-\theta\, t_2}$$

and therefore we have the sufficient statistic $\boldsymbol{t} = (m,\ \sum_{i=1}^{m} x_i) : \Omega_1 \times \cdots \times \Omega_m \longrightarrow \mathcal{R}^{k(m)=2}$

(2) Consider the Normal model $X \sim N(x|\mu, \sigma)$ and the iid experiment $e(m)$ again with $\boldsymbol{x} = \{x_1, \ldots, x_m\}$. The likelihood function is:

$$p(\boldsymbol{x}|\mu, \sigma) \propto \sigma^{-m} \exp\left\{-\frac{1}{2\sigma^2}\sum_{i=1}^{m}(x_i - \mu)^2\right\} = \sigma^{-t_1} \exp\left\{-\frac{1}{2\sigma^2}(t_3 - 2\,\mu\, t_2 + \mu^2\, t_1)\right\}$$

and $\boldsymbol{t} = (m,\ \sum_{i=1}^{m} x_i,\ \sum_{i=1}^{m} x_i^2) : \Omega_1 \times \cdots \times \Omega_m \longrightarrow \mathcal{R}^{k(m)=3}$ a sufficient statistic. Usually we shall consider $\boldsymbol{t} = \{m,\ \overline{x}, s^2\}$ with

$$\overline{x} = \frac{1}{m}\sum_{i=1}^{m} x_i \qquad \text{and} \qquad s^2 = \frac{1}{m}\sum_{i=1}^{m}(x_i - \overline{x})^2$$

the sample mean and the sample variance. Inferences on the parameters μ and σ will depend on $\boldsymbol{t}$ and all other aspects of the data are irrelevant.

(3) Consider the Uniform model $X \sim Un(x|0, \theta)$ and the iid sampling $\{x_1, x_2, \ldots, x_m\}$. Then $\boldsymbol{t} = (m,\ \max\{x_i,\ i = 1, \ldots m\}) : \Omega_1 \times \cdots \times \Omega_m \longrightarrow \mathcal{R}^{k(m)=2}$ is a sufficient statistic for θ.

2.5 Exponential Family

A probability density $p(x|\boldsymbol{\theta})$, with $x \in \Omega_X$ and $\boldsymbol{\theta} \in \Theta \subseteq \mathcal{R}^k$ belongs to the *k-parameter exponential family* if it has the form:

$$p(x|\boldsymbol{\theta}) = f(x)\, g(\boldsymbol{\theta}) \exp\left\{\sum_{i=1}^{k} c_i\, \phi_i(\boldsymbol{\theta})\, h_i(x)\right\}$$

with

$$g(\boldsymbol{\theta})^{-1} = \int_{\Omega_x} f(x) \prod_{i=1}^{k} \exp\{c_i\, \phi_i(\boldsymbol{\theta})\, h_i(x)\}\, dx \;\leq \infty$$

The family is called *regular* if supp$\{X\}$ is independent of $\boldsymbol{\theta}$; *irregular* otherwise.

If $\boldsymbol{x} = \{x_1, x_2, \ldots, x_n\}$ is an exchangeable random sampling from the k-parameter regular exponential family, then

$$p(\boldsymbol{x}|\boldsymbol{\theta}) = \left[\prod_{i=1}^{n} f(x_i)\right] [g(\boldsymbol{\theta})]^n \exp\left\{\sum_{i=1}^{k} c_i\, \phi_i(\boldsymbol{\theta}) \left(\sum_{j=1}^{n} h_i(x_j)\right)\right\}$$

and therefore $\boldsymbol{t}(\boldsymbol{x}) = \left\{n, \sum_{i=1}^{n} h_1(x_i), \ldots \sum_{i=1}^{n} h_k(x_i)\right\}$ will be a set of *sufficient statistics.*

Example 2.2 Several distributions of interest, like Poisson and Binomial, belong to the exponential family:

(1) Poisson $Po(n|\mu)$: $P(n|\mu) = \dfrac{e^{-\mu}\, \mu^n}{\Gamma(n+1)} = \dfrac{e^{-(\mu - n \ln \mu)}}{\Gamma(n+1)}$

(2) Binomial $Bi(n|N, \theta)$: $P(n|N, \theta) = \binom{N}{n} \theta^n\, (1-\theta)^{N-n} = \binom{N}{n} e^{n \ln\theta + (N-n)\ln(1-\theta)}$

However, the Cauchy $Ca(x|\alpha, \beta)$ distribution, for instance, does not because

$$p(x_1, \ldots, x_m|\alpha, \beta) \propto \prod_{i=1}^{n} \left(1 + \beta(x_i - \alpha)^2\right)^{-1} = \exp\left\{\sum_{i=1}^{m} \log(1 + \beta(x_i - \alpha)^2)\right\}$$

can not be expressed as the exponential family form. In consequence, there are no sufficient *minimal* statistics (in other words $\boldsymbol{t} = \{n, x_1, \ldots, x_n\}$ is the sufficient statistic) and we will have to work with the whole sample.

2.6 Prior Functions

In the *Bayes rule*, $p(\boldsymbol{\theta}|\boldsymbol{x}) \propto p(\boldsymbol{x}|\boldsymbol{\theta})\, p(\boldsymbol{\theta})$, the *prior function* $p(\boldsymbol{\theta})$ represents the knowledge (*degree of credibility*) that we have about the parameters before the experiment is done and it is a necessary element to obtain the *posterior density* $p(\boldsymbol{\theta}|\boldsymbol{x})$ from which we shall make inferences. If we have faithful information on them before we do the experiment, it is reasonable to incorporate that in the specification of the prior density (*informative prior*) so the new data will provide additional information that will update and improve our knowledge. The specific form of the prior can be motivated, for instance, by the results obtained in previous experiments. However, it is usual that before we do the experiment, either we have a vague knowledge of the parameters compared to what we expect to get from the experiment or simply we do not want to include previous results to perform an independent analysis. In this case, all the new information will be contained in the likelihood function $p(\boldsymbol{x}|\boldsymbol{\theta})$ of the experiment and the prior density (*non-informative prior*) will be merely a mathematical element needed for the inferential process. Being this the case, we expect that the whole weight of the inferences rests on the likelihood and the prior function has the smallest possible influence on them. To learn something from the experiment it is then desirable to have a situation like the one shown in Fig. 2.1 where the posterior distribution $p(\boldsymbol{\theta}|\boldsymbol{x})$ is dominated by the likelihood function. Otherwise, the experiment will provide little information compared to the one we had before and, unless our previous knowledge is based on suspicious observations, it will be wise to design a better experiment.

A considerable amount of effort has been put to obtain reasonable *non-informative priors* that can be used as a standard reference function for the Bayes rule. Clearly, *non-informative* is somewhat misleading because we are never in a state of absolute ignorance about the parameters and the specification of a mathematical model for the process assumes some knowledge about them (masses and life-times take non-negative real values, probabilities have support on [0, 1],...). On the other hand, it doesn't make sense to think about a function that represents ignorance in a formal and objective way so *knowing little a priory* is relative to what we may expect to learn from the experiment. Whatever prior we use will certainly have some effect on the posterior inferences and, in some cases, it would be wise to consider a reasonable set of them to see what is the effect.

The ultimate task of this section is to present the most usual approaches to derive a non-informative prior function to be used as a standard reference that contains little information about the parameters compared to what we expect to get from the experiment.[4] In many cases, these priors will not be Lebesgue integrable (*improper functions*) and, obviously, can not be considered as probability density functions that quantify any knowledge on the parameters (although, with little rigor, sometimes we still talk about prior *densities*). If one is reluctant to use them right the way one can, for instance, define them on a sufficiently large compact support that contains the region where the likelihood is dominant. However, since

[4]For a comprehensive discussion see [5].

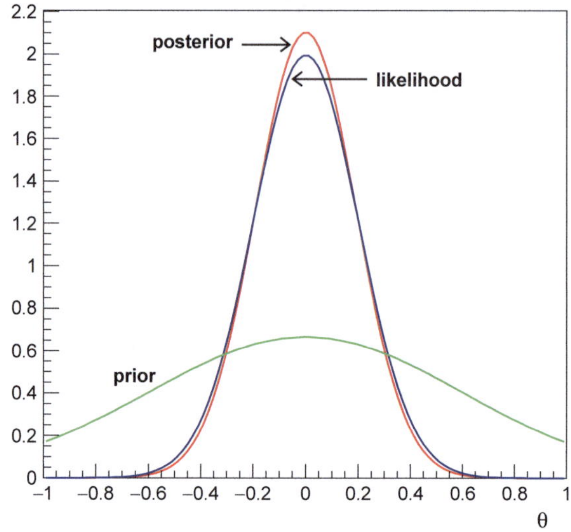

Fig. 2.1 Prior, likelihood and posterior as function of the parameter θ. In this case, the prior is a smooth function and the posterior is dominated by the likelihood

$$p(\boldsymbol{\theta}|\boldsymbol{x})\,d\boldsymbol{\theta} \;\propto\; p(\boldsymbol{x}|\boldsymbol{\theta})\,p(\boldsymbol{\theta})\,d\boldsymbol{\theta} = p(\boldsymbol{x}|\boldsymbol{\theta})\,d\mu(\boldsymbol{\theta})$$

in most cases it will be sufficient to consider them simply as what they really are: a measure. In any case, what is mandatory is that the posterior is a well defined proper density.

2.6.1 *Principle of Insufficient Reason*

The *Principle of Insufficient Reason*[5] dates back to J. Bernoulli and P.S. Laplace and, originally, it states that if we have n exclusive and exhaustive hypothesis and there is no special reason to prefer one over the other, it is reasonable to consider them equally likely and assign a prior probability $1/n$ to each of them. This certainly sounds reasonable and the idea was right the way extended to parameters taking countable possible values and to those with continuous support that, in case of compact sets, becomes a uniform density. It was extensively used by P.S. Laplace and T. Bayes, being he the first to use a uniform prior density for making inferences on the parameter of a Binomial distribution, and is usually referred to as the *"Bayes-Laplace Postulate"*. However, a uniform prior density is obviously not invariant under reparameterizations. If prior to the experiment we have a very vague knowledge about the parameter $\theta \in [a, b]$, we certainly have a vague knowledge about $\phi = 1/\theta$ or $\zeta = \log\theta$ and a uniform distribution for θ:

[5]Apparently, *"Insufficient Reason"* was coined by Laplace in reference to the Leibniz's *Principle of Sufficient Reason* stating essentially that every fact has a sufficient reason for why it is the way it is and not other way.

$$\pi(\theta)\,d\theta = \frac{1}{b-a}\,d\theta$$

implies that:

$$\pi(\phi)\,d\phi = \frac{1}{\phi^2}\,d\phi \qquad \text{and} \qquad \pi(\zeta)\,d\zeta = e^{\zeta}\,d\zeta$$

Shouldn't we take as well a uniform density for ϕ or ζ?

Nevertheless, we shall see that a uniform density, that is far from representing ignorance on a parameter, may be a reasonable choice in many cases even though, if the support of the parameter is infinite, it is an improper function.

2.6.2 Parameters of Position and Scale

An important class of parameters we are interested in are those of position and scale. Let's treat them separately and leave for a forthcoming section the argument behind that. Start with a random quantity $X \sim p(x|\mu)$ with μ a *location parameter*. The density has the form $p(x|\mu) = f(x-\mu)$ so, taking a prior function $\pi(\mu)$ we can write

$$p(x,\mu)\,dx\,d\mu = [p(x|\mu)\,dx]\;[\pi(\mu)\,d\mu] = [f(x-\mu)\,dx]\;[\pi(\mu)\,d\mu]$$

Now, consider random quantity $X' = X + a$ with $a \in \mathcal{R}$ a known value. Defining the new parameter $\mu' = \mu + a$ we have

$$p(x',\mu')\,dx'\,d\mu' = \left[p(x'|\mu')\,dx'\right]\left[\pi'(\mu')\,d\mu'\right] = \left[f(x'-\mu')\,dx'\right]\left[\pi(\mu'-a)\,d\mu'\right]$$

In both cases the models have the same structure so making inferences on μ from the sample $\{x_1, x_1, \ldots, x_n\}$ is formally equivalent to making inferences on μ' from the shifted sample $\{x'_1, x'_2, \ldots, x'_n\}$. Since we have the same prior degree of knowledge on μ and μ', it is reasonable to take the same functional form for $\pi(\cdot)$ and $\pi'(\cdot)$ so:

$$\pi(\mu'-a)\,d\mu' == \pi(\mu')\,d\mu' \qquad \forall a \in \mathcal{R}$$

and, in consequence:

$$\pi(\mu) = \text{constant}$$

If θ is a *scale parameter*, the model has the form $p(x|\theta) = \theta f(x\theta)$ so taking a prior function $\pi(\theta)$ we have that

$$p(x,\theta)\,dx\,d\theta = [p(x|\theta)\,dx]\;[\pi(\theta)\,d\theta] = [\theta f(x\theta)\,dx]\;[\pi(\theta)\,d\theta]$$

For the scaled random quantity $X' = a\,X$ with $a \in \mathcal{R}^+$ known, we have that:

$$p(x',\theta')\,dx'\,d\theta' = \left[p(x'|\theta')\,dx'\right]\left[\pi'(\theta')\,d\theta'\right] = \left[\theta' f(x'\theta')\,dx'\right]\left[\pi(a\theta')\,ad\theta\right]$$

where we have defined the new parameter $\theta' = \theta/a$. Following the same argument as before, it is sound to assume the same functional form for $\pi(\cdot)$ and $\pi'(\cdot)$ so:

$$\pi(a\theta')\,a\,d\theta' = \pi(\theta')\,d\theta' \qquad \forall a \in \mathcal{R}$$

and, in consequence:

$$\pi(\theta) = \frac{1}{\theta}$$

Both prior functions are improper so they may be explicited as

$$\pi(\mu,\theta) \propto \frac{1}{\theta}\,\mathbf{1}_{\Theta}(\theta)\,\mathbf{1}_{M}(\mu)$$

with Θ, M an appropriate sequence of compact sets or considered as prior measures provided that the posterior densities are well defined. Let's see some examples.

Example 2.3 (*The Exponential Distribution*) Consider the sequence of independent observations $\{x_1, x_2, \ldots, x_n\}$ of the random quantity $X \sim Ex(x|\theta)$ drawn under the same conditions. The joint density is

$$p(x_1, x_2, \ldots, x_n|\theta) = \theta^n\,e^{-\theta(x_1 + x_2 + \cdots x_n)}$$

The statistic $t = n^{-1}\sum_{i=1}^{n} x_i$ is sufficient for θ and is distributed as

$$p(t|\theta) = \frac{(n\theta)^n}{\Gamma(n)}\,t^{n-1}\exp\{-n\theta t\}$$

It is clear that θ is a scale parameter so we shall take the prior function $\pi(\theta) = 1/\theta$. Note that if we make the change $z = \log t$ and $\phi = \log\theta$ we have that

$$p(z|\phi) = \frac{n^n}{\Gamma(n)}\exp\{n\left((\phi + z) - e^{\phi+z}\right)\}$$

In this parameterization, ϕ is a position parameter and therefore $\pi(\phi) = \text{const}$ in consistency with $\pi(\theta)$. Then, we have the proper posterior for inferences:

$$p(\theta|t,n) = \frac{(nt)^n}{\Gamma(n)}\exp\{-nt\theta\}\,\theta^{n-1}; \qquad \theta > 0$$

Consider now the sequence of compact sets $C_k = [1/k, k]$ covering R^+ as $k \to \infty$. Then, with support on C_k we have the proper prior density

$$\pi_k(\theta) = \frac{1}{2\,\log k}\,\frac{1}{\theta}\mathbf{1}_{C_k}(\theta)$$

and the sequence of posteriors:

$$p_k(\theta|t, n) = \frac{(nt)^n}{\gamma(n, ntk) - \gamma(n, nt/k)}\exp\{-nt\theta\}\,\theta^{n-1}\,\mathbf{1}_{C_k}(\theta)$$

with $\gamma(a, x)$ the Incomplete Gamma Function. It is clear that

$$\lim_{k\to\infty} p_k(\theta|t, n) = p(\theta|t, n)$$

Example 2.4 (*The Uniform Distribution*) Consider the random quantity $X \sim Un(x|0, \theta)$ and the independent sampling $\{x_1, x_2, \ldots, x_n\}$. To draw inferences on θ, the statistics $x_M = \max\{x_1, x_2, \ldots, x_n\}$ is sufficient and is distributed as (show that):

$$p(x_M|\theta) = n\,\frac{x_M^{n-1}}{\theta^n}\mathbf{1}_{[0,\theta]}(x_M)$$

As in the previous case, θ is a scale parameter and with the change $t_M = \log x_M$, $\phi = \log\theta$ is a position parameter. Then, we shall take $\pi(\theta) \propto \theta^{-1}$ and get the posterior density (Pareto):

$$p(\theta|x_M, n) = n\,\frac{x_M^n}{\theta^{n+1}}\,\mathbf{1}_{[x_M,\infty)}(\theta)$$

Example 2.5 (*The one-dimensional Normal Distribution*) Consider the random quantity $X \sim N(x|\mu, \sigma)$ and the experiment $e(n)$ that provides the independent and exchangeable sequence $\boldsymbol{x} = \{x_1, x_2, \ldots, x_n\}$ of observations. The likelihood function will then be:

$$p(\boldsymbol{x}|\mu, \sigma) = \prod_{i=1}^{n} p(x_i|\mu, \sigma) \propto \frac{1}{\sigma^n}\exp\left\{-\frac{1}{2\sigma^2}\sum_{i=1}^{n}(x_i - \mu)^2\right\}$$

There is a three-dimensional sufficient statistic $\boldsymbol{t} = \{n, \overline{x}, s^2\}$ where

$$\overline{x} = \frac{1}{n}\sum_{i=1}^{n} x_i \qquad \text{and} \qquad s^2 = \frac{1}{n}\sum_{i=1}^{n}(x_i - \overline{x})^2$$

so we can write

$$p(\boldsymbol{x}|\mu,\sigma) \propto \frac{1}{\sigma^n} \exp\left\{-\frac{n}{2\sigma^2}\,(s^2 + (\overline{x}-\mu)^2)\right\}$$

In this case we have both position and scale parameters so we take $\pi(\mu,\sigma) = \pi(\mu)\pi(\sigma) = \sigma^{-1}$ and get the proper posterior

$$p(\mu,\sigma|\boldsymbol{x}) \propto p(\boldsymbol{x}|\mu,\sigma)\,\pi(\mu,\sigma) \propto \frac{1}{\sigma^{n+1}} \exp\left\{-\frac{n}{2\sigma^2}\,\left[s^2 + (\overline{x}-\mu)^2\right]\right\}$$

- **Marginal posterior density of** σ: Integrating the parameter $\mu \in \mathcal{R}$ we have that:

$$p(\sigma|\boldsymbol{x}) = \int_{-\infty}^{+\infty} p(\mu,\sigma|\boldsymbol{x})\,d\mu \propto \sigma^{-n} \exp\left\{-\frac{n\,s^2}{2\sigma^2}\right\} \mathbf{1}_{(0,\infty)}(\sigma)$$

and therefore, the random quantity

$$Z = \frac{n\,s^2}{\sigma^2} \sim \chi^2(z|n-1)$$

- **Marginal posterior density of** μ: Integrating the parameter $\sigma \in [0,\infty)$ we have that:

$$p(\mu|\boldsymbol{x}) = \int_0^{+\infty} p(\mu,\sigma|\boldsymbol{x})\,d\sigma \propto \left(1 + \frac{(\mu-\overline{x})^2}{s^2}\right)^{-n/2} \mathbf{1}_{(-\infty,\infty)}(\mu)$$

so the random quantity

$$T = \frac{\sqrt{n-1}(\mu-\overline{x})}{s} \sim St(t|n-1)$$

It is clear that $p(\mu,\sigma|\boldsymbol{x}) \neq p(\mu|\boldsymbol{x})\,p(\sigma|\boldsymbol{x})$ and, in consequence, are not independent.

- **Distribution of** μ **conditioned to** σ: Since $p(\mu,\sigma|\boldsymbol{x}) = p(\mu|\sigma,\boldsymbol{x})\,p(\sigma|\boldsymbol{x})$ we have that

$$p(\mu|\sigma,\boldsymbol{x}) \propto \frac{1}{\sigma} \exp\{-\frac{n}{2\sigma^2}\,(\mu-\overline{x})^2\}$$

so $\mu|\sigma \sim N(\mu|\overline{x},\sigma/\sqrt{n})$.

Example 2.6 (*Contrast of parameters of Normal Densities*) Consider two independent random quantities $X_1 \sim N(x_1,|\mu_1,\sigma_1)$ and $X_2 \sim N(x_2,|\mu_2,\sigma_2)$ and the random samplings $\boldsymbol{x}_1 = \{x_{11}, x_{12}, \ldots, x_{1n_1}\}$ and $\boldsymbol{x}_2 = \{x_{21}, x_{22}, \ldots, x_{2n_2}\}$ of sizes n_1 and n_2 under the usual conditions. From the considerations of the previous example, we can write

$$p(\boldsymbol{x}_i|\mu_i, \sigma_i) \propto \frac{1}{\sigma_i^{n_i}} \exp\left\{-\frac{n_i}{2\sigma_i^2}\left(s_i^2 + (\overline{x_i} - \mu_i)^2\right)\right\}; \qquad i = 1, 2$$

Clearly, (μ_1, μ_2) are position parameters and (σ_1, σ_2) scale parameters so, in principle, we shall take the improper prior function

$$\pi(\mu_1, \sigma_1, \mu_2, \sigma_2) = \pi(\mu_1)\pi(\mu_2)\pi(\sigma_1)\pi(\sigma_2) \propto \frac{1}{\sigma_1\,\sigma_2}$$

However, if we have know that both distributions have the same variance, then we may set $\sigma = \sigma_1 = \sigma_2$ and, in this case, the prior function will be

$$\pi(\mu_1, \mu_2, \sigma) = \pi(\mu_1)\pi(\mu_2)\pi(\sigma) \propto \frac{1}{\sigma}$$

Let's analyze both cases.

• **Marginal Distribution of** σ_1 **and** σ_2: In this case we assume that $\sigma_1 \neq \sigma_2$ and we shall take the prior $\pi(\mu_1, \sigma_1, \mu_2, \sigma_2) \propto (\sigma_1\sigma_2)^{-1}$. Integrating μ_1 and μ_2 we get:

$$p(\sigma_1, \sigma_2|\boldsymbol{x}_1, \boldsymbol{x}_2) = p(\sigma_1, |\boldsymbol{x}_1)p(\sigma_2, |\boldsymbol{x}_2) \propto \sigma_1^{-n_1}\,\sigma_2^{-n_2} \exp\left\{-\frac{1}{2}\left(\frac{n_1 s_1^2}{\sigma_1^2} + \frac{n_2 s_2^2}{\sigma_2^2}\right)\right\}$$

Now, if we define the new random quantities

$$Z = \frac{s_2^2}{w^2\,s_1^2} = \frac{(\sigma_1/s_1)^2}{(\sigma_2/s_2)^2} \qquad \text{and} \qquad W = \frac{n_1 s_1^2}{{\sigma_1}^2}$$

both with support in $(0, +\infty)$, and integrate the last we get we get that Z follows a Snedecor Distribution $Sn(z|n_2 - 1, n_1 - 1)$ whose density is

$$p(z|\boldsymbol{x}_1, \boldsymbol{x}_2) = \frac{(\nu_1/\nu_2)^{\nu_1/2}}{\text{Be}\,(\nu_1/2, \nu_2/2)}\, z^{(\nu_1/2)-1} \left(1 + \frac{\nu_1}{\nu_2}\,z\right)^{-(\nu_1+\nu_2)/2} \mathbf{1}_{(0,\infty)}(z).$$

• **Marginal Distribution of** μ_1 **and** μ_2: In this case, it is different whether we assume that, although unknown, the variances are the same or not. In the first case, we set $\sigma_1 = \sigma_2 = \sigma$ and take the reference prior $\pi(\mu_1, \mu_2, \sigma) = \sigma^{-1}$. Defining

$$A = n_1\left[s_1^2 + (\overline{x}_1 - \mu_1)^2\right] + n_2\left[s_2^2 + (\overline{x}_2 - \mu_2)^2\right]$$

we can write

$$p(\mu_1, \mu_2, \sigma|\boldsymbol{x}, \boldsymbol{y}) \propto \frac{1}{\sigma^{n_1+n_2+1}} \exp\left\{-\frac{1}{2} A/\sigma^2\right\}$$

It is left as an exercise to show that if we make the transformation

$$w = \mu_1 - \mu_2 \in (-\infty, +\infty); \quad u = \mu_2 \in (-\infty, +\infty) \quad \text{and} \quad z = \sigma^{-2} \in (0, +\infty)$$

and integrate the last two, we get

$$p(w|\boldsymbol{x}_1, \boldsymbol{x}_2) \propto \left(1 + \frac{n_1 n_2}{n_1 + n_2} \frac{[(\overline{x}_1 - \overline{x}_2) - w]^2}{n_1 s_1^2 + n_2 s_2^2}\right)^{-(n_1+n_2-1)/2}$$

Introducing the more usual terminology

$$s^2 = \frac{n_1 s_1^2 + n_2 s_2^2}{n_1 + n_2 - 2}$$

we have that

$$p(w|\boldsymbol{x}_1, \boldsymbol{x}_2) \propto \left(1 + \frac{n_1 n_2}{n_1 + n_2} \frac{[w - (\overline{x}_1 - \overline{x}_2)]^2}{s^2 (n_1 + n_2 - 2)}\right)^{-[(n_1+n_2-2)+1]/2}$$

and therefore the random quantity

$$T = \frac{(\mu_1 - \mu_2) - (\overline{x}_1 - \overline{x}_2)}{s\,(1/n_1 + 1/n_2)^{1/2}}$$

follows a Student's Distribution $St(t|\nu)$ with $\nu = n_1 + n_2 - 2$ degrees of freedom.

Let's see now the case where we can not assume that the variances are equal. Taking the prior reference function $\pi(\mu_1, \mu_2, \sigma_1, \sigma_2) = (\sigma_1 \sigma_2)^{-1}$ we get

$$p(\mu_1, \mu_2, \sigma_1, \sigma_2|\boldsymbol{x}_1, \boldsymbol{x}_2) \propto \sigma_1^{-(n_1+1)} \sigma_2^{-(n_2+1)} \exp\left\{-\frac{1}{2} \sum_{i=1}^{2} \frac{s_i^2 + (\overline{x}_i - \mu_i)^2}{\sigma_i^2/n_i}\right\}$$

After the appropriate integrations (left as exercise), defining $w = \mu_1 - \mu_2$ and $u = \mu_2$ we end up with the density

$$p(w, u|\boldsymbol{x}_1, \boldsymbol{x}_2) \propto \left(1 + \frac{(\overline{x}_1 - w - u)^2}{s_1^2}\right)^{-n_1/2} \left(1 + \frac{(\overline{x}_2 - u)^2}{s_2^2}\right)^{-n_2/2}$$

where integral over $u \in \mathcal{R}$ can not be expressed in a simple way. The density

$$p(w|\boldsymbol{x}_1, \boldsymbol{x}_2) \propto \int_{-\infty}^{+\infty} p(w, u|\boldsymbol{x}_1, \boldsymbol{x}_2)\, du$$

is called the *Behrens-Fisher Distribution*. Thus, to make statements on the difference of Normal means, we should analyze first the sample variances and decide how shall we treat them.

2.6.3 Covariance Under Reparameterizations

The question of how to establish a reasonable criteria to obtain a prior for a given model $p(\boldsymbol{x}|\boldsymbol{\theta})$ that can be used as a standard reference function was studied by Harold Jeffreys [6] in the mid XX century. The rationale behind the argument is that if we have the model $p(\boldsymbol{x}|\boldsymbol{\theta})$ with $\boldsymbol{\theta} \in \Omega_\theta \subseteq R^n$ and make a reparameterizations $\phi = \phi(\boldsymbol{\theta})$ with $\phi(\cdot)$ a one-to-one differentiable function, the statements we make about $\boldsymbol{\theta}$ should be consistent with those we make about ϕ and, in consequence, priors should be related by

$$\pi_\theta(\boldsymbol{\theta})d\boldsymbol{\theta} = \pi_\phi(\phi(\boldsymbol{\theta}))\left|\det\left[\frac{\partial\phi_i(\theta)}{\partial\theta_j}\right]\right| d\boldsymbol{\theta}$$

Now, assume that the Fisher's matrix (see Sect. 4.5)

$$\boldsymbol{I}_{ij}(\boldsymbol{\theta}) = E_X\left[\frac{\partial \log p(\boldsymbol{x}|\boldsymbol{\theta})}{\partial\theta_i}\frac{\partial \log p(\boldsymbol{x}|\boldsymbol{\theta})}{\partial\theta_j}\right]$$

exists for this model. Under a differentiable one-to-one transformation $\phi = \phi(\boldsymbol{\theta})$ we have that

$$\boldsymbol{I}_{ij}(\phi) = \frac{\partial\theta_k}{\partial\phi_i}\frac{\partial\theta_l}{\partial\phi_j}\boldsymbol{I}_{kl}(\boldsymbol{\theta})$$

so it behaves as a covariant symmetric tensor of second order (left as exercise). Then, since

$$\det\left[\boldsymbol{I}(\phi)\right] = \left|\det\left[\frac{\partial\theta_i}{\partial\phi_j}\right]\right|^2 \det\left[\boldsymbol{I}(\boldsymbol{\theta})\right]$$

Jeffreys proposed to consider the prior

$$\pi(\boldsymbol{\theta}) \propto \left[\det[\boldsymbol{I}(\boldsymbol{\theta})]\right]^{1/2}$$

In fact, if we consider the parameter space as a Riemannian manifold (see Sect. 4.7) the Fisher's matrix is the metric tensor (Fisher-Rao metric) and this is just the invariant volume element. Intuitively, if we make a transformation such that at a particular value $\phi_0 = \phi(\theta_0)$ the Fisher's tensor is constant and diagonal, the metric in a neighborhood of ϕ_0 is Euclidean and we have location parameters for which a constant prior is appropriate and therefore

$$\pi(\phi)d\phi \propto d\phi = \left[\det\left[\boldsymbol{I}(\boldsymbol{\theta})\right]^{1/2}\right] d\boldsymbol{\theta} = \pi(\boldsymbol{\theta})d\boldsymbol{\theta}$$

It should be pointed out that there may be other priors that are also invariant under reparameterizations and that, as usual, we talk loosely about *prior densities* although they usually are improper functions.

For one-dimensional parameter, the density function expressed in terms of

$$\phi \sim \int [\boldsymbol{I}(\theta)]^{1/2} \, d\theta$$

may be reasonably well approximated by a Normal density (at least in the parametric region where the likelihood is dominant) because $\boldsymbol{I}(\phi)$ is constant (see Sect. 4.5) and then, due to translation invariance, a constant prior for ϕ is justified. Let' see some examples.

Example 2.7 (The Binomial Distribution) Consider the random quantity $X \sim Bi(x|\theta, n)$:

$$p(x|n,\theta) = \binom{n}{x} \theta^k \, (1-\theta)^{n-k}; \qquad n, k \in N_0; \; k \leq n$$

with $0 < \theta < 1$. Since $E[X] = n\theta$ we have that:

$$\boldsymbol{I}(\theta) = E_X\left[\left(-\frac{\partial^2 \log p(x|n,\theta)}{\partial\theta^2}\right)\right] = \frac{n}{\theta\,(1-\theta)}$$

so the Jeffreys prior (proper in this case) for the parameter θ is

$$\pi(\theta) \propto [\theta\,(1-\theta)]^{-1/2}$$

and the posterior density will therefore be

$$p(\theta|k,n) \propto \theta^{k-1/2}\,(1-\theta)^{n-k-1/2}$$

that is; a $Be(x|k + 1/2, n - k + 1/2)$ distribution. Since

$$\phi = \int \frac{d\theta}{\sqrt{\theta\,(1-\theta)}} = 2\,\text{asin}\,(\theta^{1/2})$$

we have that $\theta = \sin^2\phi/2$ and, parameterized in terms of ϕ, $\boldsymbol{I}(\phi)$ is constant so the distribution "*looks*" more Normal (see Fig. 2.2).

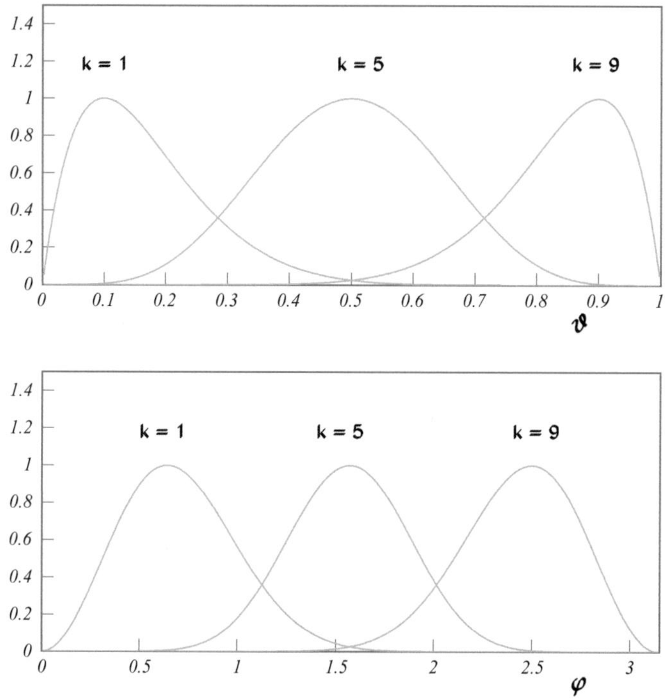

Fig. 2.2 Dependence of the likelihood function with the parameter θ (*upper*) and with $\phi = 2\,\mathrm{asin}(\theta^{1/2})$ (*lower*) for a Binomial process with $n = 10$ and $k = 1, 5$ and 9

Example 2.8 (*The Poisson Distribution*) Consider the random quantity $X \sim Po(x|\mu)$:

$$p(x|\mu) = e^{-\mu}\,\frac{\mu^x}{\Gamma(x+1)}; \qquad x \in N;\; \mu \in R^+$$

Then, since $E[X] = \mu$ we have

$$\boldsymbol{I}(\mu) = E_X\left[\left(-\frac{\partial^2 \log p(x|\mu)}{\partial \mu^2}\right)\right] = \frac{1}{\mu}$$

so we shall take as *prior* (improper):

$$\pi(\mu) = [\boldsymbol{I}(\mu)]^{1/2} = \mu^{-1/2}$$

and make inferences on μ from the proper posterior density

$$p(\mu|x) \propto e^{-\mu}\,\mu^{x-1/2}$$

that is, a $Ga(x|1, x+1/2)$ distribution.

Example 2.9 (*The Pareto Distribution*) Consider the random quantity $X \sim Pa(x|\theta, x_0)$ with $x_0 \in R^+$ known and density

$$p(x|\theta, x_0) = \frac{\theta}{x_0} \left(\frac{x_0}{x}\right)^{\theta+1} \mathbf{1}_{(x_0,\infty)}(x); \qquad \theta \in R^+$$

Then,

$$\boldsymbol{I}(\theta) = E_X\left[\left(-\frac{\partial^2 \log p(x|\theta, x_0)}{\partial \theta^2}\right)\right] = \frac{1}{\theta^2}$$

so we shall take as *prior* (improper):

$$\pi(\theta) \propto [\boldsymbol{I}(\mu)]^{1/2} = \theta^{-1}$$

and make inferences from the posterior density (proper)

$$p(\theta|x, x_0) = x^{-\theta} \log x$$

Note that if we make the transformation $t = \log x$, the density becomes

$$p(t|\theta, x_0) = \theta\, x_0^{\theta}\, e^{-\theta t} \mathbf{1}_{(\log x_0,\infty)}(t)$$

for which θ is a scale parameter and, from previous considerations, we should take $\pi(\theta) \propto \theta^{-1}$ in consistency with Jeffreys's prior.

Example 2.10 (*The Gamma Distribution*) Consider the random quantity $X \sim Ga(x|\alpha, \beta)$ with $\alpha, \beta \in R^+$ and density

$$p(x|\alpha, \beta) = \frac{\alpha^{\beta}}{\Gamma(\beta)}\, e^{-\alpha x}\, x^{\beta-1}\, \mathbf{1}_{(0,\infty)}(x)$$

Show that the Fisher's matrix is

$$\boldsymbol{I}(\alpha, \beta) = \begin{pmatrix} \beta\alpha^{-2} & -\alpha^{-1} \\ -\alpha^{-1} & \Psi'(\beta) \end{pmatrix}$$

with $\Psi'(x)$ the first derivative of the Digamma Function and, following Jeffreys' rule, we should take the prior

$$\pi(\alpha, \beta) \propto \alpha^{-1} \left[\beta\Psi'(\beta) - 1\right]^{1/2}$$

Note that α is a scale parameter so, from previous considerations, we should take $\pi(\alpha) \propto \alpha^{-1}$. Furthermore, if we consider α and β independently, we shall get

$$\pi(\alpha, \beta) = \pi(\alpha)\pi(\beta) \propto \alpha^{-1} \left[\Psi'(\beta)\right]^{1/2}$$

Example 2.11 (*The Beta Distribution*)
Show that for the $Be(x|\alpha, \beta)$ distribution with density

$$p(x|\alpha, \beta) = \frac{\Gamma(\alpha+\beta)}{\Gamma(\alpha)\Gamma(\beta)} (x^{\alpha-1} (1-x)^{\beta-1} \mathbf{1}_{[0,1]}(x); \qquad \alpha, \beta \in R^{+}$$

the Fisher's matrix is given by

$$\boldsymbol{I}(\alpha, \beta) = \begin{pmatrix} \Psi'(\alpha) - \Psi'(\alpha+\beta) & -\Psi'(\alpha+\beta) \\ -\Psi'(\alpha+\beta) & \Psi'(\beta) - \Psi'(\alpha+\beta) \end{pmatrix}$$

with $\Psi'(x)$ the first derivative of the Digamma Function.

Example 2.12 (*The Normal Distribution*)

Univariate: The Fisher's matrix is given by

$$\boldsymbol{I}(\mu, \sigma) = \begin{pmatrix} \sigma^{-2} & 0 \\ 0 & 2\sigma^{-2} \end{pmatrix}$$

so

$$\pi_1(\mu, \sigma) \propto [\det[\boldsymbol{I}(\mu, \sigma]]^{1/2} \propto \frac{1}{\sigma^2}$$

However, had we treated the two parameters independently, we should have obtained

$$\pi_2(\mu, \sigma) = \pi(\mu)\, \pi(\sigma) \propto \frac{1}{\sigma}$$

The prior $\pi_2 \propto \sigma^{-1}$ is the one we had used in Example 2.5 where the problem was treated as two one-dimensional independent problems and, as we saw:

$$T = \frac{\sqrt{n-1}(\mu - \overline{x})}{s} \sim St(t|n-1) \quad \text{and} \quad Z = \frac{n\, s^2}{\sigma^2} \sim \chi^2(z|n-1)$$

with $E[Z] = n - 1$. Had we used prior $\pi_1 \propto \sigma^{-2}$, we would have obtained that $Z \sim \chi^2(z|n)$ and therefore $E[Z] = n$. This is not reasonable. On the one hand, we know from the sampling distribution $N(x|\mu, \sigma)$ that $E[ns^2\sigma^{-2}] = n - 1$. On the other hand, we have two parameters (μ, σ) and integrate on one (σ) so the number of degrees of freedom should be $n - 1$.

Bivariate: The Fisher's matrix is given by

$$\boldsymbol{I}(\mu_1, \mu_2) = (1-\rho^2)^{-1} \begin{pmatrix} \sigma_1^{-2} & -\rho(\sigma_1\sigma_2)^{-1} \\ -\rho(\sigma_1\sigma_2)^{-1} & \sigma_2^{-2} \end{pmatrix}$$

$$\boldsymbol{I}(\sigma_1, \sigma_2, \rho) = (1-\rho^2)^{-1} \begin{pmatrix} (2-\rho^2)\sigma_1^{-2} & -\rho^2(\sigma_1\sigma_2)^{-1} & -\rho\sigma_1^{-1} \\ -\rho^2(\sigma_1\sigma_2)^{-1} & (2-\rho^2)\sigma_2^{-2} & -\rho\sigma_2^{-1} \\ -\rho\sigma_1^{-1} & -\rho\sigma_2^{-1} & (1+\rho^2)(1-\rho^2)^{-1} \end{pmatrix}$$

$$\boldsymbol{I}(\mu_1, \mu_2, \sigma_1, \sigma_2, \rho) = \begin{pmatrix} \boldsymbol{I}(\mu_1, \mu_2) & \mathbf{0} \\ \mathbf{0} & \boldsymbol{I}(\sigma_1, \sigma_2, \rho) \end{pmatrix}$$

From this,

$$\pi(\mu_1, \mu_2, \sigma_1, \sigma_2, \rho) \propto |\text{det}\boldsymbol{I}(\mu_1, \mu_2, \sigma_1, \sigma_2, \rho)|^{1/2} = \frac{1}{\sigma_1^2\sigma_2^2(1-\rho^2)^2}$$

while if we consider $\pi(\mu_1, \mu_2, \sigma_1, \sigma_2, \rho) = \pi(\mu_1, \mu_2)\pi(\sigma_1, \sigma_2, \rho)$ we get

$$\pi(\mu_1, \mu_2, \sigma_1, \sigma_2, \rho) \propto \frac{1}{\sigma_1\sigma_2(1-\rho^2)^{3/2}}$$

Problem 2.1 Show that for the density $p(\boldsymbol{x}|\boldsymbol{\theta})$; $\boldsymbol{x} \in \Omega \subseteq R^n$, the Fisher's matrix (if exists)

$$\boldsymbol{I}_{ij}(\boldsymbol{\theta}) = E_X \left[\frac{\partial \log\, p(\boldsymbol{x}|\boldsymbol{\theta})}{\partial \theta_i} \frac{\partial \log\, p(\boldsymbol{x}|\boldsymbol{\theta})}{\partial \theta_j} \right]$$

transforms under a differentiable one-to-one transformation $\boldsymbol{\phi} = \boldsymbol{\phi}(\boldsymbol{\theta})$ as a covariant symmetric tensor of second order; that is

$$\boldsymbol{I}_{ij}(\boldsymbol{\phi}) = \frac{\partial \theta_k}{\partial \phi_i} \frac{\partial \theta_l}{\partial \phi_j} \boldsymbol{I}_{kl}(\boldsymbol{\theta})$$

Problem 2.2 Show that for $X \sim Po(x|\mu + b)$ with $b \in R^+$ known (Poisson model with known background), we have that $\boldsymbol{I}(\mu) = (\mu + b)^{-1}$ and therefore the posterior (proper) is given by:

$$p(\mu|x, b) \propto e^{-(\mu+b)}\,(\mu + b)^{x-1/2}$$

Problem 2.3 Show that for the one parameter mixture model $p(x|\lambda) = \lambda p_1(x) + (1-\lambda)p_2(x)$ with $p_1(x) \neq p_2(x)$ properly normalized and $\lambda \in (0, 1)$,

$$\boldsymbol{I}(\lambda) = \frac{1}{\lambda(1-\lambda)} \left\{ 1 - \int_{-\infty}^{\infty} \frac{p_1(x) p_2(x)}{p(x|\lambda)} dx \right\}$$

When $p_1(x)$ and $p_2(x)$ are *"well separated"*, the integral is $<< 1$ and therefore $\boldsymbol{I}(\lambda) \sim [\lambda(1-\lambda)]^{-1}$. On the other hand, when they *"get closer"* we can write $p_2(x) = p_1(x) + \eta(x)$ with $\int_{-\infty}^{\infty} \eta(x) dx = 0$ and, after a Taylor expansion for $|\eta(x)| << 1$ get to first order that

$$\boldsymbol{I}(\lambda) \simeq \int_{-\infty}^{\infty} \frac{(p_1(x) - p_2(x))^2}{p_1(x)} dx + \cdots$$

independent of λ. Thus, for this problem it will be sound to consider the prior $\pi(\lambda|a, b) = Be(\lambda|a, b)$ with parameters between $(1/2, 1/2)$ and $(1, 1)$.

2.6.4 Invariance Under a Group of Transformations

Some times, we may be interested to provide the prior with invariance under some transformations of the parameters (or a subset of them) considered of interest for the problem at hand. As we have stated, from a formal point of view the prior can be treated as an absolute continuous measure with respect to Lebesgue so $p(\boldsymbol{\theta}|\boldsymbol{x})\, d\boldsymbol{\theta} \propto p(\boldsymbol{x}|\boldsymbol{\theta})\, \pi(\boldsymbol{\theta})\, d\boldsymbol{\theta} = p(\boldsymbol{x}|\boldsymbol{\theta})\, d\mu(\boldsymbol{\theta})$. Now, consider the probability space (Ω, B, μ) and a measurable homeomorphism $T : \Omega \rightarrow \Omega$. A measure μ on the Borel algebra B would be invariant by the mapping T if for any $A \subset B$, we have that $\mu(T^{-1}(A)) = \mu(A)$. We know, for instance, that there is a unique measure λ on R^n that is invariant under translations and such that for the unit cube $\lambda([0, 1]^n) = 1$: the Lebesgue measure (in fact, it could have been defined that way). This is consistent with the constant prior specified already for position parameters. The Lebesgue measure is also the unique measure in R^n that is invariant under the rotation group $SO(n)$ (see Problem 2.5). Thus, when expressed in spherical polar coordinates, it would be reasonable for the spherical surface S^{n-1} the rotation invariant prior

$$d\mu(\boldsymbol{\phi}) = \prod_{k=1}^{n-1} (\sin\phi_k)^{(n-1)-k}\, d\phi_k$$

with $\phi_{n-1} \in [0, 2\pi)$ and $\phi_j \in [0, \pi]$ for the rest. We shall use this prior function in a later problem.

In other cases, the group of invariance is suggested by the model

$$M : \{p(\boldsymbol{x}|\boldsymbol{\theta}),\ \boldsymbol{x} \in \Omega_X,\ \boldsymbol{\theta} \in \Omega_\Theta\}$$

in the sense that we can make a transformation of the random quantity $\boldsymbol{X} \rightarrow \boldsymbol{X}'$ and absorb the change in a redefinition of the parameters $\boldsymbol{\theta} \rightarrow \boldsymbol{\theta}'$ such that the expression of the probability density remains unchanged. Consider a group of transformations[6] G that acts

$$\text{on the Sample Space:} \quad x \rightarrow x' = g \circ x; \qquad g \in G;\, x, x' \in \Omega_X$$

$$\text{on the Parametric Space:} \; \theta \rightarrow \theta' = g \circ \theta; \qquad g \in G;\, \theta, \theta' \in \Omega_\Theta$$

The model M is said to be invariant under G if $\forall g \in G$ and $\forall \theta \in \Omega_\Theta$ the random quantity $X' = g \circ X$ is distributed as $p(x'|\theta') \equiv p(g \circ x | g \circ \theta)$. Therefore, transformations of data under G will make no difference on the inferences if we assign consistent "prior beliefs" to the original and transformed parameters. Note that the action of the group on the sample and parameter spaces will, in general, be different. The essential point is that, as Alfred Haar showed in 1933, for the action of the group G of transformations there is an invariant measure μ (*Haar measure*; [8]) such that

$$\int_{\Omega_X} f(g \circ x) d\mu(x) = \int_{\Omega_X} f(x') d\mu(x')$$

for any Lebesgue integrable function $f(x)$ on Ω_X. Shortly after, it was shown (Von Neumann (1934); Weil and Cartan (1940)) that this measure is unique up to a multiplicative constant. In our case, the function will be $p(\cdot|\theta)\mathbf{1}_\Theta(\theta)$ and the invariant measure we are looking for is $d\mu(\theta) \propto \pi(\theta) d\theta$. Furthermore, since the group may be non-abelian, we shall consider the action on the right and on the left of the parameter space. Thus, we shall have:

$$\int_\Theta p(\cdot|g \circ \theta)\, \pi_L(\theta)\, d\theta = \int_\Theta p(\cdot|\theta')\, \pi_L(\theta')\, d\theta'$$

if the group acts on the left and

$$\int_\Theta p(\cdot|\theta \circ g)\, \pi_R(\theta)\, d\theta = \int_\Theta p(\cdot|\theta')\, \pi_R(\theta')\, d\theta'$$

if the action is on the right. Then, we should start by identifying the group of transformations under which the model is invariant (if any; in many cases, either there is no invariance or at least not obvious) work in the parameter space. The most interesting cases for us are:

---

[6]In this context, the use of Transformation Groups arguments was pioneered by E.T. Jaynes [7].

$$\text{Affine Transformations: } x \to x' = g \circ x = a + b\,x$$

$$\text{Matrix Transformations: } x \to x' = g \circ x = R\,x$$

Translations and scale transformations are a particular case of the first and rotations of the second. Let's start with the location and scale parameters; that is, a density

$$p(x|\mu, \sigma)\,dx = \frac{1}{\sigma}\,f\left(\frac{x-\mu}{\sigma}\right)\,dx$$

the Affine group $G = \{g \equiv (a, b);\, a \in R;\, b \in R^+\}$ so $x' = g \circ x = a + bx$ and the model will be invariant if

$$(\mu', \sigma') = g \circ (\mu, \sigma) = (a, b) \circ (\mu, \sigma) = (a + b\mu, b\sigma)$$

Now,

$$\begin{aligned}
\int p(\cdot|\mu', \sigma')\,\pi_L(\mu', \sigma')\,d\mu'\,d\sigma' &= \int p(\cdot|g \circ (\mu, \sigma)\,\pi_L(\mu, \sigma)\,d\mu\,d\sigma = \\
= \int p(\cdot|\mu', \sigma')\,&\left\{\pi_L[g^{-1}(\mu', \sigma')]\,J(\mu', \sigma'; \mu, \sigma)\right\}\,d\mu'\,d\sigma' = \\
&= \int p(\cdot|\mu', \sigma')\,\left\{\pi_L\left(\frac{\mu'-a}{b}, \frac{\sigma'}{b}\right)\,\frac{1}{b^2}\right\}\,d\mu'\,d\sigma'
\end{aligned}$$

and this should hold for all $(a, b) \in R \times R^+$ so, in consequence:

$$d\mu_L(\mu, \sigma) = \pi_L(\mu, \sigma)\,d\mu\,d\sigma \propto \frac{1}{\sigma^2}\,d\mu\,d\sigma$$

However, the group of Affine Transformations is non-abelian so if we study the action on the left, there is no reason why we should not consider also the action on the right. Since

$$(\mu', \sigma') = (\mu, \sigma) \circ g = (\mu, \sigma) \circ (a, b) = (\mu + a\sigma, b\sigma)$$

the same reasoning leads to (left as exercise):

$$d\mu_R(\mu, \sigma) = \pi_R(\mu, \sigma)\,d\mu\,d\sigma \propto \frac{1}{\sigma}\,d\mu\,d\sigma$$

The first one (π_L) is the one we obtain using Jeffrey's rule in two dimensions while π_R is the one we get for position and scale parameters or Jeffrey's rule treating both parameters independently; that is, as two one-dimensional problems instead a one two-dimensional problem. Thus, although from the invariance point of view there is no reason why one should prefer one over the other, the right invariant Haar prior gives more consistent results. In fact ([9, 10]), a necessary and sufficient condition for a sequence of posteriors based on proper priors to converge in probability to an invariant posterior is that the prior is the right Haar measure.

Problem 2.4 As a remainder, given a measure space $(\Omega, \mathcal{B}, \mu)$ a mapping $T : \Omega \longrightarrow \Omega$ is measurable if $T^{-1}(A) \in \mathcal{B}$ for all $A \in \mathcal{B}$ and the measure μ is invariant under T if $\mu(T^{-1}(A)) = \mu(A)$ for all $A \in \mathcal{B}$. Show that the measure $d\mu(\theta) = [\theta(1-\theta)]^{-1/2}\, d\theta$ is invariant under the mapping $T : [0, 1] \rightarrow [0, 1]$ such that $T : \theta \rightarrow \theta' = T(\theta) = 4\theta(1-\theta)$. This is the Jeffrey's prior for the Binomial model $Bi(x|N, \theta)$.

Problem 2.5 Consider the n-dimensional spherical surface S_n of unit radius, $\boldsymbol{x} \in S_n$ and the transformation $\boldsymbol{x}' = \boldsymbol{R}\boldsymbol{x} \in S_n$ where $\boldsymbol{R} \in SO(n)$. Show that the Haar invariant measure is the Lebesgue measure on the sphere.

Hint: Recall that $\boldsymbol{R}$ is an orthogonal matrix so $\boldsymbol{R}^t = \boldsymbol{R}^{-1}$; that $|\det \boldsymbol{R}| = 1$ so $J(\boldsymbol{x}'; \boldsymbol{x}) = |\partial \boldsymbol{x}/\partial \boldsymbol{x}')| = |\partial \boldsymbol{R}^{-1}\boldsymbol{x}'/\partial \boldsymbol{x}')| = |\det \boldsymbol{R}| = 1$ and that $\boldsymbol{x}''\boldsymbol{x}' = \boldsymbol{x}^t\boldsymbol{x} = 1$.

Example 2.12 (*Bivariate Normal Distribution*) Let $\boldsymbol{X} = (X_1, X_2) \sim N(\boldsymbol{x}|\boldsymbol{0}, \phi)$ with $\phi = \{\sigma_1, \sigma_2, \rho\}$; that is:

$$p(\boldsymbol{x}|\phi) = (2\pi)^{-1}\, |\det[\boldsymbol{\Sigma}]|^{-1/2} \exp\left\{-\frac{1}{2}\left(\boldsymbol{x}^t\, \boldsymbol{\Sigma}^{-1}\, \boldsymbol{x}\right)\right\}$$

with the covariance matrix

$$\boldsymbol{\Sigma} = \begin{pmatrix} \sigma_1^2 & \rho\sigma_1\sigma_2 \\ \rho\sigma_1\sigma_2 & \sigma_2^2 \end{pmatrix} \quad \text{and} \quad \det[\boldsymbol{\Sigma}] = \sigma_1^2\sigma_2^2(1-\rho)^2$$

Using the Cholesky decomposition we can express $\boldsymbol{\Sigma}^{-1}$ as the product of two lower (or upper) triangular matrices:

$$\boldsymbol{\Sigma}^{-1} = \frac{1}{\det[\boldsymbol{\Sigma}]}\begin{pmatrix} \sigma_2^2 & -\rho\sigma_1\sigma_2 \\ -\rho\sigma_1\sigma_2 & \sigma_1^2 \end{pmatrix} = \boldsymbol{A}^t\boldsymbol{A} \quad \text{with} \quad \boldsymbol{A} = \begin{pmatrix} \frac{1}{\sigma_1} & 0 \\ \frac{-\rho}{\sigma_1\sqrt{1-\rho^2}} & \frac{1}{\sigma_2\sqrt{1-\rho^2}} \end{pmatrix}$$

For the action on the left:

$$\boldsymbol{M} = \boldsymbol{T} = \begin{pmatrix} a & 0 \\ b & c \end{pmatrix};\, a, b > 0 \quad \longrightarrow \quad J(\boldsymbol{A}'; \boldsymbol{A}) = a^2 c$$

and, in consequence

$$\pi(aa'_{11}, aa'_{21} + ba'_{22}, ca'_{22})\, ac^2 = \pi(a'_{11}, a'_{21}, a'_{22}) \longrightarrow \pi(a'_{11}, a'_{21}, a'_{22}) \propto \frac{1}{{a'_{11}}^2 a'_{22}}$$

and $\det[\boldsymbol{\Sigma}] = (\det[\boldsymbol{\Sigma}^{-1}])^{-1} = (\det[\boldsymbol{A}])^{-2}$. Thus, in the new parameterization $\boldsymbol{\theta} = \{a_{11}, a_{21}, a_{22}\}$

$$p(\boldsymbol{x}|\boldsymbol{\theta}) = (2\pi)^{-1}\,|\text{det}[\boldsymbol{A}]|\,\exp\left\{-\frac{1}{2}\left(\boldsymbol{x}^t\,\boldsymbol{A}^t\boldsymbol{A}\,\boldsymbol{x}\right)\right\}$$

Consider now the group of lower triangular $2x2$ matrices

$$G_l = \{\boldsymbol{T}\,\in\, LT_{2x2};\quad T_{ii} > 0\}$$

Since $\boldsymbol{T}^{-1} \in G_l$, inserting the identity matrix $\boldsymbol{I} = \boldsymbol{T}\boldsymbol{T}^{-1} = \boldsymbol{T}^{-1}\boldsymbol{T}$ we have: action

On the Left	On the Right
$\boldsymbol{T}\circ\boldsymbol{x}\to\boldsymbol{T}\boldsymbol{x} = \boldsymbol{x}'$	$\boldsymbol{x}\circ\boldsymbol{T}\to\boldsymbol{T}^{-1}\boldsymbol{x} = \boldsymbol{x}'$
$\left[\boldsymbol{x}^t\,(\boldsymbol{T}^t(\boldsymbol{T}^t)^{-1})\,\boldsymbol{A}^t\boldsymbol{A}\,(\boldsymbol{T}^{-1}\boldsymbol{T})\,\boldsymbol{x}\right]$	$\left[\boldsymbol{x}^t\,((\boldsymbol{T}^t)^{-1}\boldsymbol{T}^t)\,\boldsymbol{A}^t\boldsymbol{A}\,(\boldsymbol{T}\boldsymbol{T}^{-1})\,\boldsymbol{x}\right]$
$\boldsymbol{M} = \boldsymbol{T}$	$\boldsymbol{M} = \boldsymbol{T}^{-1}$

Then

$$\boldsymbol{M}\boldsymbol{x} = \boldsymbol{x}';\quad \boldsymbol{x} = \boldsymbol{M}^{-1}\boldsymbol{x}';\quad \boldsymbol{x}'^t = \boldsymbol{x}^t\boldsymbol{M}^t\quad\text{and}\quad d\boldsymbol{x} = \frac{1}{|\text{det}[\boldsymbol{M}]|}d\boldsymbol{x}'$$

so

$$p(\boldsymbol{x}'|\boldsymbol{\theta}) = (2\pi)^{-1}\,\frac{|\text{det}[\boldsymbol{A}]|}{|\text{det}[\boldsymbol{M}]|}\,\exp\left\{-\frac{1}{2}\left(\boldsymbol{x}'^t\,(\boldsymbol{A}\boldsymbol{M}^{-1})^t(\boldsymbol{A}\boldsymbol{M}^{-1})\,\boldsymbol{x}'\right)\right\}$$

and the model is invariant under G_l if the action on the parameter space is

$$G_l:\boldsymbol{A}\longrightarrow\boldsymbol{A}' = \boldsymbol{A}\boldsymbol{M}^{-1};\quad \boldsymbol{A} = \boldsymbol{A}'\boldsymbol{M};\quad \text{det}[\boldsymbol{A}] = \text{det}[\boldsymbol{A}']\,\text{det}[\boldsymbol{M}]$$

so

$$p(\boldsymbol{x}'|\boldsymbol{\theta}') = (2\pi)^{-1}\,|\text{det}[\boldsymbol{A}']|\,\exp\left\{-\frac{1}{2}\left(\boldsymbol{x}'^t\,\boldsymbol{A}'^t\boldsymbol{A}'\,\boldsymbol{x}'\right)\right\}$$

Then, the Haar equation reads

$$\int_\Theta p(\bullet|\boldsymbol{A}')\,\pi(\boldsymbol{A}')\,d\boldsymbol{A}' = \int_\Theta p(\bullet|g\circ\boldsymbol{A})\,\pi(\boldsymbol{A})\,d\boldsymbol{A} = \int_\Theta p(\bullet|\boldsymbol{A}')\,\pi(\boldsymbol{A}'\boldsymbol{M})\,J(\boldsymbol{A}';\,\boldsymbol{A})\,d\boldsymbol{A}$$

and, in consequence, $\forall\boldsymbol{M}\in G$

$$\pi(\boldsymbol{A}'\boldsymbol{M})\,J(\boldsymbol{A}';\,\boldsymbol{A})\,da'_{11}da'_{21}da'_{22} = \pi(\boldsymbol{A}')\,da'_{11}da'_{21}da'_{22}$$

For the action on the left:

$$\boldsymbol{M} = \boldsymbol{T} = \begin{pmatrix} a & 0 \\ b & c \end{pmatrix}; a, b > 0 \quad \longrightarrow \quad J(\boldsymbol{A}'; \boldsymbol{A}) = a^2 c$$

and, in consequence

$$\pi(aa'_{11}, aa'_{21} + ba'_{22}, ca'_{22})\, a^2 c = \pi(a'_{11}, a'_{21}, a'_{22}) \longrightarrow \pi(a'_{11}, a'_{21}, a'_{22}) \propto \frac{1}{{a'_{11}}^2 a'_{22}}$$

For the action on the right:

$$\boldsymbol{M} = \boldsymbol{T}^{-1} = \begin{pmatrix} a^{-1} & 0 \\ -b(ac)^{-1} & c^{-1} \end{pmatrix} \quad \longrightarrow \quad J(\boldsymbol{A}'; \boldsymbol{A}) = (ac^2)^{-1}$$

and, in consequence

$$\pi\left(\frac{a'_{11}}{a}, \frac{ca'_{21} - ba'_{22}}{ac}, \frac{a'_{22}}{c}\right) \frac{1}{ac^2} = \pi(a'_{11}, a'_{21}, a'_{22}) \longrightarrow \pi(a'_{11}, a'_{21}, a'_{22}) \propto \frac{1}{a'_{11}{a'_{22}}^2}$$

In terms of the parameters of interest $\{\sigma_1, \sigma_2, \rho\}$, since

$$da_{11} da_{21} da_{22} = \frac{1}{\sigma_1^2 \sigma_2^2 (1 - \rho^2)^2}\, d\sigma_1 d\sigma d\rho$$

we have finally that for invariance under G_l:

$$\pi_L^l(\sigma_1, \sigma_2, \rho) = \frac{1}{\sigma_1 \sigma_2 (1 - \rho^2)^{3/2}} \quad \text{and} \quad \pi_R^l(\sigma_1, \sigma_2, \rho) = \frac{1}{\sigma_2^2 (1 - \rho^2)}$$

The same analysis with decomposition in upper triangular matrices leads to

$$\pi_L^u(\sigma_1, \sigma_2, \rho) = \frac{1}{\sigma_1 \sigma_2 (1 - \rho^2)^{3/2}} \quad \text{and} \quad \pi_R^u(\sigma_1, \sigma_2, \rho) = \frac{1}{\sigma_1^2 (1 - \rho^2)}$$

As we see, in both cases the left Haar invariant prior coincides with Jeffrey's prior when $\{\mu_1, \mu_2\}$ and $\{\sigma_1, \sigma_2, \rho\}$ are decoupled.

At this point, one may be tempted to use a right Haar invariant prior where the two parameters σ_1 and σ_2 are treated on equal footing

$$\pi(\sigma_1, \sigma_2, \rho) = \frac{1}{\sigma_1 \sigma_2 (1 - \rho^2)}$$

Under this prior, since the sample correlation

$$r = \frac{\sum_i (x_{1i} - \overline{x}_1)(x_{2i} - \overline{x}_2)}{\left(\sum_i (x_{1i} - \overline{x}_1)^2 \sum_i (x_{2i} - \overline{x}_2)^2\right)^{1/2}}$$

is a sufficient statistics for ρ, we have that the posterior for inferences on the correlation coefficient will be

$$p(\rho|\boldsymbol{x}) \propto (1-\rho^2)^{(n-3)/2}\, F(n-1, n-1, n-1/2; (1+r\rho)/2)$$

with $F(a, b, c; z)$ the Hypergeometric Function.

Example 2.13 If $\theta \in \Theta \longrightarrow g\circ\theta = \phi(\theta) = \theta' \in \Theta$ with $\phi(\theta)$ is a one-to-one differentiable mapping, then

$$\begin{aligned}\int_\Theta p(\bullet|\theta')d\mu(\theta) &= \int_\Theta p(\bullet|\theta')\pi(\theta)d\theta = \int_\Theta p(\bullet|\theta')\pi(\phi^{-1}(\theta'))\left|\frac{\partial\phi^{-1}(\theta')}{\partial\theta'}\right| d\theta' = \\ &= \int_\Theta p(\bullet|\theta')\pi(\theta')d\theta' = \int_\Theta p(\bullet|\theta')d\mu(\theta')\end{aligned}$$

and therefore, Jeffreys' prior defines a Haar invariant measure.

2.6.5 Conjugated Distributions

In as much as possible, we would like to consider reference priors $\pi(\theta|a, b, \ldots)$ versatile enough such that by varying some of the parameters $a, b, \ldots$ we get diverse forms to analyze the effect on the final results and, on the other hand, to simplify the evaluation of integrals like:

$$p(x) = \int p(x|\theta)\cdot p(\theta)\, d\theta \qquad \text{and} \qquad p(y|x) = \int p(y|\theta)\cdot p(\theta|x)\, d\theta$$

This leads us to consider as reference priors the *Conjugated Distributions* [11].

Let $\mathcal{S}$ be a class of sampling distributions $p(x|\theta)$ and $\mathcal{P}$ the class of prior densities for the parameter θ. If

$$p(\theta|x) \in \mathcal{P} \qquad \text{for all} \qquad p(x|\theta) \in \mathcal{S} \;\; \text{and} \;\; p(\theta) \in \mathcal{P}$$

we say that the class $\mathcal{P}$ is conjugated to $\mathcal{S}$. We are mainly interested in the class of priors $\mathcal{P}$ that have the same functional form as the likelihood. In this case, since both the prior density and the posterior belong to the same family of distributions, we say that they are *closed under sampling*. It should be stressed that the criteria for taking conjugated reference priors is eminently practical and, in many cases, they do not exist. In fact, only the *exponential family* of distributions has conjugated prior densities. Thus, if $\boldsymbol{x} = \{x_1, x_2, \ldots, x_n\}$ is an exchangeable random sampling from the k-parameter regular exponential family, then

$$p(\boldsymbol{x}|\boldsymbol{\theta}) = f(\boldsymbol{x})\, g(\boldsymbol{\theta}) \exp\left\{\sum_{j=1}^{k} c_j\, \phi_j(\boldsymbol{\theta}) \left(\sum_{i=1}^{n} h_j(x_i)\right)\right\}$$

and the *conjugated prior* will have the form:

$$\pi(\boldsymbol{\theta}|\boldsymbol{\tau}) = \frac{1}{K(\boldsymbol{\tau})}\, [g(\boldsymbol{\theta})]^{\tau_0} \exp\left\{\sum_{j=1}^{k} c_j\, \phi_j(\boldsymbol{\theta})\, \tau_j\right\}$$

where $\boldsymbol{\theta} \in \Theta$, $\boldsymbol{\tau} = \{\tau_0, \tau_1, \ldots, \tau_k\}$ the *hyperparameters* and $K(\boldsymbol{\tau}) < \infty$ the normalization factor so $\int_{\Theta} \pi(\boldsymbol{\theta}|\boldsymbol{\tau}) d\boldsymbol{\theta} = 1$. Then, the general scheme will be[7]:

(1) Choose the class of priors $\pi(\boldsymbol{\theta}|\boldsymbol{\tau})$ that reflect the structure of the model;
(2) Choose a prior function $\pi(\boldsymbol{\tau})$ for the *hyperparameters*;
(3) Express the posterior density as $p(\boldsymbol{\theta}, \boldsymbol{\tau}|\boldsymbol{x}) \propto p(\boldsymbol{x}|\boldsymbol{\theta})\pi(\boldsymbol{\theta}|\boldsymbol{\tau})\pi(\boldsymbol{\tau})$;
(4) Marginalize for the parameters of interest:

$$p(\boldsymbol{\theta}|\boldsymbol{x}) \propto \int_{\Phi} p(\boldsymbol{x}|\boldsymbol{\theta})\pi(\boldsymbol{\theta}|\boldsymbol{\tau})\pi(\boldsymbol{\tau}) d\boldsymbol{\tau}$$

or, if desired, get the conditional density

$$p(\boldsymbol{\theta}|\boldsymbol{x}, \boldsymbol{\tau}) = \frac{p(\boldsymbol{x}, \boldsymbol{\theta}, \boldsymbol{\tau})}{p(\boldsymbol{x}, \boldsymbol{\tau})} = \frac{p(\boldsymbol{x}|\boldsymbol{\theta})\pi(\boldsymbol{\theta}|\boldsymbol{\tau})}{p(\boldsymbol{x}|\boldsymbol{\tau})}$$

The obvious question that arises is how do we choose the prior $\pi(\phi)$ for the hyperparameters. Besides *reasonableness*, we may consider two approaches. Integrating the parameters $\boldsymbol{\theta}$ of interest, we get

$$p(\boldsymbol{\tau}, \boldsymbol{x}) = \pi(\boldsymbol{\tau}) \int_{\Theta} p(\boldsymbol{x}|\boldsymbol{\theta})\pi(\boldsymbol{\theta}|\boldsymbol{\tau})\, d\boldsymbol{\theta} = \pi(\boldsymbol{\tau})\, p(\boldsymbol{x}|\boldsymbol{\tau})$$

so we may use any of the procedures under discussion to take $\pi(\boldsymbol{\tau})$ as the prior for the model $p(\boldsymbol{x}|\boldsymbol{\tau})$ and then obtain

$$\pi(\boldsymbol{\theta}) = \int_{\Omega_{\tau}} \pi(\boldsymbol{\theta}|\boldsymbol{\tau})\, \pi(\boldsymbol{\tau})\, d\boldsymbol{\tau}$$

[7]We can go an step upwards and assign a prior to the hyperparameters with hyper-hyperparameters,…

The beauty of Bayes rule but not very practical in complicated situations. A second approach, more ugly and practical, is the so called *Empirical Method* where we assign numeric values to the hyperparameters suggested by $p(\boldsymbol{x}|\boldsymbol{\tau})$ (for instance, moments, maximum-likelihood estimation,...); that is, setting, in a distributional sense, $\pi(\boldsymbol{\tau}) = \delta_{\boldsymbol{\tau}^\star}$ so $\langle \pi(\boldsymbol{\tau}), p(\boldsymbol{\theta}, \boldsymbol{x}, \boldsymbol{\tau})\rangle = p(\boldsymbol{\theta}, \boldsymbol{x}, \boldsymbol{\tau}^\star)$. Thus,

$$p(\boldsymbol{\theta}|\boldsymbol{x}, \boldsymbol{\tau}^\star) \propto p(\boldsymbol{x}|\boldsymbol{\theta})\pi(\boldsymbol{\theta}|\boldsymbol{\tau}^\star)$$

Obviously, fixing the hyperparameters assumes a perfect knowledge of them and does not allow for variations but the procedure may be useful to guess at least were to go.

Last, it may happen that a single conjugated prior does not represent sufficiently well our beliefs. In this case, we may consider a k-mixture of conjugated priors

$$\pi(\boldsymbol{\theta}|\boldsymbol{\tau}_1, \ldots, \boldsymbol{\tau}_k) = \sum_{i=1}^{k} w_i\, \pi(\boldsymbol{\theta}|\boldsymbol{\tau}_i)$$

In fact [12], any prior density for a model that belongs to the exponential family can be approximated arbitrarily close by a mixture of conjugated priors.

Example 2.14 Let's see the conjugated prior distributions for some models:

• **Poisson model** $Po(n|\mu)$: Writing

$$p(n|\mu) = \frac{e^{-\mu}\, \mu^n}{\Gamma(n+1)} = \frac{e^{-(\mu - n\log\mu)}}{\Gamma(n+1)}$$

it is clear that the Poisson distribution belongs to the exponential family and the conjugated prior density for the parameter μ is

$$\pi(\mu|\tau_1, \tau_2) \propto e^{-\tau_1\mu + \tau_2 \log\mu} \propto Ga(\mu|\tau_1, \tau_2)$$

If we set a prior $\pi(\tau_1, \tau_2)$ for the hyperparameters we can write

$$p(n, \mu, \tau_1, \tau_2)\; p(n|\mu)\; \pi(\mu|\tau_1, \tau_2) = \pi(\tau_1, \tau_2)$$

and integrating μ:

$$p(n, \tau_1, \tau_2) = \left[\frac{\Gamma(n+\tau_2)}{\Gamma(\tau_1)}\, \frac{\tau_1^{\tau_2}}{(1+\tau_1)^{n+\tau_2}}\right] \pi(\tau_1, \tau_2) = p(n|\tau_1, \tau_2)\, \pi(\tau_1, \tau_2)$$

• **Binomial model** $Bi(n|N, \theta)$: Writing

$$P(n|N, \theta) = \binom{N}{n}\, \theta^n\, (1-\theta)^{N-n} = \binom{N}{n}\, e^{n\log\theta + (N-n)\log(1-\theta)}$$

it is clear that it belong to the exponential family and the conjugated prior density for the parameter θ will be:

$$\pi(\theta|\tau_1, \tau_2) = Be(\tau|\tau_1, \tau_2)$$

- **Multinomial model** Let $\boldsymbol{X} = (X_1, X_2, \ldots, X_k) \sim Mn(\boldsymbol{x}|\boldsymbol{\theta})$; that is:

$$\boldsymbol{X} \sim p(\boldsymbol{x}|\boldsymbol{\theta}) = \Gamma(n+1) \prod_{i=1}^{k} \frac{\theta_i^{x_i}}{\Gamma(x_i+1)} \qquad \begin{cases} X_i \in N, & \sum_{i=1}^{k} X_i = n \\ \theta_i \in [0, 1], & \sum_{i=1}^{k} \theta_i = 1 \end{cases}$$

The Dirichlet distribution $Di(\boldsymbol{\theta}|\boldsymbol{\alpha})$:

$$\pi(\boldsymbol{\theta}|\boldsymbol{\alpha}) = D(\boldsymbol{\alpha}) \prod_{i=1}^{k} \theta_i^{\alpha_i - 1} \qquad \begin{cases} \boldsymbol{\alpha} = (\alpha_1, \alpha_2, \ldots, \alpha_k), \quad \alpha_i > 0, \quad \sum_{i=1}^{k} \alpha_i = \alpha_0 \\ D(\boldsymbol{\alpha}) = \Gamma(\alpha_0) \left[\prod_{i=1}^{k} \Gamma(\alpha_i)\right]^{-1} \end{cases}$$

is the natural conjugated prior for this model. It is a degenerated distribution in the sense that

$$\pi(\boldsymbol{\theta}|\boldsymbol{\alpha}) = D(\boldsymbol{\alpha}) \left[\prod_{i=1}^{k-1} \theta_i^{\alpha_i - 1}\right] \left[1 - \sum_{i=1}^{k-1} \theta_i\right]^{\alpha_k - 1}$$

The posterior density will then be $\boldsymbol{\theta} \sim Di(\boldsymbol{\theta}|\boldsymbol{x} + \boldsymbol{\alpha})$ with

$$E[\theta_i] = \frac{x_i + \alpha_i}{n + \alpha_0} \qquad \text{and} \qquad V[\theta_i, \theta_j] = \frac{E[\theta_i](\delta_{ij} - E[\theta_j])}{n + \alpha_0 + 1}$$

The parameters $\boldsymbol{\alpha}$ of the Dirichlet distribution $Di(\boldsymbol{\theta}|\boldsymbol{\alpha})$ determine the expected values $E[\theta_i] = \alpha_i/\alpha_0$. In practice, it is more convenient to control also the variances and use the *Generalized Dirichlet Distribution* $GDi(\boldsymbol{\theta}|\boldsymbol{\alpha}, \boldsymbol{\beta})$:

$$\pi(\boldsymbol{\theta}|\boldsymbol{\alpha}, \boldsymbol{\beta}) = \prod_{i=1}^{k-1} \frac{\Gamma(\alpha_i + \beta_i)}{\Gamma(\alpha_i)\Gamma(\beta_i)} \theta_i^{\alpha_i - 1} \left[1 - \sum_{j=1}^{i} \theta_j\right]^{\gamma_i}$$

where:

$$0 < \theta_i < 1, \quad \sum_{i=1}^{k-1} \theta_i < 1, \quad \theta_n = 1 - \sum_{i=1}^{k-1} \theta_i$$

$$\alpha_i > 0, \quad \beta_i > 0, \quad \text{and} \quad \gamma_i \begin{cases} \beta_i - \alpha_{i+1} - \beta_{i+1}; & i = 1, 2, \ldots, k-2 \\ \beta_{k-1} - 1; & i = k-1 \end{cases}$$

When $\beta_i = \alpha_{i+1} + \beta_{i+1}$ it becomes the Dirichlet distribution. For this prior we have that

$$E[\theta_i] = \frac{\alpha_i}{\alpha_i + \beta_i} S_i \quad \text{and} \quad V[\theta_i, \theta_j] = E[\theta_j] \left(\frac{\alpha_i + \delta_{ij}}{\alpha_i + \beta_i + 1} T_i - E[\theta_i] \right)$$

where

$$S_i = \prod_{j=1}^{i-1} \frac{\beta_j}{\alpha_j + \beta_j} \quad \text{and} \quad T_i = \prod_{j=1}^{i-1} \frac{\beta_j + 1}{\alpha_j + \beta_j + 1}$$

with $S_1 = T_1 = 1$ and we can have control over the prior means and variances.

2.6.6 Probability Matching Priors

A pragmatic criteria is that of *probability matching priors* for which the one sided credible intervals derived from the posterior distribution coincide, to a certain level of accuracy, with those derived by the classical approach. This condition leads to a differential equation for the prior distribution [13, 14]. We shall illustrate in the following lines the rationale behind for the simple one parameter case assuming that the needed regularity conditions are satisfied.

Consider then a random quantity $X \sim p(x|\theta)$ and an iid sampling $\boldsymbol{x} = \{x_1, x_2, \ldots, x_n\}$ with θ the parameter of interest. The classical approach for inferences is based on the likelihood

$$p(\boldsymbol{x}|\theta) = p(x_1, x_2, \ldots x_n|\theta) = \prod_{i=1}^{n} p(x_i|\theta)$$

and goes through the following reasoning:

(1) Assumes that the parameter θ has the *true* but unknown value θ_0 so the sample is actually drawn from $p(x|\theta_0)$;

(2) Find the estimator $\theta_m(\boldsymbol{x})$ of θ_0 as the value of θ that maximizes the likelihood; that is:

$$\theta_m = \max_{\theta}\{p(\boldsymbol{x}|\theta)\} \quad \longrightarrow \quad \left(\frac{\partial \ln p(\boldsymbol{x}|\theta)}{\partial \theta} \right)_{\theta_m} = 0$$

(3) Given the model $X \sim p(x|\theta_0)$, after the appropriate change of variables get the distribution

$$p(\theta_m|\theta_0)$$

of the random quantity $\theta_m(X_1, X_2, \dots X_n)$ and draw inferences from it.

The Bayesian inferential process considers a prior distribution $\pi(\theta)$ and draws inferences on θ from the posterior distribution of the quantity of interest

$$p(\theta|\boldsymbol{x}) \propto p(\boldsymbol{x}|\theta)\,\pi(\theta)$$

Let's start with the Bayesian and expand the term on the right around θ_m. On the one hand:

$$\ln\frac{p(\boldsymbol{x}|\theta)}{p(\boldsymbol{x}|\theta_m)} = \frac{1}{2!}\left(\frac{\partial^2 \ln p(\boldsymbol{x}|\theta)}{\partial\theta^2}\right)_{\theta_m}(\theta-\theta_m)^2 + \frac{1}{3!}\left(\frac{\partial^3 \ln p(\boldsymbol{x}|\theta)}{\partial\theta^3}\right)_{\theta_m}(\theta-\theta_m)^3 + \cdots$$

Now,

$$-\frac{1}{n}\frac{\partial^2 \ln p(\boldsymbol{x}|\theta)}{\partial\theta^2} = \frac{1}{n}\sum_{i=1}^{n}\frac{\partial^2(-\ln p(x_i|\theta))}{\partial\theta^2} \;\xrightarrow{n\to\infty}\; \mathrm{E}_X\left[\frac{\partial^2(-\ln p(x|\theta))}{\partial\theta^2}\right] = I(\theta)$$

so we can substitute:

$$\left(\frac{\partial^2 \ln p(\boldsymbol{x}|\theta)}{\partial\theta^2}\right)_{\theta_m} = -n\,I(\theta_m) \quad \text{and} \quad \left(\frac{\partial^3 \ln p(\boldsymbol{x}|\theta)}{\partial\theta^3}\right)_{\theta_m} = -n\left(\frac{\partial I(\theta)}{\partial\theta}\right)_{\theta_m}$$

to get

$$p(\boldsymbol{x}|\theta) = e^{\ln p(\boldsymbol{x}|\theta)} \propto e^{-\frac{nI(\theta_m)}{2}(\theta-\theta_m)^2}\left(1 - \frac{n}{3!}\left(\frac{\partial I(\theta)}{\partial\theta}\right)_{\theta_m}(\theta-\theta_m)^3 + \cdots\right)$$

On the other hand:

$$\pi(\theta) = \pi(\theta_m)\left(1 + \left(\frac{1}{\pi(\theta)}\frac{\partial\pi(\theta)}{\partial\theta}\right)_{\theta_m}(\theta-\theta_m) + \cdots\right)$$

so If we define the random quantity $T = \sqrt{nI(\theta_m)}(\theta-\theta_m)$ and consider that

$$I^{-3/2}(\theta)\,\frac{\partial I(\theta)}{\partial\theta} = -2\,\frac{\partial I^{-1/2}}{\partial\theta}$$

we get finally:

$$p(t|\boldsymbol{x}) = \frac{\exp(-t^2/2)}{\sqrt{2\pi}}\left(1 + \frac{1}{\sqrt{n}}\left[\left(\frac{I^{-1/2}(\theta)}{\pi(\theta)}\frac{\partial\pi(\theta)}{\partial\theta}\right)_{\theta_m} t + \frac{1}{3}\left(\frac{\partial I^{-1/2}}{\partial\theta}\right)_{\theta_m} t^3\right] + O\left(\frac{1}{n}\right)\right)$$

Let's now find

$$P(T \leq z|\boldsymbol{x}) = \int_{-\infty}^{z} p(t|\boldsymbol{x})dt$$

Defining

$$Z(x) = \frac{1}{\sqrt{2\pi}} e^{-x^2/2} \quad \text{and} \quad P(x) = \int_{-\infty}^{x} Z(t)dt$$

and considering that

$$\int_{-\infty}^{z} Z(t)\, t\, dt = -Z(z) \quad \text{and} \quad \int_{-\infty}^{z} Z(t)\, t^3\, dt = -Z(z)\, (z^2+2)$$

it is straight forward to get:

$$P(T \leq z|\boldsymbol{x}) = P(z) - \frac{Z(z)}{\sqrt{n}} \left[\left(\frac{I^{-1/2}(\theta)}{\pi(\theta)} \frac{\partial \pi(\theta)}{\partial \theta} \right)_{\theta_m} + \frac{z^2+2}{3} \left(\frac{\partial I^{-1/2}}{\partial \theta} \right)_{\theta_m} \right] O\left(\frac{1}{n}\right)$$

From this probability distribution, we can infer what the classical approach will get. Since he will draw inferences from $p(\boldsymbol{x}|\theta_0)$, we can take a sequence of proper priors $\pi_k(\theta|\theta_0)$ for $k = 1, 2, \ldots$ that induce a sequence of distributions such that

$$\lim_{k\to\infty} \langle \pi_k(\theta|\theta_0),\, p(\boldsymbol{x}|\theta) \rangle = p(\boldsymbol{x}|\theta_0)$$

In Distributional sense, the sequence of distributions generated by

$$\pi_k(\theta|\theta_0) = \frac{k}{2}\, \mathbf{1}_{[\theta_0 - 1/k, \theta_0 + 1/k]}; \qquad k = 1, 2, \ldots$$

converge to the Delta distribution δ_{θ_0} and, from distributional derivatives, as $k \to \infty$,

$$\langle \frac{d}{d\theta} \pi_k(\theta|\theta_0),\, I^{-1/2}(\theta) \rangle = -\langle \pi_k(\theta|\theta_0),\, \frac{d}{d\theta} I^{-1/2}(\theta) \rangle \simeq - \left(\frac{\partial I^{-1/2}(\theta))}{\partial \theta} \right)_{\theta_0}$$

But $\theta_0 = \theta_m + O(1/\sqrt{n})$ so, for a sequence of priors that shrink to $\theta_0 \simeq \theta_m$,

$$P(T \leq z|\boldsymbol{x}) = P(z) - \frac{Z(z)}{\sqrt{n}} \left[\frac{z^2+1}{3} \left(\frac{\partial I^{-1/2}}{\partial \theta} \right)_{\theta_m} \right] + O\left(\frac{1}{n}\right)$$

For terms of order $O(1/\sqrt{n})$ in both expressions of $P(T \leq z|\boldsymbol{x})$ to be the same, we need that:

$$\left(\frac{1}{\sqrt{I(\theta)}} \frac{1}{\pi(\theta)} \frac{\partial \pi(\theta)}{\partial \theta}\right)_{\theta_m} = -\left(\frac{\partial I^{-1/2}}{\partial \theta}\right)_{\theta_m}$$

and therefore

$$\pi(\theta) = I^{1/2}(\theta)$$

that is, Jeffrey's prior. In the case of n-dimensional parameters, the reasoning goes along the same lines but the expressions and the development become much more lengthy and messy and we refer to the literature.

The procedure for a first order *probability matching prior* [15, 16] starts from the likelihood

$$p(x_1, x_2, \ldots x_n | \theta_1, \theta_2, \ldots \theta_p)$$

and then:

(1) Get the Fisher's matrix $\boldsymbol{I}(\theta_1, \theta_2, \ldots \theta_p)$ and the inverse $\boldsymbol{I}^{-1}(\theta_1, \theta_2, \ldots \theta_p)$;
(2) Suppose we are interested in the parameter $t = t(\theta_1, \theta_2, \ldots \theta_p)$ a twice continuous and differentiable function of the parameters. Define the column vector

$$\nabla_t = \left(\frac{\partial t}{\partial \theta_1}, \frac{\partial t}{\partial \theta_2}, \ldots, \frac{\partial t}{\partial \theta_p}\right)^T$$

(3) Define the column vector

$$\eta = \frac{\boldsymbol{I}^{-1} \nabla_t}{(\nabla_t^T \boldsymbol{I}^{-1} \nabla_t)^{1/2}} \qquad \text{so that} \qquad \eta^T \boldsymbol{I} \eta = 1$$

(4) The probability matching prior for the parameter $t = t(\boldsymbol{\theta})$ in terms of $\theta_1, \theta_2, \ldots \theta_p$ is given by the equation:

$$\sum_{k=1}^{p} \frac{\partial}{\partial \theta_k} [\eta_k(\boldsymbol{\theta})\, \pi(\boldsymbol{\theta})] = 0$$

Any solution $\pi(\theta_1, \theta_2, \ldots \theta_p)$ will do the job.
(5) Introduce $t = t(\boldsymbol{\theta})$ in this expression, say, for instance $\theta_1 = \theta_1(t, \theta_2, \ldots \theta_p)$, and the corresponding Jacobian $J(t, \theta_2, \ldots \theta_p)$. Then we get the prior for the parameter t of interest and the nuisance parameters $\theta_2, \ldots \theta_p$ that, eventually, will be integrated out.

Example 2.15 Consider two independent random quantities X_1 and X_2 such that

$$P(X_i = n_k) = Po(n_k | \mu_i).$$

We are interested in the parameter $t = \mu_1/\mu_2$ so setting $\mu = \mu_2$ we have the ordered parameterization $\{t, \mu\}$. The joint probability is

$$P(n_1, n_2|\mu_1, \mu_2) = P(n_1|\mu_1)\, P(n_2|\mu_2) = e^{-(\mu_1 + \mu_2)} \frac{\mu_1^{n_1} \mu_2^{n_2}}{\Gamma(n_1 + 1)\Gamma(n_2 + 1)}$$

from which we get the Fisher's matrix

$$\mathbf{I}(\mu_1, \mu_2) = \begin{pmatrix} 1/\mu_1 & 0 \\ 0 & 1/\mu_2 \end{pmatrix} \qquad \text{and} \qquad \mathbf{I}^{-1}(\mu_1, \mu_2) = \begin{pmatrix} \mu_1 & 0 \\ 0 & \mu_2 \end{pmatrix}$$

We are interested in the parameter $t = \mu_1/\mu_2$, a twice continuous and differentiable function of the parameters, so

$$\nabla_t(\mu_1, \mu_2) = \left(\frac{\partial t}{\partial \mu_1}, \frac{\partial t}{\partial \mu_2} \right)^T = \left(\mu_2^{-1}, -\mu_1 \mu_2^{-2} \right)^T = \begin{pmatrix} \mu_2^{-1} \\ -\mu_1 \mu_2^{-2} \end{pmatrix}$$

Therefore:

$$\mathbf{I}^{-1}\, \nabla_t = \begin{pmatrix} \mu_1 \mu_2^{-1} \\ -\mu_1 \mu_2^{-1} \end{pmatrix} \qquad S = \nabla_t^T\, \mathbf{I}^{-1}\, \nabla_t = \frac{\mu_1(\mu_1 + \mu_2)}{\mu_2^3}$$

$$\eta = \frac{\mathbf{I}^{-1}\, \nabla_t}{(\nabla_t^T \mathbf{I}^{-1}\, \nabla_t)^{1/2}} = \begin{pmatrix} (\mu_1 \mu_2)^{1/2} (\mu_1 + \mu_2)^{-1/2} \\ -(\mu_1 \mu_2)^{1/2} (\mu_1 + \mu_2)^{-1/2} \end{pmatrix}$$

so that $\eta^T \mathbf{I} \eta = 1$. The probability matching prior for the parameter $t = \mu_1/\mu_2$ in terms of μ_1 and μ_2 is given by the equation:

$$\sum_{k=1}^{2} \frac{\partial}{\partial \mu_k} \left[\eta_k(\mu)\, \pi(\mu) \right] = 0$$

so, if $f(\mu_1, \mu_2) = (\mu_1 \mu_2)^{1/2} (\mu_1 + \mu_2)^{-1/2}$, we have to solve

$$\frac{\partial}{\partial \mu_1} f(\mu_1, \mu_2)\, \pi(\mu_1, \mu_2) = \frac{\partial}{\partial \mu_2} f(\mu_1, \mu_2)\, \pi(\mu_1, \mu_2)$$

Any solution will do so:

$$\pi(\mu_1, \mu_2) \propto f^{-1}(\mu_1, \mu_2) = \frac{\sqrt{\mu_1 + \mu_2}}{\sqrt{\mu_1 \mu_2}}$$

Substituting $\mu_1 = t\mu_2$ and including the Jacobian $J = \mu_2$ we have finally:

$$\pi(t, \mu_2) \propto \sqrt{\mu_2}\sqrt{\frac{1+t}{t}}$$

The posterior density will be:

$$p(t, \mu_2|n_1, n_2) \propto p(n_1, n_2|t, \mu_2)\pi(t, \mu_2) \propto e^{-\mu_2(1+t)} t^{n_1-1/2}\,(1+t)^{1/2}\,\mu_2^{n+3/2-1}$$

and, integrating the nuisance parameter $\mu_2 \in [0, \infty)$, we get the posterior density:

$$p(t|n_1, n_2) = N\,\frac{t^{n_1-1/2}}{(1+t)^{n+1}}$$

with $N^{-1} = B(n_1 + 1/2, n_2 + 1/2)$.

Example 2.16 (*Gamma distribution*) Show that for $Ga(x|\alpha, \beta)$:

$$p(x|\alpha, \beta) = \frac{\alpha^{\beta}}{\Gamma(\beta)}\,e^{-\alpha x}\,x^{\beta-1}\,\mathbf{1}_{(0,\infty)}(x)$$

the probability matching prior for the ordering

- $\{\beta, \alpha\}$ is $\pi(\alpha, \beta) = \beta^{-1/2}\left[\alpha^{-1}\sqrt{\beta\Psi'(\beta)-1}\right]$
- $\{\alpha, \beta\}$ is $\pi(\alpha, \beta) = \left[\alpha^{-1}\sqrt{\Psi'(\beta)}\right]\sqrt{\beta\Psi'(\beta)-1}$

to be compared with Jeffrey's prior $\pi_2^J(\alpha, \beta) = \alpha^{-1}\sqrt{\beta\Psi'(\beta)-1}$ and Jeffrey's prior when both parameters are treated individually $\pi_{1+1}^J(\alpha, \beta) = \alpha^{-1}\sqrt{\Psi'(\beta)}$

Example 2.17 (*Bivariate Normal Distribution*)

For the ordered parameterization ρ, σ_1, σ_2: the Fisher's matrix (see Example 2.12) is:

$$\boldsymbol{I}(\rho, \sigma_1, \sigma_2) = (1-\rho^2)^{-1}\begin{pmatrix} (1+\rho^2)(1-\rho^2)^{-1} & -\rho\sigma_1^{-1} & -\rho\sigma_2^{-1} \\ -\rho\sigma_1^{-1} & (2-\rho^2)\sigma_1^{-2} & -\rho^2(\sigma_1\sigma_2)^{-1} \\ -\rho\sigma_2^{-1} & -\rho^2(\sigma_1\sigma_2)^{-1} & (2-\rho^2)\sigma_2^{-2} \end{pmatrix}$$

and the inverse:

$$\boldsymbol{I}^{-1}(\rho, \sigma_1, \sigma_2) = \frac{1}{2}\begin{pmatrix} 2(1-\rho^2)^2 & \sigma_1\rho(1-\rho^2) & \sigma_2\rho(1-\rho^2) \\ \sigma_1\rho(1-\rho^2) & \sigma_1^2 & \rho^2\sigma_1\sigma_2 \\ \sigma_2\rho(1-\rho^2) & \rho^2\sigma_1\sigma_2 & \sigma_2^2 \end{pmatrix}$$

Then

$$\frac{2}{\rho}\frac{\partial}{\partial\rho}[\pi(1-\rho^2)] + \frac{\partial}{\partial\sigma_1}[\pi\sigma_1] + \frac{\partial}{\partial\sigma_2}[\pi\sigma_2] = 0$$

for which

$$\pi(\sigma_1, \sigma_2, \rho) = \frac{1}{\sigma_1 \sigma_2 (1 - \rho^2)}$$

is a solution.

Problem 2.6 Consider

$$X \sim p(x|a, b, \sigma) = \frac{\sinh[\sigma(b-a)]}{2(b-a)} \frac{1}{\cosh[\sigma(x-a)]\cosh[\sigma(b-x)]} \mathbf{1}_{(-\infty,\infty)}(x)$$

where $a < b \in \mathcal{R}$ and $\sigma \in (0, \infty)$. Show that

$$E[X] = \frac{b+a}{2} \qquad \text{and} \qquad V[x] = \frac{(b-a)^2}{12} + \frac{\pi^2}{12\sigma^2}$$

and that, for known $\sigma >>$, the probability matching prior for a and b tends to $\pi_{pm}(a, b) \sim (b-a)^{-1/2}$. Show also that, under the same limit, $\pi_{pm}(\theta) \sim \theta^{-1/2}$ for $(a, b) = (-\theta, \theta)$ and $(a, b) = (0, \theta)$. Since $p(x|a, b, \sigma) \stackrel{\sigma>>}{\rightarrow} Un(x|a, b)$ discuss in this last case what is the difference with the Example 2.4.

2.6.7 *Reference Analysis*

The expected amount of information (*Expected Mutual Information*) on the parameter θ provided by k independent observations of the model $p(\boldsymbol{x}|\theta)$ relative to the prior knowledge on θ described by $\pi(\theta)$ is

$$I\,[e(k), \pi(\theta)] = \int_{\Theta} \pi(\theta)\, d\theta \int_{\Omega_X} p(z_k|\theta)\, \log\, \frac{p(\theta|z_k)}{\pi(\theta)} dz_k$$

where $z_k = \{\boldsymbol{x}_1, \ldots, \boldsymbol{x}_k\}$. If $\lim_{k\rightarrow\infty} I\,[e(k), \pi(\theta)]$ exists, it will quantify the maximum amount of information that we could obtain on θ from experiments described by this model relative to the prior knowledge $\pi(\theta)$. The central idea of the *reference analysis* [4, 17] is to take as *reference prior for the model* $p(\boldsymbol{x}|\theta)$ that which maximizes the maximum amount of information we may get so it will be the *less informative* for this model. From Calculus of Variations, if we introduce the prior $\pi^\star(\theta) = \pi(\theta) + \epsilon\eta(\theta)$ with $\pi(\theta)$ an extremal of the expected information $I\,[e(k), \pi(\theta)]$ and $\eta(\theta)$ such that

$$\int_{\Theta} \pi(\theta) d\theta = \int_{\Theta} \pi^\star(\theta) d\theta = 1 \qquad \longrightarrow \qquad \int_{\Theta} \eta(\theta) d\theta = 0$$

it is easy to see (left as exercise) that

$$\pi(\theta) \propto \exp\left\{\int_{\Omega_X} p(z_k|\theta)\, \log\, p(\theta|z_k)\, dz_k\right\} = f_k(\theta)$$

This is a nice but complicated implicit equation because, on the one hand, $f_k(\theta)$ depends on $\pi(\theta)$ through the posterior $p(\theta|z_k)$ and, on the other hand, the limit $k \rightarrow \infty$ is usually divergent (intuitively, the more precision we want for θ, the more information is needed and to know the actual value from the experiment requires an infinite amount of information). This can be circumvented regularizing the expression as

$$\pi(\theta) \propto \pi(\theta_0) \lim_{k\rightarrow\infty} \frac{f_k(\theta)}{f_k(\theta_0)}$$

with θ_0 any interior point of Θ (we are used to that in particle physics!). Let's see some examples.

Example 2.18 Consider again the exponential model for which $t = n^{-1}\sum_{i=1}^{n} x_i$ is sufficient for θ and distributed as

$$p(t|\theta) = \frac{(n\theta)^n}{\Gamma(n)}\, t^{n-1} \exp\{-n\theta t\}$$

Taking $\pi(\theta) = \mathbf{1}_{(0,\infty)}(\theta)$ we have the proper posterior

$$\pi^{\star}(\theta|t) = \frac{(nt)^{n+1}}{\Gamma(n+1)}\, \exp\{-n\theta t\}\, \theta^n$$

Then $\log \pi^{\star}(\theta|t) = -(n\theta)t + n\, \log\theta + (n+1)\log t + g_1(n)$ and

$$f_n(\theta) = \exp\left\{\int_{\Omega_X} p(t|\theta)\, \log\, \pi^{\star}(\theta|t)\, dt\right\} = \frac{g_2(n)}{\theta} \quad\longrightarrow\quad \pi(\theta) \propto \pi(\theta_0) \lim_{n\rightarrow\infty} \frac{f_n(\theta)}{f_n(\theta_0)} \propto \frac{1}{\theta}$$

Example 2.19 Prior functions depend on the particular model we are treating. To learn about a parameter, we can do different experimental designs that respond to different models and, even though the parameter is the same, they may have different priors. For instance, we may be interested in the *acceptance*; the probability to accept an event under some conditions. For this, we can generate for instance a sample of N observed events and see how many (x) pass the conditions. This experimental design corresponds to a Binomial distribution

$$p(x|N,\theta) = \binom{N}{x}\, \theta^x\, (1-\theta)^{N-x}$$

with $x = \{0, 1, \ldots, N\}$. For this model, the reference prior (also Jeffrey's and PM) is $\pi(\theta) = \theta^{-1/2}(1-\theta)^{-1/2}$ and the posterior $\theta \sim Be(\theta|x + 1/2, N - x + 1/2)$. Conversely, we can generate events until r are accepted and see how many (x) have we generated. This experimental design corresponds to a Negative Binomial distribution

$$p(x|r, \theta) = \binom{x-1}{r-1} \theta^r (1-\theta)^{x-r}$$

where $x = r, r+1, \ldots$ and $r \geq 1$. For this model, the reference prior (Jeffrey's and PM too) is $\pi(\theta) = \theta^{-1}(1-\theta)^{-1/2}$ and the posterior $\theta \sim Be(\theta|r, x - r + 1/2)$.

Problem 2.7 Consider

(1) $X \sim Po(x|\theta) = \exp\{-\theta\} \frac{\theta^x}{\Gamma(x+1)}$ and the experiment $e(k) \stackrel{iid}{\to} \{x_1, x_2, \ldots x_k\}$. Take $\pi^\star(\theta) = \mathbf{1}_{(0,\infty)}(\theta)$, and show that

$$\pi(\theta) \propto \pi(\theta_0) \lim_{k\to\infty} \frac{f_k(\theta)}{f_k(\theta_0)} \propto \theta^{-1/2}$$

(2) $X \sim Bi(x|N, \theta) = \binom{N}{x} \theta^x (1-\theta)^{N-x}$ and the experiment $e(k) \stackrel{iid}{\to} \{x_1, x_2, \ldots x_k\}$. Take $\pi^\star(\theta) \propto \theta^{a-1}(1-\theta)^{b-1}\mathbf{1}_{(0,1)}(\theta)$ with $a, b > 0$ and show that

$$\pi(\theta) \propto \pi(\theta_0) \lim_{k\to\infty} \frac{f_k(\theta)}{f_k(\theta_0)} \propto \theta^{-1/2}(1-\theta)^{-1/2}$$

(Hint: For (1) and (2) consider the Taylor expansion of $\log\Gamma(z, \cdot)$ around $E[z]$ and the asymptotic behavior of the Polygamma Function $\Psi^{n)}(z) = a_n z^{-n} + a_{n+1} z^{-(n+1)} + \cdots$).

(3) $X \sim Un(x|0, \theta)$ and the iid sample $\{x_1, x_2, \ldots x_k\}$. For inferences on θ, show that $f_k = \theta^{-1} g(k)$ and in consequence the posterior is Pareto $Pa(\theta|x_M, n)$ with $x_M = \max\{x_1, x_2, \ldots x_k\}$ the sufficient statistic.

A very useful constructive theorem to obtain the *reference prior* is given in [18]. First, a *permissible prior* for the model $p(\boldsymbol{x}|\theta)$ is defined as a strictly positive function $\pi(\theta)$ such that it renders a proper posterior; that is,

$$\forall x \in \Omega_X \qquad \int_\Theta p(\boldsymbol{x}|\theta)\, \pi(\theta)\, d\theta < \infty$$

and that for some approximating sequence $\Theta_k \subset \Theta$; $\lim_{k\to\infty} \Theta_k = \Theta$, the sequence of posteriors $p_k(\theta|\boldsymbol{x}) \propto p(\boldsymbol{x}|\theta)\pi_k(\theta)$ converges logarithmically to $p(\theta|\boldsymbol{x}) \propto p(\boldsymbol{x}|\theta)\pi(\theta)$. Then, the *reference prior* is just a *permissible prior* that maximizes the maximum amount of information the experiment can provide for the parameter. The constructive procedure for a one-dimensional parameter consists on:

(1) Take $\pi^{\star}(\theta)$ as a continuous strictly positive function such that the corresponding posterior

$$\pi^{\star}(\theta|z_k) = \frac{p(z_k|\theta)\,\pi^{\star}(\theta)}{\int_{\Theta} p(z_k|\theta)\,\pi(\theta)\,d\theta}$$

is proper and asymptotically consistent. $\pi^{\star}(\theta)$ is arbitrary so it can be taken for convenience to simplify the integrals.

(2) Obtain

$$f_k^{\star}(\theta) = \exp\left\{\int_{\Omega_X} p(z_k|\theta)\,\log\,\pi^{\star}(\theta|z_k)\,dz_k\right\} \qquad \text{and} \qquad h_k(\theta;\theta_0) = \frac{f_k^{\star}(\theta)}{f_k^{\star}(\theta_0)}$$

for any interior point $\theta_0 \in \Theta$;

(3) If

(3.1) each $f_k^{\star}(\theta)$ is continuous;
(3.2) for any fixed θ and large k, is $h_k(\theta;\theta_0)$ is either monotonic in k or bounded from above by $h(\theta)$ that is integrable on any compact set;
(3.3) $\pi(\theta) = \lim_{k\to\infty} h_k(\theta;\theta_0)$ is a *permissible prior function*

then $\pi(\theta)$ is a reference prior for the model $p(\boldsymbol{x}|\theta)$. It is important to note that there is no requirement on the existence of the Fisher's information $\boldsymbol{I}(\theta)$. If it exists, a simple Taylor expansion of the densities shows that for a one-dimensional parameter $\pi(\theta) = [\boldsymbol{I}(\theta)]^{1/2}$ in consistency with Jeffrey's proposal. Usually, the last is easier to evaluate but not always as we shall see.

In many cases supp(θ) is unbounded and the prior $\pi(\theta)$ is not a propper density. As we have seen this is not a problem as long as the posterior $p(\theta|z_k) \propto p(z_k|\theta)\pi(\theta)$ is propper although, in any case, one can proceed *"more formally"* considering a sequence of proper priors $\pi_m(\theta)$ defined on a sequence of compact sets $\Theta_m \subset \Theta$ such that $\lim_{m\to\infty} \Theta_m = \Theta$ and taking the limit of the corresponding sequence of posteriors $p_m(\theta|z_k) \propto p(z_k|\theta)\pi_m(\theta)$. Usually simple sequences as for example $\Theta_m = [1/m, m]$; $\lim_{m\to\infty} \Theta_m = (0,\infty)$, or $\Theta_m = [-m, m]$; $\lim_{m\to\infty} \Theta_m = (-\infty,\infty)$ will suffice.

When the parameter $\boldsymbol{\theta}$ is n-dimensional, the procedure is more laborious. First, one starts [4] arranging the parameters in decreasing order of importance $\{\theta_1, \theta_2, \ldots, \theta_n\}$ (as we did for the Probability Matching Priors) and then follow the previous scheme to obtain the conditional prior functions

$$\pi(\theta_n|\theta_1,\theta_2,\ldots,\theta_{n-1})\,\pi(\theta_{n-1}|\theta_1,\theta_2,\ldots,\theta_{n-2})\,\ldots\,\pi(\theta_2|\theta_1)\,\pi(\theta_1)$$

For instance in the case of two parameters and the ordered parameterization $\{\theta,\lambda\}$:

(1) Get the conditional $\pi(\lambda|\theta)$ as the reference prior for λ keeping θ fixed;

(2) Find the marginal model

$$p(\boldsymbol{x}|\theta) = \int_{\Phi} p(\boldsymbol{x}|\theta, \lambda)\, \pi(\lambda|\theta)\, d\lambda$$

(3) Get the reference prior $\pi(\theta)$ from the marginal model $p(\boldsymbol{x}|\theta)$

Then $\pi(\theta, \lambda) \propto \pi(\lambda|\theta)\pi(\theta)$. This is fine if $\pi(\lambda|\theta)$ and $\pi(\theta)$ are propper functions; seldom the case. Otherwise one has to define the appropriate sequence of compact sets observing, among other things, that this has to be done for the full parameter space and usually the limits depend on the parameters. Suppose that we have the sequence $\Theta_i \times \Lambda_i \overset{i\to\infty}{\longrightarrow} \Theta \times \Lambda$. Then:

(1) Obtain $\pi_i(\lambda|\theta)$:

$$\pi_i^\star(\lambda|\theta)\mathbf{1}_{\Lambda_i}(\lambda) \longrightarrow \pi_i^\star(\lambda|\theta, z_k) = \frac{p(z_k|\theta, \lambda)\pi_i^\star(\lambda|\theta)}{\int_{\Lambda_i} p(z_k|\theta, \lambda)\pi_i^\star(\lambda|\theta)\, d\lambda} \longrightarrow \pi_i(\lambda|\theta) = \lim_{k\to\infty} \frac{f_k^\star(\lambda|\Lambda_i, \theta, \ldots)}{f_k^\star(\lambda_0|\Lambda_i, \theta, \ldots)}$$

(2) Get the marginal density $p_i(\boldsymbol{x}|\theta)$:

$$p_i(\boldsymbol{x}|\theta) = \int_{\Lambda_i} p(\boldsymbol{x}|\theta, \lambda)\, \pi_i(\lambda|\theta)\, d\lambda$$

(3) Determine $\pi_i(\theta)$:

$$\pi_i^\star(\theta)\mathbf{1}_{\Theta_i}(\theta) \longrightarrow \pi_i^\star(\theta|z_k) = \frac{p_i(z_k|\theta)\pi_i^\star(\theta)}{\int_{\Theta_i} p_i(z_k|\theta)\pi_i^\star(\theta)\, d\theta} \longrightarrow \pi_i(\theta) = \lim_{k\to\infty} \frac{f_k^\star(\theta|\Theta_i, \Lambda_i, \ldots)}{f_k^\star(\theta_0|\Theta_i, \Lambda_i, \ldots)}$$

(4) The reference prior for the ordered parameterization $\{\theta, \lambda\}$ will be:

$$\pi(\theta, \lambda) = \lim_{i\to\infty} \frac{\pi_i(\lambda|\theta)\, \pi_i(\theta)}{\pi_i(\lambda_0|\theta_0)\, \pi_i(\theta_0)}$$

In the case of two parameters, if Λ is independent of θ the Fisher's matrix usually exists and, if $\boldsymbol{I}(\theta, \lambda)$ and $\boldsymbol{S}(\theta, \lambda) = \boldsymbol{I}^{-1}(\theta, \lambda)$ are such that:

$$\boldsymbol{I}_{22}(\theta, \lambda) = a_1^2(\theta)\, b_1^2(\lambda) \qquad \text{and} \qquad \boldsymbol{S}_{11}(\theta, \lambda) = a_0^{-2}(\theta)\, b_0^{-2}(\lambda)$$

then [19] $\pi(\theta, \lambda) = \pi(\lambda|\theta)\pi(\theta) = a_0(\theta)\, b_1(\lambda)$ is a permissible prior even if the conditional reference priors are not proper. The reference priors are usually *probability matching priors*.

Example 2.20 A simple example is the Multinomial distribution $\boldsymbol{X} \sim Mn(\boldsymbol{x}|\boldsymbol{\theta})$ with $\dim\boldsymbol{X} = k+1$ and probability

$$p(\boldsymbol{x}|\theta) \propto \theta_1^{x_1}\,\theta_2^{x_2}\,\ldots\,\theta_k^{x_k}\,(1-\delta_k)^{x_{k+1}}; \qquad \delta_k = \sum_{j=1}^{k}\theta_j$$

Consider the ordered parameterization $\{\theta_1, \theta_2, \ldots, \theta_k\}$. Then

$$\pi(\theta_1, \theta_2, \ldots, \theta_k) = \pi(\theta_k|\theta_{k-1}, \theta_{k-2}\ldots\theta_2, \theta_1)\,\pi(\theta_{k-1}|\theta_{k-2}\ldots\theta_2, \theta_1)\,\ldots\,\pi(\theta_2|\theta_1)\,\pi(\theta_1)$$

In this case, all the conditional densities are proper

$$\pi(\theta_m|\theta_{m-1}, \ldots\,\theta_1) \propto \theta_m^{-1/2}\,(1-\delta_m)^{-1/2}$$

and therefore

$$\pi(\theta_1, \theta_2, \ldots\,\theta_k) \propto \prod_{i=1}^{k}\theta_i^{-1/2}\,(1-\delta_i)^{-1/2}$$

The posterior density will be then

$$p(\boldsymbol{\theta}|\boldsymbol{x}) \propto \left[\prod_{i=1}^{k}\theta_i^{x_i-1/2}\,(1-\delta_i)^{-1/2}\right](1-\delta_k)^{x_{k+1}}$$

Example 2.21 Consider again the case of two independent Poisson distributed random quantities X_1 and X_2 with joint density

$$P(n_1, n_2|\mu_1, \mu_2) = P(n_1|\mu_1)\,P(n_2|\mu_2) = e^{-(\mu_1+\mu_2)}\frac{\mu_1^{n_1}\mu_2^{n_2}}{\Gamma(n_1+1)\Gamma(n_2+1)}$$

We are interested in the parameter $\theta = \mu_1/\mu_2$ so setting $\mu = \mu_2$ we have the ordered parameterization $\{\theta, \mu\}$ and:

$$P(n_1, n_2|\theta, \mu) = e^{-\mu(1+\theta)}\frac{\theta^{n_1}\mu^{n}}{\Gamma(n_1+1)\Gamma(n_2+1)}$$

where $n = n_1 + n_2$. Since $E[X_1] = \mu_1 = \theta\mu$ and $E[X_2] = \mu_2 = \mu$ the Fisher's matrix and its inverse will be

$$\mathbf{I} = \begin{pmatrix} \mu/\theta & 1 \\ 1 & (1+\theta)/\mu \end{pmatrix}; \qquad \det(\mathbf{I}) = \theta^{-1} \quad \text{and} \quad \mathbf{S} = \mathbf{I}^{-1} = \begin{pmatrix} \theta(1+\theta)/\mu & -\theta \\ -\theta & \mu \end{pmatrix}$$

Therefore

$$S_{11} = \theta(1+\theta)/\mu \qquad \text{and} \qquad F_{22} = (1+\theta)/\mu$$

and, in consequence:

$$\pi(\theta)\, f_1(\mu) \propto S_{11}^{-1/2} = \frac{\sqrt{\mu}}{\sqrt{\theta(1+\theta)}} \qquad\qquad \pi(\mu|\theta)\, f_2(\theta) \propto F_{22}^{1/2} = \frac{\sqrt{1+\theta}}{\sqrt{\mu}}$$

Thus, we have for the ordered parameterization $\{\theta, \mu\}$ the reference prior:

$$\pi(\theta, \mu) = \pi(\mu|\theta)\, \pi(\theta) \propto \frac{1}{\sqrt{\mu\theta(1+\theta)}}$$

and the posterior density will be:

$$p(\theta, \mu|n_1, n_2) \;\propto\; \exp\{-\mu(1+\theta)\}\; \theta^{n_1-1/2}\,(1+\theta)^{-1/2}\,\mu^{n-1/2}$$

and, integrating the nuisance parameter $\mu \in [0, \infty)$ we get finally

$$p(\theta|n_1, n_2) = N\, \frac{\theta^{n_1-1/2}}{(1+\theta)^{n+1}}$$

with $\theta = \mu_1/\mu_2$, $n = n_1 + n_2$ and $N^{-1} = B(n_1 + 1/2, n_2 + 1/2)$. The distribution function will be:

$$P(\theta|n_1, n_2) = \int_0^{\theta} p(\theta'|n_1, n_2)\, d\theta' = I(\theta/(1+\theta); n_1 + 1/2, n_2 + 1/2)$$

with $I(x; a, b)$ the Incomplete Beta Function and the moments, when they exist;

$$E[\theta^m] = \frac{\Gamma(n_1 + 1/2 + m)\,\Gamma(n_2 + 1/2 - m)}{\Gamma(n_1 + 1/2)\,\Gamma(n_2 + 1/2)}$$

It is interesting to look at the problem from a different point of view. Consider again the ordered parameterization $\{\theta, \lambda\}$ with $\theta = \mu_1/\mu_2$ but now, the nuisance parameter is $\lambda = \mu_1 + \mu_2$. The likelihood will be:

$$P(n_1, n_2|\theta, \lambda) = \frac{1}{\Gamma(n_1+1)\Gamma(n_2+1)} e^{-\lambda}\,\lambda^n\, \frac{\theta^{n_1}}{(1+\theta)^n}$$

The domains are $\Theta = (0, \infty)$ and $\Lambda = (0, \infty)$, independent. Thus, no need to specify the prior for λ since

$$p(\theta|n_1, n_2) \propto \pi(\theta)\, \frac{\theta^{n_1}}{(1+\theta)^n} \int_{\Lambda} e^{-\lambda}\,\lambda^n \pi(\lambda) d\lambda \propto \frac{\theta^{n_1}}{(1+\theta)^n}\, \pi(\theta)$$

In this case we have that

$$I(\theta) \propto \frac{1}{\theta\,(1+\theta)^2} \quad \longrightarrow \quad \pi(\theta) = \frac{1}{\theta^{1/2}\,(1+\theta)}$$

and, in consequence,

$$p(\theta|n_1, n_2) = N\,\frac{\theta^{n_1-1/2}}{(1+\theta)^{n+1}}$$

Problem 2.8 Show that the reference prior for the Pareto distribution $Pa(x|\theta, x_0)$ (see Example 2.9) is $\pi(\theta, x_0) \propto (\theta x_0)^{-1}$ and that for an iid sample $\boldsymbol{x} = \{x_1, \ldots, x_n\}$, if $x_m = \min\{x_i\}_{i=1}^n$ and $a = \sum_{i=1}^n \ln(x_i/x_m)$ the posterior

$$p(\theta, x_0|\boldsymbol{x}) = \frac{n a^{n-1}}{x_m \Gamma(n-1)}\, e^{-a\theta}\,\theta^{n-1} \left(\frac{x_0}{x_m}\right)^{n\theta-1} \mathbf{1}_{(0,\infty)}(\theta)\mathbf{1}_{(0,x_m)}(x_0)$$

is proper for a sample size $n > 1$. Obtain the marginal densities

$$p(\theta|\boldsymbol{x}) = \frac{a^{n-1}}{\Gamma(n-1)}\, e^{-a\theta}\,\theta^{n-2}\mathbf{1}_{(0,\infty)}(\theta) \qquad \text{and}$$

$$p(x_0|\boldsymbol{x}) = \frac{n(n-1)}{a}\, x_0^{-1} \left[1 + \frac{n}{a}\ln\left(\frac{x_m}{x_0}\right)\right]^{-n} \mathbf{1}_{(0,x_m)}(x_0)$$

and show that for large n (see Sect. 2.10.2) $E[\theta] \simeq n a^{-1}$ and $E[x_0] \simeq x_m$.

Problem 2.9 Show that for the shifted Pareto distribution (Lomax distribution):

$$p(x|\theta, x_0) = \frac{\theta}{x_0}\left(\frac{x_0}{x+x_0}\right)^{\theta+1} \mathbf{1}_{(0,\infty)}(x); \qquad \theta, x_0 \in R^+$$

the reference prior for the ordered parameterization $\{\theta, x_0\}$ is $\pi_r(\theta, x_0) \propto (x_0\theta(\theta+1))^{-1}$ and for $\{x_0, \theta\}$ is $\pi_r(x_0, \theta) \propto (x_0\theta)^{-1}$. Show that the first one is a first order probability matching prior while the second is not. In fact, show that for $\{x_0, \theta\}$, $\pi_{pm}(x_0, \theta) \propto (x_0\theta^{3/2}\sqrt{\theta+2})^{-1}$ is a matching prior and that for both orderings the Jeffrey's prior is $\pi_J(\theta, x_0) \propto (x_0(\theta+1)\sqrt{\theta(\theta+2)})^{-1}$.

Problem 2.10 Show that for the Weibull distribution

$$p(x|\alpha, \beta) = \alpha\beta\, x^{\beta-1} \exp\left\{-\alpha x^{\beta}\right\} \mathbf{1}_{(0,\infty)}(x)$$

with $\alpha, \beta > 0$, the reference prior functions are

$$\pi_r(\beta, \alpha) = (\alpha\beta)^{-1} \qquad \text{and} \qquad \pi_r(\alpha, \beta) = \left(\alpha\beta\sqrt{\zeta(2) + (\psi(2) - \ln\alpha)^2}\right)^{-1}$$

for the ordered parameterizations $\{\beta, \alpha\}$ and $\{\alpha, \beta\}$ respectively being $\zeta(2) = \pi^2/6$ the Riemann Zeta Function and $\psi(2) = 1 - \gamma$ the Digamma Function.

2.7 Hierarchical Structures

In many circumstances, even though the experimental observations respond to the same phenomena it is not always possible to consider the full set of observations as an exchangeable sequence but rather exchangeability within subgroups of observations. As stated earlier, this may be the case when the results come from different experiments or when, within the same experiment, data taking conditions (acceptances, efficiencies,…) change from run to run. A similar situation holds, for instance, for the results of responses under a drug performed at different hospitals when the underlying conditions of the population vary between zones, countries,… In general, we shall have different groups of observations

$$\begin{aligned}
\boldsymbol{x}_1 &= \{x_{11}, x_{21}, \ldots, x_{n_1 1}\} \\
&\vdots \\
\boldsymbol{x}_j &= \{x_{1j}, x_{2j}, \ldots, x_{n_j j}\} \\
&\vdots \\
\boldsymbol{x}_J &= \{x_{1J}, x_{2J}, \ldots, x_{n_J J}\}
\end{aligned}$$

from J experiments $e_1(n_1), e_2(n_2), \ldots, e_J(n_J)$. Within each sample $\boldsymbol{x}_j$, we can consider that exchangeability holds and also for the sets of observations $\{\boldsymbol{x}_1, \boldsymbol{x}_2, \ldots, \boldsymbol{x}_J\}$ In this case, it is appropriate to consider *hierarchical structures.*

Let's suppose that for each experiment $e(j)$ the observations are drawn from the model

$$p(\boldsymbol{x}_j|\boldsymbol{\theta}_j); \qquad j = 1, 2, \ldots, J$$

Since the experiments are independent we assume that the parameters of the sequence $\{\boldsymbol{\theta}_1, \boldsymbol{\theta}_2, \ldots, \boldsymbol{\theta}_J\}$ are exchangeable and that, although different, they can be assumed to have a common origin since they respond to the same phenomena. Thus,we can set

$$p(\boldsymbol{\theta}_1, \boldsymbol{\theta}_2, \ldots, \boldsymbol{\theta}_J|\phi) = \prod_{i=1}^{J} p(\boldsymbol{\theta}_i|\phi)$$

Fig. 2.3 Structure of the hierarchical model

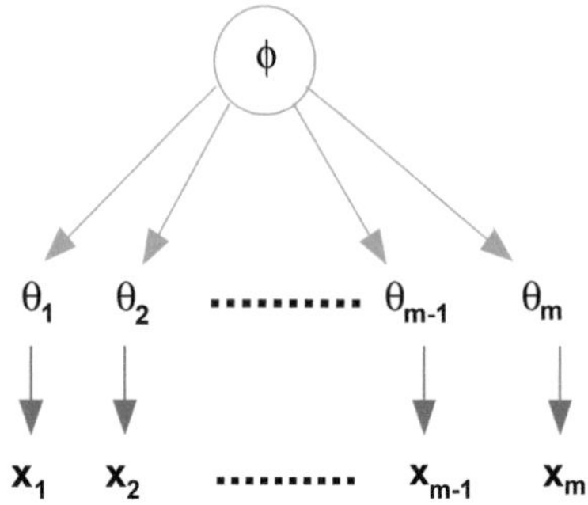

with ϕ the *hyperparameters* for which we take a prior $\pi(\phi)$. Then we have the structure (Fig. 2.3.)

$$p(\boldsymbol{x}_1, \ldots, \boldsymbol{x}_J, \boldsymbol{\theta}_1, \ldots, \boldsymbol{\theta}_J, \phi) = \pi(\phi) \prod_{i=1}^{J} p(\boldsymbol{x}_i|\boldsymbol{\theta}_i)\, \pi(\boldsymbol{\theta}_i|\phi)$$

This structure can be repeated sequentially if we consider appropriate to assign a prior $\pi(\phi|\boldsymbol{\tau})$ to the hyperparameters ϕ so that

$$p(\boldsymbol{x}, \boldsymbol{\theta}, \phi, \boldsymbol{\tau}) = p(\boldsymbol{x}|\boldsymbol{\theta})\, \pi(\boldsymbol{\theta}|\phi)\, \pi(\phi|\boldsymbol{\tau})\, \pi(\boldsymbol{\tau})$$

Now, consider the model $p(\boldsymbol{x}, \boldsymbol{\theta}, \phi)$. We may be interested in $\boldsymbol{\theta}$, in the hyperparameters ϕ or in both. In general we shall need the conditional densities:

- $p(\phi|\boldsymbol{x}) \propto p(\phi) \int p(\boldsymbol{x}|\boldsymbol{\theta})\, p(\boldsymbol{\theta}|\phi)\, d\boldsymbol{\theta}$
- $p(\boldsymbol{\theta}|\boldsymbol{x}, \phi) = \dfrac{p(\boldsymbol{\theta}, \boldsymbol{x}, \phi)}{p(\boldsymbol{x}, \phi)}$ and
- $p(\boldsymbol{\theta}|\boldsymbol{x}) = \dfrac{p(\boldsymbol{x}|\boldsymbol{\theta})}{p(\boldsymbol{x})}\, p(\boldsymbol{\theta}) = \dfrac{p(\boldsymbol{x}|\boldsymbol{\theta})}{p(\boldsymbol{x})} \int p(\boldsymbol{\theta}|\phi)\, p(\phi)\, d\phi$

that can be expressed as

$$p(\boldsymbol{\theta}|\boldsymbol{x}) = \int \frac{p(\boldsymbol{x}, \boldsymbol{\theta}, \phi)}{p(\boldsymbol{x})}\, d\phi = \int p(\boldsymbol{\theta}|\boldsymbol{x}, \phi)\, p(\phi|\boldsymbol{x})\, d\phi$$

and, since

$$p(\boldsymbol{\theta}|\boldsymbol{x}) = p(\boldsymbol{x}|\boldsymbol{\theta}) \int p(\boldsymbol{\theta}|\phi)\, \frac{p(\phi|\boldsymbol{x})}{p(\boldsymbol{x}|\phi)}\, d\phi$$

we can finally write

$$\frac{p(\boldsymbol{\theta}, \boldsymbol{\phi})}{p(\boldsymbol{x})} = p(\boldsymbol{\theta}|\boldsymbol{\phi})\,\frac{p(\boldsymbol{\phi})}{p(\boldsymbol{x})} = p(\boldsymbol{\theta}|\boldsymbol{\phi})\,\frac{p(\boldsymbol{\phi}|\boldsymbol{x})}{p(\boldsymbol{x}|\boldsymbol{\phi})}$$

In general, this conditional densities have complicated expressions and we shall use Monte Carlo methods to proceed (see Gibbs Sampling, Example 3.15, in Chap. 3).

It is important to note that if the prior distributions are not proper we can have improper marginal and posterior densities that obviously have no meaning in the inferential process. Usually, conditional densities are better behaved but, in any case, we have to check that this is so. In general, the better behaved is the likelihood the wildest behavior we can accept for the prior functions. We can also used prior distributions that are a mixture of proper distributions:

$$p(\boldsymbol{\theta}|\boldsymbol{\phi}) = \sum_i w_i\, p_i(\boldsymbol{\theta}|\boldsymbol{\phi})$$

with $w_i \geq 0$ and $\sum w_i = 1$ so that the combination is convex and we assure that it is proper density or, extending this to a continuous mixture:

$$p(\boldsymbol{\theta}|\boldsymbol{\phi}) = \int w(\boldsymbol{\sigma})\, p(\boldsymbol{\theta}|\boldsymbol{\phi}, \boldsymbol{\sigma})\, d\boldsymbol{\sigma}.$$

2.8 Priors for Discrete Parameters

So far we have discussed parameters with continuous support but in some cases it is either finite or countable. If the parameter of interest can take only a finite set of n possible values, the reasonable option for an *uninformative prior* is a Discrete Uniform Probability $P(X = x_i) = 1/n$. In fact, it is shown in Sect. 4.2 that maximizing the expected information provided by the experiment with the normalization constraint (i.e. the probability distribution for which the *prior* knowledge is minimal) drives to $P(X = x_i) = 1/n$ in accordance with the *Principle of Insufficient Reason*.

Even though finite discrete parameter spaces are either the most usual case we shall have to deal with or, at least, a sufficiently good approximation for the real situation, it may happen that a non-informative prior is not the most appropriate (see Example 2.22). On the other hand, if the parameter takes values on a countable set the problem is more involved. A possible way out is to devise a hierarchical structure in which we assign the discrete parameter θ a prior $\pi(\theta|\boldsymbol{\lambda})$ with $\boldsymbol{\lambda}$ a set of continuous hyperparameters. Then, since

$$p(\boldsymbol{x}, \boldsymbol{\lambda}) = \sum_{\theta\in\Theta} p(\boldsymbol{x}|\theta)\pi(\theta|\boldsymbol{\lambda})\,\pi(\boldsymbol{\lambda}) = p(\boldsymbol{x}|\boldsymbol{\lambda})\,\pi(\boldsymbol{\lambda})$$

we get the prior $\pi(\boldsymbol{\lambda})$ by any of the previous procedures for continuous parameters with the model $p(\boldsymbol{x}|\boldsymbol{\lambda})$ and obtain

$$\pi(\boldsymbol{\theta}) \propto \int_{\Lambda} \pi(\theta|\boldsymbol{\lambda})\, \pi(\boldsymbol{\lambda})\, d\boldsymbol{\lambda}$$

Different procedures are presented and discussed in [20].

Example 2.22 The absolute value of the electric charge (Z) of a particle is to be determined from the number of photons observed by a Cherenkov Counter. We know from test beam studies and Monte Carlo simulations that the number of observed photons n_γ produced by a particle of charge Z is well described by a Poisson distribution with parameter $\mu = n_0 Z^2$; that is

$$P(n_\gamma|n_0, Z) = e^{-n_0 Z^2}\, \frac{(n_0 Z^2)^{n_\gamma}}{\Gamma(n_\gamma + 1)}$$

so $E[n_\gamma|Z = 1] = n_0$. First, by physics considerations Z has a finite support $\Omega_Z = \{1, 2, \ldots, n\}$. Second, we know *a priory* that not all incoming nuclei are equally likely so a *non-informative* prior may not be the best choice. In any case, a discrete uniform prior will give the posterior:

$$P(Z = k|n_\gamma, n_0, n) = \frac{e^{-n_0 k^2}\, k^{2n_\gamma}}{\sum_{k=1}^{n} e^{-n_0 k^2}\, k^{2n_\gamma}}.$$

2.9 Constrains on Parameters and Priors

Consider a parametric model $p(\boldsymbol{x}|\boldsymbol{\theta})$ and the prior $\pi_0(\boldsymbol{\theta})$. Now we have some information on the parameters that we want to include in the prior. Typically we shall have say k constraints of the form

$$\int_{\Theta} g_i(\boldsymbol{\theta})\, \pi(\boldsymbol{\theta})\, d\boldsymbol{\theta} = a_i; \qquad i = 1, \ldots, k$$

Then, we have to find the prior $\pi(\boldsymbol{\theta})$ for which $\pi_0(\boldsymbol{\theta})$ is the best approximation, in the Kullback-Leibler sense, including the constraints with the corresponding Lagrange multipliers λ_i; that is, the extremal of

$$\mathcal{F} = \int_{\Theta} \pi(\boldsymbol{\theta})\, \log \frac{\pi(\boldsymbol{\theta})}{\pi_0(\boldsymbol{\theta})}\, d\boldsymbol{\theta} + \sum_{i=1}^{k} \lambda_i \left(\int_{\Theta} g_i(\boldsymbol{\theta})\, \pi(\boldsymbol{\theta})\, d\boldsymbol{\theta} - a_i \right)$$

Again, it is left as an exercise to show that from Calculus of Variations we have the well known solution

$$\pi(\boldsymbol{\theta}) \propto \pi_0(\boldsymbol{\theta})\, \exp\left\{ \sum_{i=1}^{k} \lambda_i\, g_i(\boldsymbol{\theta}) \right\} \qquad \text{where} \qquad \lambda_i \mid \int_{\Theta} g_i(\boldsymbol{\theta})\, \pi(\boldsymbol{\theta})\, d\boldsymbol{\theta} = a_i$$

Quite frequently we are forced to include constraints on the support of the parameters: some are non-negative (masses, energies, momentum, life-times,...), some are bounded in (0, 1) ($\beta = v/c$, efficiencies, acceptances,...),... At least from a formal point of view, to account for constraints on the support is a trivial problem. Consider the model $p(\boldsymbol{x}|\boldsymbol{\theta})$ with $\boldsymbol{\theta} \in \Theta_0$ and a reference prior $\pi_0(\boldsymbol{\theta})$. Then, our inferences on $\boldsymbol{\theta}$ shall be based on the posterior

$$p(\boldsymbol{\theta}|\boldsymbol{x}) = \frac{p(\boldsymbol{x}|\boldsymbol{\theta})\,\pi_0(\boldsymbol{\theta})}{\int_{\Theta_0} p(\boldsymbol{x}|\boldsymbol{\theta})\,\pi_0(\boldsymbol{\theta})\,d\boldsymbol{\theta}}$$

Now, if we require that $\boldsymbol{\theta} \in \Theta \subset \Theta_0$ we define

$$g_1(\boldsymbol{\theta}) = \mathbf{1}_\Theta(\boldsymbol{\theta}) \longrightarrow \int_{\Theta_0} g_1(\boldsymbol{\theta})\,\pi(\boldsymbol{\theta})\,d\boldsymbol{\theta} = \int_{\Theta} \pi(\boldsymbol{\theta})\,d\boldsymbol{\theta} = 1 - \epsilon$$

$$g_2(\boldsymbol{\theta}) = \mathbf{1}_{\Theta^c}(\boldsymbol{\theta}) \longrightarrow \int_{\Theta_0} g_2(\boldsymbol{\theta})\,\pi(\boldsymbol{\theta})\,d\boldsymbol{\theta} = \int_{\Theta^c} \pi(\boldsymbol{\theta})\,d\boldsymbol{\theta} = \epsilon$$

and in the limit $\epsilon \to 0$ we have the *restricted reference prior*

$$\pi(\boldsymbol{\theta}) = \frac{\pi_0(\boldsymbol{\theta})}{\int_\Theta \pi_0(\boldsymbol{\theta})\,d\boldsymbol{\theta}}\,\mathbf{1}_\Theta(\boldsymbol{\theta})$$

as we have obviously expected. Therefore

$$p(\boldsymbol{\theta}|\boldsymbol{x}, \boldsymbol{\theta} \in \Theta) = \frac{p(x|\boldsymbol{\theta})\,\pi(\boldsymbol{\theta})}{\int_\Theta p(x|\boldsymbol{\theta})\,\pi(\boldsymbol{\theta})\,d\boldsymbol{\theta}} = \frac{p(x|\theta)\,\pi_0(\boldsymbol{\theta})}{\int_\Theta p(x|\boldsymbol{\theta})\,\pi_0(\boldsymbol{\theta})\,d\boldsymbol{\theta}}\,\mathbf{1}_\Theta(\boldsymbol{\theta})$$

that is, the same initial expression but normalized in the domain of interest Θ.

2.10 Decision Problems

Even though all the information we have on the parameters of relevance is contained in the posterior density it is interesting, as we saw in Chap. 1, to explicit some particular values that characterize the probability distribution. This certainly entails a considerable and unnecessary reduction of the available information but in the end, quoting Lord Kelvin, "*... when you cannot express it in numbers, your knowledge is of a meager and unsatisfactory kind*". In statistics, to specify a particular value of the parameter is termed *Point Estimation* and can be formulated in the framework of *Decision Theory*.

In general, *Decision Theory* studies how to choose the *optimal action* among several possible alternatives based on what has been experimentally observed. Given a particular problem, we have to explicit the set Ω_θ of the possible "*states of nature*", the set Ω_X of the possible experimental outcomes and the set Ω_A of the possible actions we can take. Imagine, for instance, that we do a test on an individual suspected to have some disease for which the medical treatment has some potentially dangerous collateral effects. Then, we have:

$$\Omega_\theta = \{\text{healthy}, \ \text{sic}\}$$

$$\Omega_X = \{\text{test positive}, \ \text{test negative}\}$$

$$\Omega_A = \{\text{apply treatment}, \ \text{do not apply treatment}\}$$

Or, for instance, a detector that provides within some accuracy the momentum (p) and the velocity (β) of charged particles. If we want to assign an hypothesis for the mass of the particle we have that $\Omega_\theta = \mathcal{R}^+$ is the set of all possible states of nature (all possible values of the mass), Ω_X the set of experimental observations (the momentum and the velocity) and Ω_A the set of all possible actions that we can take (assign one or other value for the mass). In this case, we shall take a decision based on the probability density $p(m|p, \beta)$.

Obviously, unless we are in a state of absolute certainty we can not take an action without potential losses. Based on the observed experimental outcomes, we can for instance assign the particle a mass m_1 when the *true state of nature* is $m_2 \neq m_1$ or consider that the individual is healthy when is actually sic. Thus, the first element of Decision Theory is the *Loss Function*:

$$l(a, \theta) : (\theta, a) \in \Omega_\theta \times \Omega_A \longrightarrow \mathcal{R}^+ + \{0\}$$

This is a non-negative function, defined for all $\boldsymbol{\theta} \in \Omega_\theta$ and the set of possible actions $\boldsymbol{a} \in \Omega_A$, that quantifies the *loss* associated to take the action $\boldsymbol{a}$ (decide for $\boldsymbol{a}$) when the state of nature is $\boldsymbol{\theta}$.

Obviously, we do not have a perfect knowledge of the *state of nature*; what we know comes from the observed data $\boldsymbol{x}$ and is contained in the posterior distribution $p(\boldsymbol{\theta}|\boldsymbol{x})$. Therefore, we define the *Risk Function* (*risk* associated to take the action $\boldsymbol{a}$, or decide for $\boldsymbol{a}$ when we have observed the data $\boldsymbol{x}$) as the expected value of the Loss Function:

$$R(\boldsymbol{a}|\boldsymbol{x}) = E_{\boldsymbol{\theta}}[l(\boldsymbol{a}, \boldsymbol{\theta})] = \int_{\Omega_\theta} l(\boldsymbol{a}, \boldsymbol{\theta})\, p(\boldsymbol{\theta}|\boldsymbol{x})\, d\boldsymbol{\theta}$$

Sound enough, the Bayesian decision criteria consists on taking the action $\boldsymbol{a}(\boldsymbol{x})$ (*Bayesian action*) that minimizes the risk $R(\boldsymbol{a}|\boldsymbol{x})$ (*minimum risk*); that is, that minimizes the expected loss under the posterior density function.[8] Then, we shall encounter to kinds of problems:

---

[8]The problems studied by *Decision Theory* can be addressed from the point of view of *Game Theory*. In this case, instead of *Loss Functions* one works with *Utility Functions* $u(\boldsymbol{\theta}, \boldsymbol{a})$ that, in essence, are nothing else but $u(\boldsymbol{\theta}, \boldsymbol{a}) = K - l(\boldsymbol{\theta}, \boldsymbol{a}) \geq 0$; it is just matter of personal optimism to

- *inferential problems*, where $\Omega_A = \mathcal{R}$ y $\boldsymbol{a}(\boldsymbol{x})$ is a statistic that we shall take as *estimator* of the parameter $\boldsymbol{\theta}$;
- *decision problems* (or *hypothesis testing*) where $\Omega_A = \{\text{accept, reject}\}$ or choose one among a set of hypothesis.

Obviously, the actions depend on the loss function (that we have to specify) and on the posterior density and, therefore, on the data through the model $p(\boldsymbol{x}|\boldsymbol{\theta})$ and the prior function $\pi(\boldsymbol{\theta})$. It is then possible that, for a particular model, two different loss functions drive to the same decision or that the same loss function, depending on the prior, take to different actions.

2.10.1 Hypothesis Testing

Consider the case where we have to choose between two exclusive and exhaustive hypothesis H_1 and $H_2(=H_1{}^c)$. From the data sample and our prior beliefs we have the posterior probabilities

$$P(H_i|\text{data}) = \frac{P(\text{data}|H_i)\,P(H_i)}{P(\text{data})}; \quad i = 1, 2$$

and the actions to be taken are then:

$$\begin{aligned} a_1 &: \text{action to take if we decide upon } H_1 \\ a_2 &: \text{action to take if we decide upon } H_2 \end{aligned}$$

Then, we define the loss function $l(a_i, H_j)$; $i, j = 1, 2$ as:

$$l(a_i|H_j) = \begin{cases} l_{11} = l_{22} = 0 & \text{if we make the correct choice; that is,} \\ & \text{if we take action } a_1 \text{ when the state of} \\ & \text{nature is } H_1 \text{ or } a_2 \text{ when it is } H_2; \\ & \\ l_{12} > 0 & \text{if we take action } a_1 (\text{decide upon } H_1) \\ & \text{when the state of nature is } H_2 \\ & \\ l_{21} > 0 & \text{if we take action } a_2 \text{ (decide upon } H_2) \\ & \text{when the state of nature is } H_1 \end{cases}$$

(Footnote 8 continued)
work with *"utilities"* or *"losses"*. J. Von Neumann and O. Morgenstern introduced in 1944 the idea of *expected utility* and the criteria to take as *optimal action* hat which maximizes the expected utility.

so the risk function will be:

$$R(a_i|\text{data}) = \sum_{j=1}^{2} l(a_i|H_j)\, P(H_j|\text{data})$$

that is:

$$\begin{aligned} R(a_1|\text{data}) &= l_{11}\, P(H_1|\text{data}) + l_{12}\, P(H_2|\text{data}) \\ R(a_2|\text{data}) &= l_{21}\, P(H_1|\text{data}) + l_{22}\, P(H_2|\text{data}) \end{aligned}$$

and, according to the minimum Bayesian risk, we shall choose the hypothesis H_1 (action a_1) if

$$R(a_1|\text{data}) \; < \; R(a_2|\text{data}) \quad \longrightarrow \quad P(H_1|\text{data})\,(l_{11} - l_{21}) \; < \; P(H_2|\text{data})\,(l_{22} - l_{12})$$

Since we have chosen $l_{11} = l_{22} = 0$ in this particular case, we shall take action a_1 (decide for hypothesis H_1) if:

$$\frac{P(H_1|\text{data})}{P(H_2|\text{data})} > \frac{l_{12}}{l_{21}}$$

or action a_2 (decide in favor of hypothesis H_2) if:

$$R(a_2, \text{data}) \; < \; R(a_1, \text{data}) \quad \longrightarrow \quad \frac{P(H_2|\text{data})}{P(H_1|\text{data})} > \frac{l_{21}}{l_{12}}$$

that is, we take action a_i $(i = 1, 2)$ if:

$$\frac{P(H_i|\text{data})}{P(H_j|\text{data})} = \left[\frac{P(\text{data}|H_i)}{P(\text{data}|H_j)}\right]\left[\frac{P(H_i)}{P(H_j)}\right] > \frac{l_{ij}}{l_{ji}}$$

The ratio of likelihoods

$$B_{ij} = \frac{P(\text{data}|H_i)}{P(\text{data}|H_j)}$$

is called **Bayes Factor** B_{ij} and changes our prior beliefs on the two alternative hypothesis based on the evidence we have from the data; that is, quantifies how strongly data favors one model over the other. Thus, we shall decide in favor of hypothesis H_i against H_j $(i, j = 1, 2)$ if

$$\frac{P(H_i|\text{data})}{P(H_j|\text{data})} > \frac{l_{ij}}{l_{ji}} \qquad \longrightarrow \qquad B_{ij} > \frac{P(H_j)}{P(H_i)}\frac{l_{ij}}{l_{ji}}$$

If we consider the same loss if we decide upon the wrong hypothesis whatever it be, we have $l_{12} = l_{21}$ (Zero-One Loss Function). In general, we shall be interested in testing:

(1) **Two simple hypothesis**, H_1 versus H_2, for which the models $M_i = \{X \sim p_i(x|\theta_i)\}$; $i = 1, 2$ are fully specified including the values of the parameters (that is, $\Theta_i = \{\theta_i\}$). In this case, the Bayes Factor will be given by the ratio of likelihoods

$$B_{12} = \frac{p_1(\boldsymbol{x}|\theta_1)}{p_2(\boldsymbol{x}|\theta_2)} \qquad\qquad \left(\text{usually } \frac{p(\boldsymbol{x}|\theta_1)}{p(\boldsymbol{x}|\theta_2)}\right)$$

The classical Bayes Factor is the ratio of the likelihoods for the two competing models evaluated at their respective maximums.

(2) **A simple** (H_1) **versus a composite hypothesis** H_2 for which the parameters of the model $M_2 = \{X \sim p_2(x|\theta_2)\}$ have support on Θ_2. Then we have to average the likelihood under H_2 and

$$B_{12} = \frac{p_1(\boldsymbol{x}|\theta_1)}{\int_{\Theta_2} p_2(\boldsymbol{x}|\theta)\pi_2(\theta)d\theta}$$

(3) **Two composite hypothesis**: in which the models M_1 and M_2 have parameters that are not specified by the hypothesis so

$$B_{12} = \frac{\int_{\Theta_1} p_1(\boldsymbol{x}|\theta_1)\pi_1(\theta_1)d\theta_1}{\int_{\Theta_2} p_2(\boldsymbol{x}|\theta_2)\pi_2(\theta_2)d\theta_2}$$

and, since $P(H_1|\text{data}) + P(H_2|\text{data}) = 1$, we can express the posterior probability $P(H_1|\text{data})$ as

$$P(H_1|\text{data}) = \frac{B_{12}\, P(H_1)}{P(H_2) + B_{12}\, P(H_1)}$$

Usually, we consider equal prior probabilities for the two hypothesis ($P(H_1) = P(H_2) = 1/2$) but be aware that in some cases this may not be a realistic assumption.

Bayes Factors are independent of the prior beliefs on the hypothesis ($P(H_i)$) but, when we have composite hypothesis, we average the likelihood with a prior and if it is an improper function they are not well defined. If we have prior knowledge about the parameters, we may take informative priors that are proper but this is not always the case. One possible way out is to consider sufficiently general proper priors (conjugated priors for instance) so the Bayes factors are well defined and then study what is the sensitivity for different reasonable values of the hyperparameters. A more practical and interesting approach to avoid the indeterminacy due to improper priors [21, 22] is to take a subset of the observed sample to render a proper posterior (with, for instance, reference priors) and use that as proper prior density to compute the Bayes Factor with the remaining sample. Thus, if the sample $\boldsymbol{x} = \{x_1, \ldots, x_n\}$ consists on iid observations, we may consider $\boldsymbol{x} = \{\boldsymbol{x}_1, \boldsymbol{x}_2\}$ and, with the reference prior $\pi(\boldsymbol{\theta})$, obtain the proper posterior

$$\pi(\boldsymbol{\theta}|\boldsymbol{x}_1) = \frac{p(\boldsymbol{x}_1|\boldsymbol{\theta})\,\pi(\boldsymbol{\theta})}{\int_\Theta p(\boldsymbol{x}_1|\boldsymbol{\theta})\,\pi(\boldsymbol{\theta})\,d\boldsymbol{\theta}}$$

The remaining subsample $(\boldsymbol{x}_2)$ is then used to compute the partial Bayes Factor[9]:

$$B_{12}(x_2|x_1) = \frac{\int_{\Theta_1} p_1(\boldsymbol{x}_2|\boldsymbol{\theta}_1)\,\pi_1(\boldsymbol{\theta}_1|\boldsymbol{x}_1)\,d\boldsymbol{\theta}_1}{\int_{\Theta_2} p_2(\boldsymbol{x}_2|\boldsymbol{\theta}_2)\,\pi_2(\boldsymbol{\theta}_2|\boldsymbol{x}_1)\,d\boldsymbol{\theta}_2} \qquad \left(= \frac{BF(\boldsymbol{x}_1, \boldsymbol{x}_2)}{BF(\boldsymbol{x}_1)}\right)$$

for the hypothesis testing. Berger and Pericchi propose to use the minimal amount of data needed to specify a proper prior (usually $\max\{\dim(\boldsymbol{\theta}_i)\}$) so as to leave most of the sample for the model testing and dilute the dependence on a particular election of the training sample evaluating the Bayes Factors with all possible minimal samples and choosing the truncated mean, the geometric mean or the median, less sensitive to outliers, as a characteristic value (see Example 2.24). A thorough analysis of Bayes Factors, with its caveats and advantages, is given in [23].

A different alternative to quantify the evidence in favour of a particular model that avoids the need of the prior specification and is easy to evaluate is the Schwarz criteria [24] (or "*Bayes Information Criterion (BIC)*"). The rationale is the following. Consider a sample $\boldsymbol{x} = \{x_1, \ldots, x_n\}$ and two alternative hypothesis for the models $M_i = \{p_i(x|\boldsymbol{\theta}_i); \dim(\boldsymbol{\theta}_i) = d_i\}$; $i = 1, 2$. As we can see in Sect. 4.5, under the appropriate conditions we can approximate the likelihood as

$$l(\boldsymbol{\theta}|\boldsymbol{x}) \simeq l(\widehat{\boldsymbol{\theta}}|\boldsymbol{x}) \exp\left\{-\frac{1}{2}\sum_{k=1}^{d}\sum_{m=1}^{d}(\theta_k - \widehat{\theta}_k)\left[n\boldsymbol{I}_{km}(\widehat{\boldsymbol{\theta}})\right](\theta_m - \widehat{\theta}_m)\right\}$$

so taking a uniform prior for the parameters $\boldsymbol{\theta}$, reasonable in the region where the likelihood is dominant, we can approximate

$$J(\boldsymbol{x}) = \int_\Theta p(\boldsymbol{x}|\boldsymbol{\theta})\,\pi(\boldsymbol{\theta})\,d\boldsymbol{\theta} \simeq p(\boldsymbol{x}|\widehat{\boldsymbol{\theta}})\,(2\pi/n)^{d/2}\,|\det[\boldsymbol{I}(\widehat{\boldsymbol{\theta}})]|^{-1/2}$$

and, ignoring terms that are bounded as $n \to \infty$, define the $BIC(M_i)$ for the model M_i as

$$2\ln\,J_i(\boldsymbol{x}) \simeq BIC(M_i) \equiv 2\,\ln\,p_i(\boldsymbol{x}|\widehat{\boldsymbol{\theta}}_i) - d_i\,\ln\,n$$

so:

[9]Essentially, the ratio of the predictive inferences for $\boldsymbol{x}_2$ after $\boldsymbol{x}_1$ has been observed.

$$B_{12} \simeq \frac{p_1(\boldsymbol{x}|\widehat{\boldsymbol{\theta}}_1)}{p_2(\boldsymbol{x}|\widehat{\boldsymbol{\theta}}_2)}\, n^{(d_2-d_1)/2} \quad \longrightarrow \quad \Delta_{12} = 2\ln B_{12} \simeq 2\ln\left(\frac{p_1(\boldsymbol{x}|\widehat{\boldsymbol{\theta}}_1)}{p_2(\boldsymbol{x}|\widehat{\boldsymbol{\theta}}_2)}\right) - (d_1 - d_2)\ln n$$

and therefore, larger values of $\Delta_{12} = BIC(M_1) - BIC(M_2)$ indicate a preference for the hypothesis $H_1(M_1)$ against $H_2(M_2)$ being commonly accepted that for values grater than 6 the evidence is "*strong*"[10] although, in some cases, it is worth to study the behaviour with a Monte Carlo sampling. Note that the last term penalises models with larger number of parameters and that this quantification is sound when the sample size n is much larger than the dimensions d_i of the parameters.

Example 2.23 Suppose that from the information provided by a detector we estimate the mass of an incoming particle and we want to decide upon the two exclusive and alternative hypothesis H_1 (particle of type 1) and $H_2(=H_1{}^c)$ (particle of type 2). We know from calibration data and Monte Carlo simulations that the mass distributions for both hypothesis are, to a very good approximation, Normal with means m_1 and m_2 variances σ_1^2 and σ_2^2 respectively. Then for an observed value of the mass m_0 we have:

$$B_{12} = \frac{p(m_0|H_1)}{p(m_0|H_2)} = \frac{N(m_0|m_1,\sigma_1)}{N(m_0|m_2,\sigma_2)} = \frac{\sigma_2}{\sigma_1}\exp\left\{\frac{(m_0-m_2)^2}{2\sigma_2^2} - \frac{(m_0-m_1)^2}{2\sigma_1^2}\right\}$$

Taking $(l_{12} = l_{21}; l_{11} = l_{22} = 0)$, the Bayesian decision criteria in favor of the hypothesis H_1 is:

$$B_{12} > \frac{P(H_2)}{P(H_1)} \quad \longrightarrow \quad \ln B_{12} > \ln\frac{P(H_2)}{P(H_1)}$$

Thus, we have a critical value m_c of the mass:

$$\sigma_1{}^2\,(m_c - m_2)^2 - \sigma_2{}^2\,(m_c - m_1)^2 = 2\,\sigma_1{}^2\,\sigma_2{}^2 \ln\left(\frac{P(H_2)\,\sigma_1}{P(H_1)\,\sigma_2}\right)$$

such that, if $m_0 < m_c$ we decide in favor of H_1 and for H_2 otherwise. In the case that $\sigma_1 = \sigma_2$ and $P(H_1) = P(H_2)$, then $m_c = (m_1 + m_2)/2$. This, however, may be a quite unrealistic assumption for if $P(H_1) > P(H_2)$, it may be more likely that the event is of type 1 being $B_{12} < 1$.

Example 2.24 Suppose we have an iid sample $\boldsymbol{x} = \{x_1, \ldots, x_n\}$ of size n with $X \sim= N(x|\mu, 1)$ and the two hypothesis $H_1 = \{N(x|0,1)\}$ and $H_2 = \{N(x|\mu,1);\, \mu \neq 0\}$. Let us take $\{x_i\}$ as the minimum sample and, with the usual constant prior, consider the proper posterior

$$\pi(\mu|x_i) = \frac{1}{\sqrt{2\pi}}\exp\{-(\mu - x_i)^2/2\}$$

[10] If $P(H_1) = P(H_2) = 1/2$, then $P(H_1|\text{data}) = 0.95 \longrightarrow B_{12} = 19 \longrightarrow \Delta_{12} \simeq 6$.

that we use as a prior for the rest of the sample $\boldsymbol{x}' = \{x_1, \ldots, x_{i-1}, x_{i+1}, \ldots, x_n\}$. Then

$$\frac{P(H_1|\boldsymbol{x}', x_i)}{P(H_2|\boldsymbol{x}', x_i)} = B_{12}(i)\,\frac{P(H_1)}{P(H_2)}$$

$$\text{where}\quad B_{12}(i) = \frac{p(\boldsymbol{x}'|0)}{\int_{-\infty}^{\infty} p(\boldsymbol{x}'|\mu)\pi(\mu|x_i)d\mu} = n^{1/2}\,\exp\{-(n\,\overline{x}^2 - x_i^2)/2\}$$

and $\overline{x} = n^{-1}\sum_{k=1}^{n} x_k$. To avoid the effect that a particular choice of the minimal sample ($\{x_i\}$) may have, this is evaluated for all possible minimal samples and the median (or the geometric mean) of all the $B_{12}(i)$ is taken. Since $P(H_1|\boldsymbol{x}) + P(H_2|\boldsymbol{x}) = 1$, if we assign equal prior probabilities to the two hypothesis ($P(H_1) = P(H_2) = 1/2$) we have that

$$P(H_1|\boldsymbol{x}) = \frac{B_{12}}{1 + B_{12}} = \left(1 + n^{-1/2}\,\exp\{(n\,\overline{x}^2 - \text{med}\{x_i^2\})/2\}\right)^{-1}$$

is the posterior probability that quantifies the evidence in favor of the hypothesis H_1. It is left as an exercise to compare the Bayes Factor obtained from the geometric mean with what you would get if you were to take a proper prior $\pi(\mu|\sigma) = N(\mu|0, \sigma)$.

Problem 2.11 Suppose we have n observations (independent, under the same experimental conditions,...) of energies or decay time of particles above a certain known threshold and we want to test the evidence of an exponential fall against a power law. Consider then a sample $\boldsymbol{x} = \{x_1, \ldots, x_n\}$ of observations with $\text{supp}(X) = (1, \infty)$ and the two models

$$M_1 : p_1(x|\theta) = \theta\,\exp\{-\theta(x - 1)\}\mathbf{1}_{(1,\infty)}(x) \quad \text{and} \quad M_2 : p_2(x|\alpha) = \alpha x^{-(\alpha+1)}\mathbf{1}_{(1,\infty)}(x)$$

that is, Exponential and Pareto with unknown parameters θ and α. Show that for the minimal sample $\{x_i\}$ and reference priors, the Bayes Factor $B_{12}(i)$ is given by

$$B_{12}(i) = \left(\frac{x_g\,\ln x_g}{\overline{x} - 1}\right)^n \left(\frac{x_i - 1}{x_i\,\ln x_i}\right) = \frac{p_1(\boldsymbol{x}|\widehat{\boldsymbol{\theta}})}{p_2(\boldsymbol{x}|\widehat{\boldsymbol{\alpha}})}\left(\frac{x_i - 1}{x_i\,\ln x_i}\right)$$

where $(\overline{x}, x_g)$ are the arithmetic and geometric sample means and $(\widehat{\boldsymbol{\theta}}, \widehat{\boldsymbol{\alpha}})$ the values that maximize the likelihoods and therefore

$$\text{med}\{B_{12}(i)\}_{i=1}^{n} = \left(\frac{x_g\,\ln x_g}{\overline{x} - 1}\right)^n\,\text{med}\left\{\frac{x_i - 1}{x_i\,\ln x_i}\right\}_{i=1}^{n}$$

Problem 2.12 Suppose we have two experiments $e_i(n_i)$; $i = 1, 2$ in which, out of n_i trials, x_i successes have been observed and we are interested in testing whether both treatments are different or not (*contingency tables*). If we assume Binomial models $Bi(x_i|n_i, \theta_i)$ for both experiments and the two hypothesis $H_1 : \{\theta_1 = \theta_2\}$ and $H_2 : \{\theta_1 \neq \theta_2\}$, the Bayes Factor will be

$$B_{12} = \frac{\int_{\Theta} Bi(x_1|n_1, \theta) Bi(x_2|n_2, \theta) \pi(\theta) d\theta}{\int_{\Theta_1} Bi(x_1|n_1, \theta_1) \pi(\theta_1) d\theta_1 \int_{\Theta_2} Bi(x_2|n_2, \theta) \pi(\theta_2) d\theta_2}$$

We may consider proper Beta prior densities $Be(\theta|a, b)$. In a specific pharmacological analysis, a sample of $n_1 = 52$ individuals were administered a placebo and $n_2 = 61$ were treated with an a priori beneficial drug. After the essay, positive effects were observed in $x_1 = 22$ out of the 52 and $x_2 = 41$ out of the 61 individuals. It is left as an exercise to obtain the posterior probability $P(H_2|\text{data})$ with Jeffreys' ($a = b = 1/2$) and Uniform ($a = b = 1$) priors and to determine the BIC difference Δ_{12}.

2.10.2 Point Estimation

When we have to face the problem to characterize the posterior density by a single number, the most usual *Loss Functions* are:

• **Quadratic Loss**: In the simple one-dimensional case, the Loss Function is

$$l(\theta, a) = (\theta - a)^2$$

so, minimizing the *Risk*:

$$\min \int_{\Omega_\theta} (\theta - a)^2 \, p(\theta|\mathbf{x}) \, d\theta \quad \longrightarrow \quad \int_{\Omega_\theta} (\theta - a) \, p(\theta|\mathbf{x}) \, d\theta = 0$$

and therefore $a = E[\theta]$; that is, the posterior mean.

In the $k-$dimensional case, if $\mathcal{A} = \Omega_\theta = \mathcal{R}^k$ we shall take as Loss Function

$$l(\boldsymbol{\theta}, \boldsymbol{a}) = (\boldsymbol{a} - \boldsymbol{\theta})^T \, \mathbf{H} \, (\boldsymbol{a} - \boldsymbol{\theta})$$

where $\mathbf{H}$ is a positive defined symmetric matrix. It is clear that:

$$\min \int_{\mathcal{R}^k} (\boldsymbol{a} - \boldsymbol{\theta})^T \, \mathbf{H} \, (\boldsymbol{a} - \boldsymbol{\theta}) \, p(\boldsymbol{\theta}|\boldsymbol{x}) \, d\boldsymbol{\theta} \quad \longrightarrow \quad \mathbf{H} \, \boldsymbol{a} = \mathbf{H} \, E[\boldsymbol{\theta}]$$

so, if $\mathbf{H}^{-1}$ exists, then $\boldsymbol{a} = E[\boldsymbol{\theta}]$. Thus, we have that the Bayesian estimate under a quadratic loss function is the mean of $p(\boldsymbol{\theta}|\boldsymbol{x})$ (... if exists!).

• **Linear Loss**: If $\mathcal{A} = \Omega_\theta = \mathcal{R}$, we shall take the loss function:

$$l(\theta, a) = c_1\,(a-\theta)\,\mathbf{1}_{\theta \le a} + c_2\,(\theta - a)\,\mathbf{1}_{\theta > a}$$

Then, the estimator will be such that

$$\min \int_{\Omega_\theta} l(a,\theta) p(\theta|\boldsymbol{x}) d\theta = \min \left(c_1 \int_{-\infty}^{a} (a-\theta) p(\theta|\boldsymbol{x}) d\theta + c_2 \int_{a}^{\infty} (\theta - a) p(\theta|\boldsymbol{x}) d\theta \right)$$

After derivative with respect to a we have $(c_1 + c_2)\, P(\theta \le a) - c_2 = 0$ and therefore the estimator will be the value of a such that

$$P(\theta \le a) = \frac{c_2}{c_1 + c_2}$$

In particular, if $c_1 = c_2$ then $P(\theta \le a) = 1/2$ and we shall have the median of the distribution $p(\boldsymbol{\theta}|\boldsymbol{x})$. In this case, the Loss Function can be expressed more simply as $l(\theta, a) = |\theta - a|$.

• **Zero-One Loss**: Si $\mathcal{A} = \Omega_\theta = \mathcal{R}^k$, we shall take the Loss Function

$$l(\boldsymbol{\theta}, \boldsymbol{a}) = 1 - \mathbf{1}_{\mathcal{B}_\epsilon(\boldsymbol{a})}$$

where $\mathcal{B}_\epsilon(\boldsymbol{a}) \in \Omega_\theta$ is an open ball of radius ϵ centered at $\boldsymbol{a}$. The corresponding point estimator will be:

$$\min \int_{\Omega_\theta} (1 - \mathbf{1}_{\mathcal{B}_\epsilon(\boldsymbol{a})})\, p(\boldsymbol{\theta}|\boldsymbol{x})\, d\boldsymbol{\theta} = \max \int_{\mathcal{B}_\epsilon(\boldsymbol{a})} p(\boldsymbol{\theta}|\boldsymbol{x})\, d\boldsymbol{\theta}$$

It is clear that, in the limit $\epsilon \to 0$, the Bayesian estimator for the Zero-One Loss Function will be the mode of $p(\boldsymbol{\theta}|\boldsymbol{x})$ if exists.

As explained in Chap. 1, the mode, the median and the mean can be very different if the distribution is not symmetric. Which one should we take then? Quadratic losses, for which large deviations from the *true* value are penalized quadratically, are the most common option but, even if for unimodal symmetric the three statistics coincide, it may be misleading to take this value as a characteristic number for the information we got about the parameters or even be nonsense. In the hypothetical case that the posterior is essentially the same as the likelihood (that is the case for a sufficiently smooth prior), the Zero-One Loss points to the classical estimate of the *Maximum Likelihood Method*. Other considerations of interest in Classical Statistics (like bias, consistency, minimum variance,…) have no special relevance in Bayesian inference.

Problem 2.13 (*The Uniform Distribution*) Show that for the posterior density (see Example 2.4)

$$p(\theta|x_M, n) = n\,\frac{x_M^n}{\theta^{n+1}}\,\mathbf{1}_{[x_M,\infty)}(\theta)$$

the point estimates under quadratic, linear and 0–1 loss functions are

$$\theta_{QL} = x_M\,\frac{n}{n-1}; \qquad \theta_{LL} = x_M\,2^{1/n} \quad \text{and} \qquad \theta_{01L} = x_M$$

and discuss which one you consider more reasonable.

2.11 Credible Regions

Let $p(\boldsymbol{\theta}|\boldsymbol{x})$, with $\boldsymbol{\theta} \in \Omega \subseteq \mathcal{R}^n$ be a posterior density function. A credible region with probability content $1-\alpha$ is a region of $V_\alpha \subseteq \Theta$ of the parametric space such that

$$P(\boldsymbol{\theta} \in V_\alpha) = \int_{V_\alpha} p(\boldsymbol{\theta}|\boldsymbol{x})\,d\boldsymbol{\theta} = 1-\alpha$$

Obviously, for a given probability content credible regions are not unique and a sound criteria is to specify the one that the smallest possible volume. A region C of the parametric space Ω is called *Highest Probability Region* (HPD) with probability content $1-\alpha$ if:

(1) $P(\boldsymbol{\theta} \in C) = 1-\alpha; \quad C \subseteq \Omega;$
(2) $p(\boldsymbol{\theta}_1|\cdot) \geq p(\boldsymbol{\theta}_2|\cdot)$ for all $\boldsymbol{\theta}_1 \in C$ and $\boldsymbol{\theta}_2 \notin C$ except, at most, for a subset of Ω with zero probability measure.

It is left as an exercise to show that condition (2) implies that the HPD region so defined is of minimum volume so both definitions are equivalent. Further properties that are easy to demonstrate are:

(1) If $p(\theta|\cdot)$ is *not uniform*, the HPD region with probability content $1-\alpha$ *is unique*;
(2) If $p(\boldsymbol{\theta}_1|\cdot) = p(\boldsymbol{\theta}_2|\cdot)$, then $\boldsymbol{\theta}_1$ and $\boldsymbol{\theta}_2$ are both either included or excluded of the HPD region;
(3) If $p(\boldsymbol{\theta}_1|\cdot) \neq p(\boldsymbol{\theta}_2\cdot)$, there is an HPD region for some value of $1-\alpha$ that contains one value of $\boldsymbol{\theta}$ and not the other;
(4) $C = \{\boldsymbol{\theta} \in \Theta | p(\boldsymbol{\theta}|\boldsymbol{x}) \geq k_\alpha\}$ where k_α is the largest constant for which $P(\boldsymbol{\theta} \in C) \geq \alpha$;
(5) If $\phi = f(\boldsymbol{\theta})$ is a one-to-one transformation, then
    (a) any region with probability content $1-\alpha$ for $\boldsymbol{\theta}$ will have probability content $1-\alpha$ for ϕ but…
    (b) an HPD region for $\boldsymbol{\theta}$ will not, in general, be an HPD region for ϕ unless the transformation is linear.

In general, evaluation of credible regions is a bit messy task. A simple way through is to do a Monte Carlo sampling of the posterior density and use the 4th property.

For a one-dimensional parameter, the condition that the HPD region with probability content $1 - \alpha$ has the minimum length allows to write a relation that may be useful to obtain those regions in an easier manner. Let $[\theta_1, \theta_2]$ be an interval such that

$$\int_{\theta_1}^{\theta_2} p(\theta|\cdot)\, d\theta = 1 - \alpha$$

For this to be an HPD region we have to find the extremal of the function

$$\phi(\theta_1, \theta_2, \lambda) = (\theta_2 - \theta_1) + \lambda \left(\int_{\theta_1}^{\theta_2} p(\theta|\cdot)\, d\theta - (1 - \alpha) \right)$$

Taking derivatives we get:

$$\begin{array}{cccc} \left(\dfrac{\partial\phi(\theta_1, \theta_2, \lambda)}{\partial\theta_i}\right)_{i=1,2} = 0 & \longrightarrow & p(\theta_1|\cdot) = p(\theta_2\cdot) \\ \dfrac{\partial\phi(\theta_1, \theta_2, \lambda)}{\partial\lambda} = 0 & \longrightarrow & \int_{\theta_1}^{\theta_2} p(\theta)\, d\theta = 1 - \alpha \end{array}$$

Thus, from the first two conditions we have that $p(\theta_1|\cdot) = p(\theta_2|\cdot)$ and, from the third, we know that $\theta_1 \neq \theta_2$. In the special case that the distribution is unimodal and symmetric the only possible solution is $\theta_2 = 2E[\theta] - \theta_1$.

The HPD regions are useful to summarize the information on the parameters contained in the posterior density $p(\boldsymbol{\theta}|\boldsymbol{x})$ but it should be clear that there is no justification to reject a particular value $\boldsymbol{\theta}_0$ just because is not included in the HPD region (or, in fact, in whatever confidence region) and that in some circumstances (distributions with more than one mode for instance) it may be the union of disconnected regions.

2.12 Bayesian ($\mathcal{B}$) Versus Classical ($\mathcal{F}$) Philosophy

The Bayesian philosophy aims at the right questions in a very intuitive and, at least conceptually, simple manner. However the *"classical"* (frequentist) approach to statistics, that has been very useful in scientific reasoning over the last century, is at present more widespread in the Particle Physics community and most of the stirred up controversies are originated by misinterpretations. It is worth to take a look for instance at [2]. Let's see how a simple problem is attacked by the two schools. "We" are $\mathcal{B}$, "they" are $\mathcal{F}$.

Suppose we want to estimate the life-time of a particle. We both *"assume"* an exponential model $X \sim Ex(x|1/\tau)$ and do an experiment $e(n)$ that provides an iid sample $\boldsymbol{x} = \{x_1, x_2, \ldots, x_n\}$. In this case there is a sufficient statistic $\boldsymbol{t} = (n, \overline{x})$ with $\overline{x}$ the sample mean so let's define the random quantity

$$X = \frac{1}{n}\sum_{i=1}^{n} X_i \sim p(x|n,\tau) = \left(\frac{n}{\tau}\right)^n \frac{1}{\Gamma(n)} \exp\left\{-nx\tau^{-1}\right\} x^{n-1}\,\mathbf{1}_{(0,\infty)}(x)$$

What can we say about the parameter of interest τ?

$\mathcal{F}$ will start by finding the *estimator* (statistic) $\widehat{\tau}$ that maximizes the likelihood (MLE). In this case it is clear that $\widehat{\tau} = \overline{x}$, the sample mean. We may ask about the rationale behind because, apparently, there is no serious mathematical reasoning that justifies this procedure. $\mathcal{F}$ will respond that, in a certain sense, even for us this should be a reasonable way because if we have a smooth prior function, the posterior is dominated by the likelihood and one possible point estimator is the mode of the posterior. Beside that, he will argue that maximizing the likelihood renders an estimator that often has "good" properties like unbiasedness, invariance under monotonous one-to-one transformations, consistency (convergence in probability), smallest variance within the class of unbiased estimators (Cramèr-Rao bound), approximately well known distribution,... We may question some of them (unbiased estimators are not always the best option and invariance... well, if the transformation is not linear usually the MLE is biased), argue that the others hold in the asymptotic limit,... Anyway; for this particular case one has that:

$$E[\widehat{\tau}] = \tau \qquad \text{and} \qquad V[\widehat{\tau}] = \frac{\tau^2}{n}$$

and $\mathcal{F}$ will claim that *"if you repeat the experiment"* many times under the same conditions, you will get a sequence of estimators $\{\widehat{\tau}_1, \widehat{\tau}_2, \ldots\}$ that eventually will cluster around the life-time τ. Fine but we shall point out that, first, although desirable we usually do not repeat the experiments (and under the same conditions is even more rare) so we have just one observed sample ($\boldsymbol{x} \to \overline{x} = \widehat{\tau}$) from $e(n)$. Second, *"if you repeat the experiment you will get"* is a free and unnecessary hypothesis. You do not know what you will get, among other things, because the model we are considering may not be the way nature behaves. Besides that, it is quite unpleasant that inferences on the life-time depend upon what you think you will get if you do what you know you are not going to do. And third, that this is in any case a nice sampling property of the estimator $\hat{\tau}$ but eventually we are interested in τ so, What can we say about it?

For us, the answer is clear. Being τ a scale parameter we write the posterior density function

$$p(\tau|n,\overline{x}) = \frac{(n\,\overline{x})^n}{\Gamma(n)} \exp\left\{-n\,\overline{x}\tau^{-1}\right\} \tau^{-(n+1)}\,\mathbf{1}_{(0,\infty)}(\tau)$$

for the *degree of belief* we have on the parameter and easily get for instance:

$$E[\tau^k] = (n\overline{x})^k \frac{\Gamma(n-k)}{\Gamma(n)} \quad \longrightarrow \quad E[\tau] = \overline{x}\frac{n}{n-1}; \quad V[\tau] = \overline{x}^2 \frac{n^2}{(n-1)^2(n-2)}; \ldots$$

Cleaner and simpler impossible.

To bound the life-time, $\mathcal{F}$ proceeds with the determination of the *Confidence Intervals*. The classical procedure was introduced by J. Neyman in 1933 and rests on establishing, for an specified probability content, the domain of the random quantity (usually a statistic) as function of the possible values the parameters may take. Consider a one dimensional parameter θ and the model $X \sim p(x|\theta)$. Given a desired probability content $\beta \in [0, 1]$, he determines the interval $[x_1, x_2] \subset \Omega_X$ such that

$$P(X \in [x_1, x_2]) = \int_{x_1}^{x_2} p(x|\theta)\, dx = \beta$$

for a particular fixed value of θ. Thus, for each possible value of θ he has one interval $[x_1 = f_1(\theta; \beta), x_1 = f_2(\theta; \beta)] \subset \Omega_X$ and the sequence of those intervals gives a band in the $\Omega_\theta \times \Omega_X$ region of the real plane. As for the *Credible Regions*, these intervals are not uniquely determined so one usually adds the condition:

$$(1) \quad \int_{-\infty}^{x_1} p(x|\theta)\, dx = \int_{x_2}^{\infty} p(x|\theta)\, dx = \frac{1-\beta}{2} \qquad \text{or}$$

$$(2) \quad \int_{x_1}^{\theta} p(x|\theta)\, dx = \int_{\theta}^{x_2} p(x|\theta)\, dx = \frac{\beta}{2}$$

or, less often, (3) chooses the interval with smallest size. Now, for an invertible mapping $x_i \longrightarrow f_i(\theta)$ one can write

$$\beta = P(f_1(\theta) \leq X \leq f_2(\theta)) = P(f_2^{-1}(X) \leq \theta \leq f_1^{-1}(X))$$

and get the random interval $[f_2^{-1}(X), f_1^{-1}(X)]$ that contains the given value of θ with probability β. Thus, for each possible value that X may take he will get an interval $[f_2^{-1}(X), f_1^{-1}(X)]$ on the θ axis and a particular experimental observation $\{x\}$ will single out one of them. This is the *Confidence Interval* that the frequentist analyst will quote. Let's continue with the life-time example and take, for illustration, $n = 50$ and $\beta = 0.68$. The bands $[x_1 = f_1(\tau), x_2 = f_2(\tau)]$ in the (τ, X) plane, in this case obtained with the third prescription, are shown in Fig. 2.4(1). They are essentially straight lines so $P[X \in (0.847\tau, 1.126\tau)] = 0.68$. This is a correct statement, but doesn't say anything about τ so he inverts that and gets $0.89\, X < \tau < 1.18\, X$ in such a way that an observed value $\{x\}$ singles out an interval in the vertical τ axis. We, Bayesians, will argue this does not mean that τ has a 0.68 chance to lie in this interval and the frequentist will certainly agree on that. In fact, this is not an admissible question for him because in the classical philosophy τ is a number, unknown but a *fixed* number. If he repeats the experiment τ will not change; it is the interval that will be different because x will change. They are *random intervals* and what the 68% means is just that if he repeats the experiment a large number N of times, he will end up with N intervals of which ~68% will contain the true value τ whatever it is. But the experiment is done only once so: Does the interval derived from this observation contain τ or not? We don't know, we have no idea if it does contain τ,

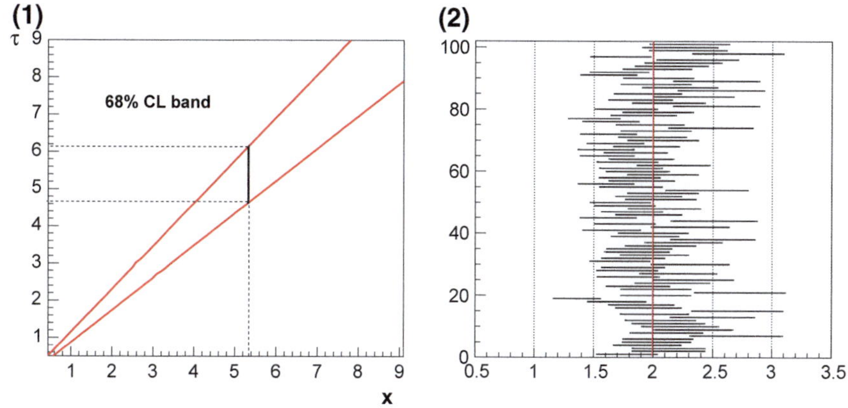

Fig. 2.4 (**1**) 68% confidence level bands in the (τ, X) plane. (**2**) 68% confidence intervals obtained for 100 repetitions of the experiment

if it does not and how far is the unknown true value. Figure 2.4(2) shows the 68% confidence intervals obtained after 100 repetitions of the experiment for $\tau = 2$ and 67 of them did contain the true value. But when the experiment is done once, he picks up one of those intervals and has a 68% chance that the one chosen contains the true value. We $\mathcal{B}$ shall proceed in a different manner. After integration of the posterior density we get the HPD interval $P\,[\tau \in (0.85\,x, 1.13\,x)] = 0.68$; almost the same but with a direct interpretation in terms of what we are interested in. Thus, both have an absolutely different philosophy:
$\mathcal{F}$: "Given a particular value of the parameters of interest, How likely is the observed data?"
$\mathcal{B}$: "Having observed this data, What can we say about the parameters of interest?"
... and the probability if the causes, as Poincare said, is the most important from the point of view of scientific applications.

In many circumstances we are also interested in one-sided intervals. That is for instance the case when the data is consistent with the hypothesis $H : \{\theta = \theta_0\}$ and we want to give an upper bound on θ so that $P(\theta \in (-\infty, \theta_\beta] = \beta$. The frequentist rationale is the same: obtain the interval $[-\infty, x_2] \subset \Omega_X$ such that

$$P(X \leq x_2) = \int_{-\infty}^{x_2} p(x|\theta)\,dx = \beta$$

where $x_2 = f_2(\theta)$; in this case without ambiguity. For the random interval $(-\infty, f_2^{-1}(X))$ $\mathcal{F}$ has that

$$P\left(\theta < f_2^{-1}(X)\right) = 1 - P\left(\theta \geq f_2^{-1}(X)\right) = 1 - \beta$$

so, for a probability content α (say 0.95), one should set $\beta = 1 - \alpha$ (=0.05). Now, consider for instance the example of the anisotropy is cosmic rays discussed in the last

Sect. 2.13.3. For a dipole moment (details are unimportant now) we have a statistic

$$X \sim p(x|\theta, 1/2) = \frac{\exp\{-\theta^2/2\}}{\sqrt{2\pi}\theta} \exp\{-x/2\}\sinh(\theta\sqrt{x})\, \mathbf{1}_{(0,\infty)}(x)$$

where the parameter θ is the dipole coefficient multiplied by a factor that is irrelevant for the example. It is argued in Sect. 2.13.3 that the reasonable prior for this model is $\pi(\theta) = \text{constant}$ so we have the posterior

$$p(\theta|x, 1/2) = \frac{\sqrt{2}}{\sqrt{\pi x}\, M(1/2, 3/2, x/2)} \exp\{-\theta^2/2\}\, \theta^{-1} \sinh(\theta\sqrt{x})\, \mathbf{1}_{(0,\infty)}(\theta)$$

with $M(a, b, z)$ the Kummer's Confluent Hypergeometric Function. In fact, θ has a compact support but since the observed values of X are consistent with $H_0 : \{\theta = 0\}$ and the sample size is very large [AMS13],[11] $p(\theta|x, 1/2)$ is concentrated in a small interval $(0, \epsilon)$ and it is easier for the evaluations to extend the domain to $\mathcal{R}^+$ without any effect on the results. Then we, Bayesians, shall derive the one-sided upper credible region $[0, \theta_{0.95}(x)]$ with $\alpha = 95\%$ probability content as simply as:

$$\int_0^{\theta_{0.95}} p(\theta|x, 1/2)\, d\theta = \alpha = 0.95$$

This upper bound shown as function of x in Fig. 2.5 under "*Bayes*" (red line). Neyman's construction is also straight forward. From

$$\int_0^{x_2} p(x|\theta, 1/2)\, dx = 1 - \alpha = 0.05$$

(essentially a χ^2 probability for $\nu = 3$), $\mathcal{F}$ will get the upper bound shown in the same figure under "*Neyman*" (blue broken line). As you can see, they get closer as x grows but, first, there is no solution for $x \leq x_c = 0.352$. In fact, $E[X] = \theta^2 + 3$ so if the dipole moment is $\delta = 0$ $(\theta = 0)$, $E[X] = 3$ and observed values below x_c will be an unlikely fluctuation downwards (assuming of course that the model is correct) but certainly a possible experimental outcome. In fact, you can see that for values of x less than 2, even though there is a solution Neyman's upper bound is underestimated. To avoid this "little" problem, a different prescription has to be taken.

The most interesting solution is the one proposed by Feldman and Cousins [25] in which the region $\Delta_X \subset \Omega_X$ that is considered for the specified probability content is determined by the ratio of probability densities. Thus, for a given value θ_0, the interval Δ_X is such that

[11] [AMS13]: Aguilar M. et al. (2013); Phys. Rev. Lett. 110, 141102.

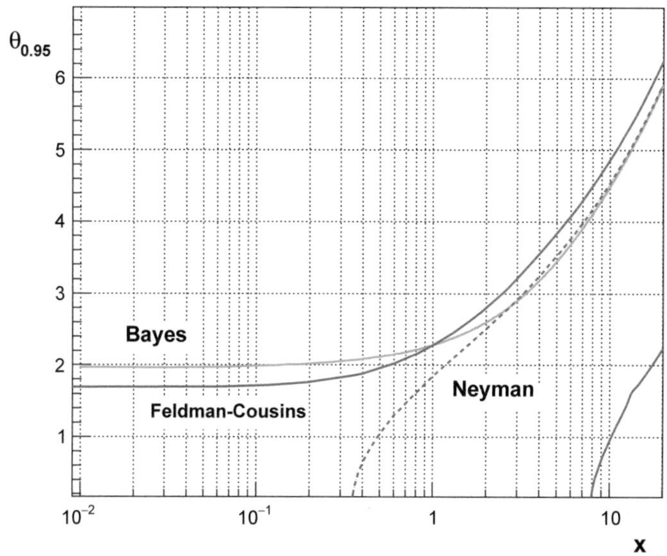

Fig. 2.5 95% upper bounds on the parameter θ following the Bayesian approach (*red*), the Neyman approach (*broken blue*) and Feldman and Cousins (*solid blue line*)

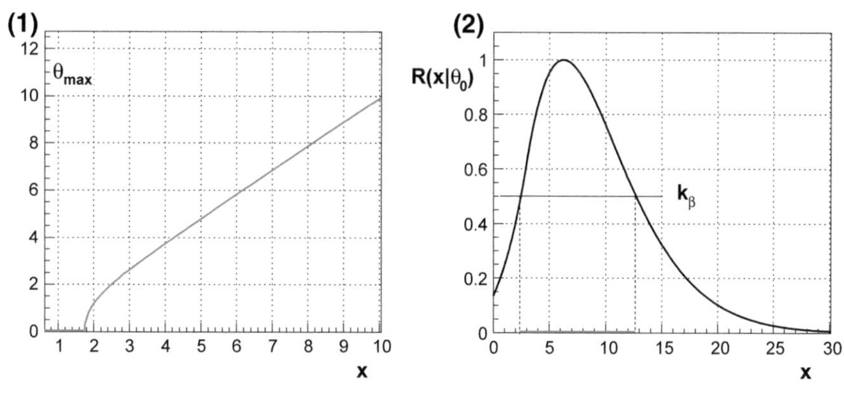

Fig. 2.6 (**1**) Dependence of θ_m with x. (**2**) Probability density ratio $R(x|\theta)$ for $\theta = 2$

$$\int_{\Delta_X} p(x|\theta_0)\,dx = \beta \qquad \text{with} \qquad R(x|\theta_0) = \frac{p(x|\theta_0)}{p(x|\theta_b)} > k_\beta;\ \forall x \in \Delta_X$$

and where θ_b is the *best* estimation of θ for a given $\{x\}$; usually the one that maximizes the likelihood (θ_m). In our case, it is given by:

$$\theta_m = \begin{cases} 0 & \text{if } x \le \sqrt{3} \\ \theta_m + \theta_m^{-1} - \sqrt{x}\coth(\theta_m\sqrt{x}) = 0 & \text{if } x > \sqrt{3} \end{cases}$$

and the dependence with x is shown in Fig. 2.6(1) ($\theta_m \simeq x$ for $x >>$). As illustration, function $R(x|\theta)$ is shown in Fig. 2.6(2) for the particular value $\theta_0 = 2$. Following this procedure,[12] the 0.95 probability content band is shown in Fig. 2.5 under *"Feldman-Cousins"* (blue line). Note that for large values of x, the confidence region becomes an interval. It is true that if we observe a large value of X, the hypothesis $H_0 : \{\delta = 0\}$ will not be favoured by the data and a different analysis will be more relevant although, by a simple modification of the ordering rule, we still can get an upper bound if desired or use the standard Neyman's procedure.

The Feldman and Cousins prescription allows to consider constrains on the parameters in a simpler way than Neyman's procedure and, as opposed to it, will always provide a region with the specified probability content. However, on the one hand, they are frequentist intervals and as such have to be interpreted. On the other hand, for discrete random quantities with image in $\{x_1, \ldots, x_k, \ldots\}$ it may not be possible to satisfy exactly the probability content equation since for the Distribution Function one has that $F(x_{k+1}) = F(x_k) + P(X = x_{k+1})$. And last, it is not straight forward to deal with nuisance parameters. Therefore, the best advice: "Be Bayesian!".

2.13 Some Worked Examples

2.13.1 Regression

Consider the exchangeable sequence $z = \{(x_1, y_1), (x_2, y_2), \ldots, (x_n, y_n)\}$ of n samplings from the two-dimensional model $N(x_i, y_i|\cdot) = N(x_i|\mu_{x_i}, \sigma^2_{x_i})N(y_i|\mu_{y_i}, \sigma^2_{y_i})$. Then

$$p(z|\cdot) \propto \exp\left\{-\frac{1}{2}\sum_{i=1}^{n}\left[\frac{(y_i - \mu_{yi})^2}{\sigma^2_{yi}} + \frac{(x_i - \mu_{xi})^2}{\sigma^2_{xi}}\right]\right\}$$

We shall assume that the precisions σ_{xi} and σ_{xi} are known and that there is a functional relation $\mu_y = f(\mu_x; \boldsymbol{\theta})$ with unknown parameters $\boldsymbol{\theta}$. Then, in terms of the new parameters of interest:

$$p(\boldsymbol{y}|\cdot) \propto \exp\left\{-\frac{1}{2}\sum_{i=1}^{n}\left[\frac{(y_i - f(\mu_{x_i}; \boldsymbol{\theta}))^2}{\sigma^2_{y_i}} + \frac{(x_i - \mu_{x_i})^2}{\sigma^2_{x_i}}\right]\right\}$$

Consider a linear relation $f(\mu_x; a, b) = a + b\mu_x$ with a, b the unknown parameters so:

[12] In most cases,a Monte Carlo simulation will simplify life.

$$p(\mathbf{z}|\cdot) \propto \exp\left\{-\frac{1}{2}\sum_{i=1}^{n}\left[\frac{(y_i - a - b\mu_{x_i})^2}{\sigma_{y_i}^2} + \frac{(x_i - \mu_{x_i})^2}{\sigma_{x_i}^2}\right]\right\}$$

and assume, in first place, that $\mu_{x_i} = x_i$ without uncertainty. Then,

$$p(\mathbf{y}|a, b) \propto \exp\left\{-\frac{1}{2}\sum_{i=1}^{n}\left[\frac{(y_i - a - bx_i)^2}{\sigma_{y_i}^2}\right]\right\}$$

There is a set of sufficient statistics for (a, b):

$$\mathbf{t} = \{t_1, t_2, t_3, t_4, t_5\} = \left\{\sum_{i=1}^{n}\frac{1}{\sigma_i^2}, \sum_{i=1}^{n}\frac{x_i^2}{\sigma_i^2}, \sum_{i=1}^{n}\frac{x_i}{\sigma_i^2}, \sum_{i=1}^{n}\frac{y_i}{\sigma_i^2}, \sum_{i=1}^{n}\frac{y_i x_i}{\sigma_i^2}\right\}$$

and, after a simple algebra, it is easy to write

$$p(\mathbf{y}|a, b) \propto \exp\left\{-\frac{1}{2(1-\rho^2)}\left[\frac{(a-a_0)^2}{\sigma_a^2} + \frac{(b-b_0)^2}{\sigma_b^2} - 2\rho\frac{(a-a_0)}{\sigma_a}\frac{(b-b_0)}{\sigma_b}\right]\right\}$$

where the new statistics $\{a_0, b_0, \sigma_a, \sigma_b, \rho\}$ are defined as:

$$a_0 = \frac{t_2 t_4 - t_3 t_5}{t_1 t_2 - t_3^2}, \qquad b_0 = \frac{t_1 t_5 - t_3 t_4}{t_1 t_2 - t_3^2}$$
$$\sigma_a^2 = \frac{t_2}{t_1 t_2 - t_3^2}, \qquad \sigma_b^2 = \frac{t_1}{t_1 t_2 - t_3^2}, \qquad \rho = -\frac{t_3}{\sqrt{t_1 t_2}}$$

Both (a, b) are position parameters so we shall take a uniform prior and in consequence

$$p(a, b|\cdot) = \frac{1}{2\pi\sigma_a\sigma_b\sqrt{1-\rho^2}} e^{\left\{-\frac{1}{2(1-\rho^2)}\left[\frac{(a-a_0)^2}{\sigma_a^2} + \frac{(b-b_0)^2}{\sigma_b^2} - 2\rho\frac{(a-a_0)}{\sigma_a}\frac{(b-b_0)}{\sigma_b}\right]\right\}}$$

This was obviously expected.

When μ_{x_i} are n unknown parameters, if we take $\pi(\mu_{xi}) = \mathbf{1}_{(0,\infty)}(\mu_{xi})$ and marginalize for (a, b) we have

$$p(a, b|\cdot) \propto \pi(a, b)\exp\left\{-\frac{1}{2}\sum_{i=1}^{n}\frac{(y_i - a - b\,x_i)^2}{\sigma_{yi}^2 + b^2\,\sigma_{xi}^2}\right\}\left\{\prod_{i=1}^{n}(\sigma_{yi}^2 + b^2\,\sigma_{xi}^2)\right\}^{-1/2}$$

In general, the expressions one gets for non-linear regression problems are complicated and setting up priors is a non-trivial task but fairly vague priors easy to deal with are usually a reasonable choice. In this case, for instance, one may consider uniform priors or normal densities $N(\cdot|0, \sigma >>)$ for both parameters (a, b) and sample the proper posterior with a Monte Carlo algorithm (Gibbs sampling will be appropriate).

The same reasoning applies if we want to consider other models or more involved relations with several explanatory variables like $\theta_i = \sum_{j=1}^{k} \alpha_j x_{ij}^{b_j}$. In counting experiments, for example, $y_i \in \mathcal{N}$ so we may be interested in a Poisson model $Po(y_i|\mu_i)$ where μ_i is parameterized as a simple log-linear form $\ln(\mu_i) = \alpha_0 + \alpha_1 x_i$ (so $\mu_i > 0$ for whatever $\alpha_0, \alpha_1 \in \mathcal{R}$). Suppose for instance that we have the sample $\{(y_i, x_i)\}_{i=1}^{n}$. Then:

$$p(\boldsymbol{y}|\alpha_1, \alpha_2, \boldsymbol{x}) \propto \prod_{i=1}^{n} \exp\{-\mu_i\}\mu_i{}^{y_i} = \exp\left\{\alpha_1 s_1 + \alpha_2 s_2 - e^{\alpha_1} \sum_{i=1}^{n} e^{\alpha_2 x_i}\right\}$$

where $s_1 = \sum_{i=1}^{n} y_i$ and $s_2 = \sum_{i=1}^{n} y_i x_i$. In this case, the Normal distribution $N(\alpha_i|a_i, \sigma_i)$ with $\sigma_i >>$ is a reasonable smooth and easy to handle proper prior density for both parameters. Thus, we get the posterior conditional densities

$$p(\alpha_i|\alpha_j, \boldsymbol{y}, \boldsymbol{x}) \propto \exp\left\{-\frac{\alpha_i^2}{2\sigma_i^2} + \alpha_i\left(\frac{a_i}{\sigma_i^2} + s_i\right) - e^{\alpha_1} \sum_{i=1}^{n} e^{\alpha_2 x_i}\right\}; \; i = 1, 2$$

that are perfectly suited for the Gibbs sampling to be discussed in Sect. 4.1 of Chap. 3.

Example 2.25 (*Proton Flux in Primary Cosmic Rays*) For energies between ~20 and ~200 GeV, the flux of protons of the primary cosmic radiation is reasonably well described by a power law $\phi(r) = c\,r^{\gamma}$ where r is the *rigidity*[13] and $\gamma = d\ln\phi/ds$, with $s = \ln r$, is the *spectral index*. At lower energies, this dependence is significantly modified by the geomagnetic cut-off and the solar wind but at higher energies, where these effects are negligible, the observations are not consistent with a single power law (Fig. 2.7(1)). One may characterize this behaviour with a simple phenomenological model where the spectral index is no longer constant but has a dependence $\gamma(s) = \alpha + \beta \tanh[a(s - s_0)]$ such that $\lim_{s\to-\infty} \gamma(s) = \gamma_1$ $(r\to 0)$ and $\lim_{s\to\infty} \gamma(s) = \gamma_2$ $(r\to +\infty)$. After integration, the flux can be expressed in terms of 5 parameters $\boldsymbol{\theta} = \{\phi_0, \gamma_1, \delta = \gamma_2 - \gamma_1, r_0, \sigma\}$ as:

$$\phi(r; \boldsymbol{\theta}) = \phi_0\, r^{\gamma_1} \left[1 + \left(\frac{r}{r_0}\right)^{\sigma}\right]^{\delta/\sigma}$$

For this example, I have used the data above 45 GeV published by the AMS experiment[14] and considered only the quoted *statistical errors* (see Fig. 2.7(1)). Last, for a better description of the flux the previous expression has been modified to account for the effect of the solar wind with the force-field approximation in consistency with

[13]The *rigidity* (r) is defined as the momentum (p) divided by the electric charge (Z) so $r = p$ for protons.

[14][AMS15]: Aguilar M. et al. (2015); PRL 114, 171103 and references therein.

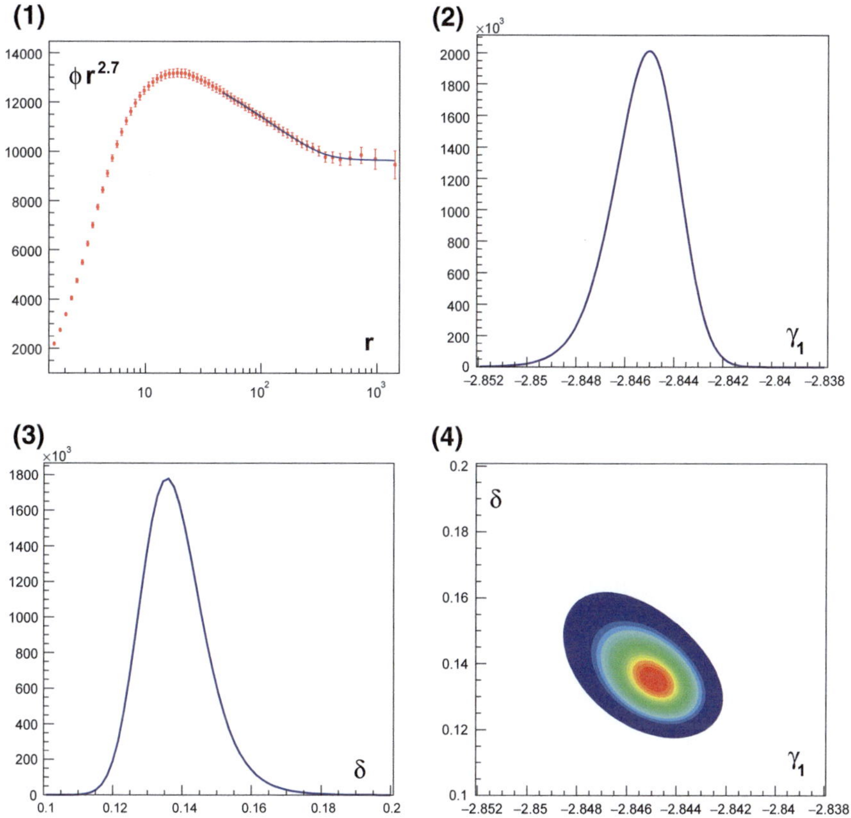

Fig. 2.7 (**1**) Observed flux multiplied by $r^{2.7}$ in $\mathrm{m}^{-2}\mathrm{sr}^{-1}\mathrm{sec}^{-1}\mathrm{GV}^{1.7}$ as given in [AMS15]; (**2**) Posterior density of the parameter γ_1 (arbitrary vertical scale); (**3**) Posterior density of the parameter $\delta = \gamma_2 - \gamma_1$ (arbitrary vertical scale); (4): Projection of the posterior density $p(\gamma_1, \delta)$

[AMS15]. This is just a technical detail, irrelevant for the purpose of the example. Then, assuming a Normal model for the observations we can write the posterior density

$$p(\boldsymbol{\theta}|\text{data}) = \pi(\boldsymbol{\theta}) \prod_{i=1}^{n} \exp\left\{-\frac{1}{2\sigma_i{}^2}\left(\phi_i - \phi(r_i; \boldsymbol{\theta})\right)^2\right\}$$

I have taken Normal priors with large variances ($\sigma_i >>$) for the parameters γ_1 and δ and restricted the support to $\mathcal{R}^+$ for $\{\phi_0, r_0, \sigma\}$. The posterior densities for the parameters γ_1 and δ are shown in Fig. 2.7(2, 3) together with the projection (Fig. 2.7(4)) that gives an idea of correlation between them. For a visual inspection, the phenomenological form of the flux is shown in Fig. 2.7(1) (blue line) overimposed to the data when the parameters are set to their expected posterior values.

2.13.2 Characterization of a Possible Source of Events

Suppose that we observe a particular region Ω of the sky during a time t and denote by λ the rate at which events from this region are produced. We take a Poisson model to describe the number of produced events: $k \sim Po(k|\lambda t)$. Now, denote by ϵ the probability to detect one event (detection area, efficiency of the detector,...). The number of observed events n from the region Ω after an exposure time t and detection probability ϵ will follow:

$$n \sim \sum_{k=n}^{\infty} Bi(k|n, \epsilon)\, Po(k|\lambda t) = Po(n|\lambda t \epsilon)$$

The approach to the problem will be the same for other counting process like, for instance, events collected from a detector for a given integrated luminosity. We suspect that the events observed in a particular region Ω_o of the sky are background events together with those from an emitting source. To determine the significance of the potential source we analyze a nearby region, Ω_b, to infer about the expected background. If after a time t_b we observe n_b events from this region with detection probability e_b then, defining $\beta = \epsilon_b t_b$ we have that

$$n_b \sim Po(n_b|\lambda_b\, \beta) = \exp\{-\beta\lambda_b\}\, \frac{(\beta\lambda_b)^{n_b}}{\Gamma(n_b + 1)}$$

At Ω_o we observe n_o events during a time t_o with a detection probability ϵ_o. Since $n_o = n_1 + n_2$ with $n_1 \sim Po(n_1|\lambda_s\alpha)$ signal events ($\alpha = \epsilon_o t_o$) and $n_2 \sim Po(n_2|\lambda_b\alpha)$ background events (assume reasonably that $e_s = e_b = e_o$ in the same region), we have that

$$n_o \sim \sum_{n_1=0}^{n_o} Po(n_1|\lambda_s\alpha) Po(n_o - n_1|\lambda_b\alpha) = Po(n_o|(\lambda_s + \lambda_b)\alpha)$$

Now, we can do several things. We can assume for instance that the overall rate form the region Ω_o is λ, write $n_o \sim Po(n_o|\alpha\lambda)$ and study the fraction λ/λ_b of the rates from the information provided by the observations in the two different regions. Then, reparameterizing the model in terms of $\theta = \lambda/\lambda_b$ and $\phi = \lambda_b$ we have

$$p(n_o, n_b|\cdot) = Po(n_o|\alpha\lambda) Po(n_b|\beta\lambda_b) \sim e^{-\beta\phi(1+\gamma\theta)}\, \theta^{n_o}\, \phi^{n_o+n_b}$$

where $\gamma = \alpha/\beta = (\epsilon_s t_s)/(\epsilon_b t_b)$. For the ordering $\{\theta, \phi\}$ we have that the Fisher's matrix and its inverse are

$$\mathbf{I}(\theta, \phi) = \begin{pmatrix} \frac{\gamma\beta\phi}{\theta} & \gamma\beta \\ \gamma\beta & \frac{\beta(1+\gamma\theta)}{\phi} \end{pmatrix} \quad \text{and} \quad \mathbf{I}^{-1}(\mu_1, \mu_2) = \begin{pmatrix} \frac{\theta(1+\gamma\theta)}{\phi\gamma\beta} & -\frac{\theta}{\beta} \\ -\frac{\theta}{\beta} & \frac{\phi}{\beta} \end{pmatrix}$$

Then

$$\pi(\theta,\phi)=\pi(\phi|\theta)\,\pi(\theta)\;\propto\;\frac{\phi^{-1/2}}{\sqrt{\theta(1+\gamma\theta)}}$$

and integrating the nuisance parameter ϕ we get finally:

$$p(\theta|n_o,n_b,\gamma)=\frac{\gamma^{n_o+1/2}}{B(n_o+1/2,n_b+1/2)}\,\frac{\theta^{n_o-1/2}}{(1+\gamma\theta)^{n_o+n_b+1}}$$

From this:

$$E[\theta^m]=\frac{1}{\gamma^m}\,\frac{\Gamma(n_o+1/2+m)\,\Gamma(n_b+1/2-m)}{\Gamma(n_o+1/2)\,\Gamma(n_b+1/2)}\quad\longrightarrow\quad E[\theta]=\frac{1}{\gamma}\,\frac{n_o+1/2}{n_b-1/2}$$

and

$$P(\theta\leq\theta_0)=\int_0^{\theta_0}p(\theta|\cdot)\,d\theta=1-IB(n_b+1/2,n_o+1/2;\,(1+\gamma\theta_0)^{-1})$$

with $IB(x,y;z)$ the Incomplete Beta Function. Had we interest in $\theta=\lambda_s/\lambda_b$, the corresponding reference prior will be

$$\pi(\theta,\phi)\;\propto\;\frac{\phi^{-1/2}}{\sqrt{(1+\theta)(\delta+\theta)}}\qquad\text{with}\qquad\delta=\frac{1+\gamma}{\gamma}$$

A different analysis can be performed to make inferences on λ_s. In this case, we may consider as an informative prior for the nuisance parameter the posterior what we had from the study of the background in the region Ω_b; that is:

$$p(\lambda_b|n_b,\,\beta)\;\propto\;\exp\{-\beta\lambda_b\}\;\lambda_b{}^{n_b-1/2}$$

and therefore:

$$p(\lambda_s|\cdot)\;\propto\;\pi(\lambda_s)\int_0^{\infty}p(n_o|\alpha(\lambda_s+\lambda_b))\,p(\lambda_b|n_b,\,\beta)\,d\lambda_b\propto\pi(\lambda_s)\,e^{-\alpha\lambda_s}\,\lambda_s^{n_o}\sum_{k=0}^{n_o}a_k\,\lambda_s{}^{-k}$$

where

$$a_k=\binom{n_o}{k}\,\frac{\Gamma(k+n_b+1/2)}{[(\alpha+\beta)]^k}$$

A reasonable choice for the prior will be a conjugated prior $\pi(\lambda_s)=Ga(\lambda_s|a,b)$ that simplifies the calculations and provides enough freedom analyze the effect of different shapes on the inferences. The same reasoning is valid if the knowledge on λ_b is represented by a different $p(\lambda_b|\cdot)$ from, say, a Monte Carlo simulation. Usual

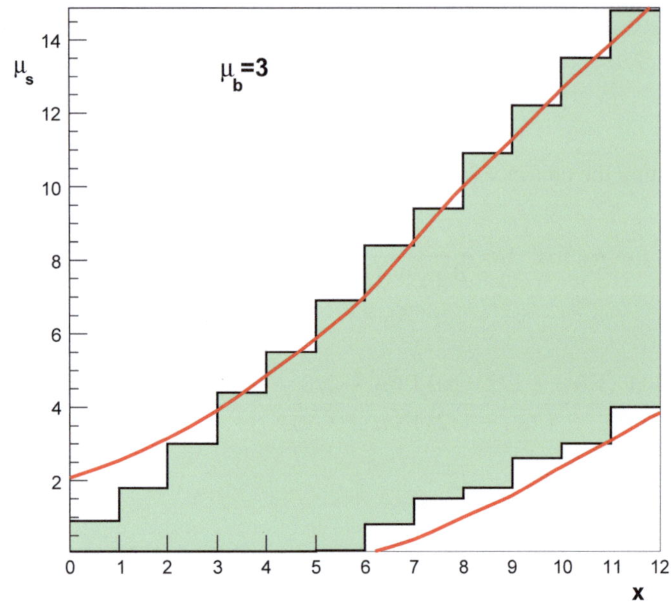

Fig. 2.8 90% Confidence Belt derived with Feldman and Cousins (*filled band*) and the Bayesian HPD region (*red lines*) for a background parameter $\mu_b = 3$

distributions in this case are the Gamma and the Normal with non-negative support. Last, it is clear that if the rate of background events is known with high accuracy then, with $\mu_i = \alpha\lambda_i$ and $\pi(\mu_s) \propto (\mu_s + \mu_b)^{-1/2}$ we have

$$p(\mu_s|\cdot) = \frac{1}{\Gamma(x+1/2, \mu_b)} \exp\{-(\mu_s + \mu_b)\} \, (\mu_s + \mu_b)^{x-1/2} \mathbf{1}_{(0,\infty)}(\mu_s)$$

As an example, we show in Fig. 2.8 the 90% HPD region obtained from the previous expression (red lines) as function of x for $\mu_b = 3$ (conditions as given in the example of [25]) and the Confidence Belt derived with the Feldman and Cousins approach (filled band). In this case, $\mu_{s,m} = \max\{0, x - \mu_b\}$ and therefore, for a given μ_s:

$$\sum_{x_1}^{x_2} Po(x|\mu_s + \mu_b) = \beta \quad \text{with} \quad R(x|\mu_s) = e^{(\mu_{s,m} - \mu_s)} \left(\frac{\mu_s + \mu_b}{\mu_{s,m} + \mu_b} \right)^x > k_\beta$$

for all $x \in [x_1, x_2]$.

Problem 2.14 In the search for a new particle, assume that the number of observed events follows a Poisson distribution with $\mu_b = 0.7$ known with enough precision from extensive Monte Carlo simulations. Consider the hypothesis $H_0 : \{\mu_s = 0\}$ and $H_1 : \{\mu_s \neq 0\}$. It is left as an exercise to obtain the Bayes Factor BF_{01} with the proper prior $\pi(\mu_s|\mu_b) = \mu_b(\mu_s + \mu_b)^{-2}$ proposed in [26], $P(H_1|n)$ and the BIC difference Δ_{01} as function of $n = 1, \ldots 7$ and decide when, based on this results, will you consider that there is evidence for a signal.

2.13.3 Anisotropies of Cosmic Rays

The angular distribution of cosmic rays in galactic coordinates is analyzed searching for possible anisotropies. A well-behaved real function $f(\theta, \phi) \in L_2(\Omega)$, with $(\theta, \phi) \in \Omega = [0, \pi]\times[0, 2\pi]$, can be expressed in the real harmonics basis as:

$$f(\theta, \phi) = \sum_{l=0}^{\infty} \sum_{m=-l}^{l} a_{lm} Y_{lm}(\theta, \phi) \qquad \text{where} \qquad a_{lm} = \int_{\Omega} f(\theta, \phi) Y_{lm}(\theta, \phi)\, d\mu;$$

$a_{lm} \in R$ and $d\mu = \sin\theta d\theta d\phi$. The convention adopted for the spherical harmonic functions is such that (*orthonormal basis*):

$$\int_{\Omega} Y_{lm}(\theta, \phi)\, Y_{l'm'}(\theta, \phi)\, d\mu = \delta_{ll'}\, \delta_{mm'} \qquad \text{and} \qquad \int_{\Omega} Y_{lm}(\theta, \phi)\, d\mu = \sqrt{4\pi}\, \delta_{l0}$$

In consequence, a probability density function $p(\theta, \phi)$ with support in Ω can be expanded as

$$p(\theta, \phi) = c_{00}\, Y_{00}(\theta, \phi) + \sum_{l=1}^{\infty} \sum_{m=-l}^{l} c_{lm}\, Y_{lm}(\theta, \phi)$$

The normalization imposes that $c_{00} = 1/\sqrt{4\pi}$ so we can write

$$p(\theta, \phi|\boldsymbol{a}) = \frac{1}{4\pi}\, (1 + a_{lm}\, Y_{lm}(\theta, \phi))$$

where $l \geq 1$,

$$a_{lm} = 4\pi c_{lm} = 4\pi \int_{\Omega} p(\theta, \phi)\, Y_{lm}(\theta, \phi)\, d\mu = 4\pi\, E_{p;\mu}[Y_{lm}(\theta, \phi)]$$

and summation over repeated indices understood. Obviously, for any $(\theta, \phi) \in \Omega$ we have that $p(\theta, \phi|\boldsymbol{a}) \geq 0$ so the set of parameters $\boldsymbol{a}$ are constrained on a compact support.

Even though we shall study the general case, we are particularly interested in the expansion up to $l = 1$ (dipole terms) so, to simplify the notation, we redefine the indices $(l, m) = \{(1, -1), (1, 0), (1, 1)\}$ as $i = \{1, 2, 3\}$ and, accordingly, the coefficients $\boldsymbol{a} = (a_{1-1}, a_{10}, a_{11})$ as $\boldsymbol{a} = (a_1, a_2, a_3)$. Thus:

$$p(\theta, \phi|\boldsymbol{a}) = \frac{1}{4\pi}\,(1 + a_1 Y_1 + a_2 Y_2 + a_3 Y_3)$$

In this case, the condition $p(\theta, \phi|\boldsymbol{a}) \geq 0$ implies that the coefficients are bounded by the sphere $a_1^2 + a_2^2 + a_3^2 \leq 4\pi/3$ and therefore, the coefficient of anisotropy

$$\delta \stackrel{def.}{=} \sqrt{\frac{3}{4\pi}}\,\left(a_1^2 + a_2^2 + a_3^2\right)^{1/2} \leq 1$$

There are no sufficient statistics for this model but the Central Limit Theorem applies and, given the large amount of data, the experimental observations can be cast in the statistic $\boldsymbol{a} = (a_1, a_2, a_3)$ such that[15]

$$p(\boldsymbol{a}|\boldsymbol{\mu}) = \prod_{i=1}^{3} N(a_i|\mu_i, \sigma_i^2)$$

with $V(a_i) = 4\pi/n$ known and with negligible correlations ($\rho_{ij} \simeq 0$).

Consider then a k-dimensional random quantity $\boldsymbol{Z} = \{Z_1, \ldots, Z_k\}$ and the distribution

$$p(z|\boldsymbol{\mu}, \boldsymbol{\sigma}) = \prod_{j=1}^{k} N(z_j|\mu_j, \sigma_j^2)$$

The interest is centered on the euclidean norm $||\boldsymbol{\mu}||$, with $\dim\{\boldsymbol{\mu}\} = k$, and its square; in particular, in

$$\delta = \sqrt{\frac{3}{4\pi}}||\boldsymbol{\mu}|| \quad \text{for } k = 3 \qquad \text{and} \qquad C_k = \frac{||\boldsymbol{\mu}||^2}{k}$$

First, let us define $X_j = Z_j/\sigma_j$ and $\rho_j = \mu_j/\sigma_j$ so $X_j \sim N(x_j|\rho_j, 1)$ and make a transformation of the parameters ρ_j to spherical coordinates:

$$\begin{aligned}
\rho_1 &= \rho\cos\phi_1 \\
\rho_2 &= \rho\sin\phi_1\cos\phi_2 \\
\rho_3 &= \rho\sin\phi_1\sin\phi_2\cos\phi_3 \\
&\;\vdots \\
\rho_{k-1} &= \rho\sin\phi_1\sin\phi_2\ldots\sin\phi_{k-2}\cos\phi_{k-1} \\
\rho_k &= \rho\sin\phi_1\sin\phi_2\ldots\sin\phi_{k-2}\sin\phi_{k-1}
\end{aligned}$$

The Fisher's matrix is the Riemann metric tensor so the square root of the determinant is the k-dimensional volume element:

---

[15]Essentially, $a_{lm} = \frac{4\pi}{n}\sum_{i=1}^{n} Y_{lm}(\theta_i, \phi_i)$ for a sample of size n.

$$dV^k = \rho^{k-1} d\rho \, dS^{k-1}$$

with

$$dS^{k-1} = \sin^{k-2}\phi_1 \, \sin^{k-3}\phi_2 \ldots \sin\phi_{k-2} \, d\phi_1 \, d\phi_2 \ldots d\phi_{k-1} = \prod_{j=1}^{k-1} \sin^{(k-1)-j} \phi_j d\phi_j$$

the $k-1$ dimensional spherical surface element, $\phi_{k-1} \in [0, 2\pi)$ and $\phi_{1,\ldots,k-2} \in [0, \pi]$. The interest we have is on the parameter ρ so we should consider the ordered parameterization $\{\rho; \boldsymbol{\phi}\}$ with $\boldsymbol{\phi} = \{\phi_1, \phi_2, \ldots, \phi_{k-1}\}$ nuisance parameters. Being ρ and ϕ_i independent for all i, we shall consider the surface element (that is, the determinant of the submatrix obtained for the angular part) as prior density (proper) for the nuisance parameters. As we have commented in Chap. 1, this is just the Lebesgue measure on the $k-1$ dimensional sphere (the Haar invariant measure under rotations) and therefore the natural choice for the prior; in other words, a uniform distribution on the $k-1$ dimensional sphere. Thus, we start integrating the angular parameters. Under the assumption that the variances σ_i^2 are all the same and considering that

$$\int_0^{\pi} e^{\pm\beta\cos\theta} \sin^{2\nu}\theta d\theta = \sqrt{\pi} \left(\frac{2}{\beta}\right)^{\nu} \Gamma(\nu + \frac{1}{2}) \, I_\nu(\beta) \qquad \text{for} \qquad Re(\nu) > -\frac{1}{2}$$

one gets $p(\phi|\text{data}) \propto p(\phi_m|\phi)\pi(\phi)$ where

$$p(\phi_m|\phi, \nu) = b \, e^{-b(\phi+\phi_m)} \left(\frac{\phi_m}{\phi}\right)^{\nu/2} I_\nu(2b\sqrt{\phi_m}\sqrt{\phi})$$

is properly normalized,

$$\nu = k/2 - 1 \, ; \qquad \phi = ||\boldsymbol{\mu}||^2; \qquad \phi_m = ||\boldsymbol{a}||^2; \qquad b = \frac{1}{2\sigma^2} = \frac{n}{8\pi}$$

and $\dim\{\boldsymbol{\mu}\} = \dim\{\boldsymbol{a}\} = k$. This is nothing else but a non-central χ^2 distribution.

From the series expansion of the Bessel functions it is easy to prove that this process is just a compound Poisson-Gamma process

$$p(\phi_m|\phi, \nu) = \sum_{k=0}^{\infty} Po(k|b\phi) \, Ga(\phi_m|b, \nu + k + 1)$$

and therefore the sampling distribution is a Gamma-weighted Poisson distribution with the parameter of interest that of the Poisson. From the Mellin Transform:

$$\mathcal{M}(s)_{\langle -\nu,\infty\rangle} = \frac{b \, e^{-b\phi}}{\Gamma(\nu+1)} \frac{\Gamma(s+\nu)}{b^s} M(s+\nu, \nu+1, b\phi)$$

with $M(a, b, z)$ the Kummer's function one can easily get the moments ($E[\phi_m^n] = M(n+1)$); in particular

$$E[\phi_m] = \phi + b^{-1}(\nu + 1) \qquad \text{and} \qquad V[\phi_m] = 2\phi b^{-1} + b^{-2}(\nu + 1)$$

Now that we have the model $p(\phi_m|\phi)$, let's go for the prior function $\pi(\phi)$ or $\pi(\delta)$. One may guess already what shall we get. The first element of the Fisher's matrix (diagonal) corresponds to the norm and is constant so it would not be surprising to get the Lebesgue measure for the norm $d\lambda(\delta) = \pi(\delta)d\delta = c\, d\delta$. As a second argument, for large sample sizes ($n >>$) we have $b >>$ so $\phi_m \sim N(\phi_m|\phi, \sigma^2 = 2\phi/b)$ and, to first order, Jeffreys' prior is $\pi(\phi) \sim \phi^{-1/2}$. From the reference analysis, if we take for instance

$$\pi^\star(\phi) = \phi^{(\nu-1)/2}$$

we end up, after some algebra, with

$$\pi(\phi) \propto \pi(\phi_0) \lim_{k\to\infty} \frac{f_k(\phi)}{f_k(\phi_0)} \propto \left(\frac{\phi_0}{\phi}\right)^{1/2} \lim_{b\longrightarrow\infty} e^{-3b(\phi-\phi_0)/2 + [I(\phi,b) - I(\phi_0,b)]}$$

where

$$I(\phi, b) = \int_0^\infty p(\phi_m|\phi) \log \frac{I_\nu(2b\sqrt{\phi\phi_m})}{I_{\nu/2}(b\phi_m/2)} d\phi_m$$

and ϕ_0 any interior point of $\Lambda(\phi) = [0, \infty)$. From the asymptotic behavior of the Bessel functions one gets

$$\pi(\phi) \propto \phi^{-1/2}$$

and therefore, $\pi(\delta) = c$. It is left as an exercise to get the same result with other priors like $\pi^\star(\phi) = c$ or $\pi^\star(\phi) = \phi^{-1/2}$.

For this problem, it is easier to derive the prior from the reference analysis. Nevertheless, the Fisher's information that can be expressed as:

$$F(\phi; \nu) = b^2 \left\{ -1 + b \frac{e^{-b\phi}}{\phi^{\nu/2+1}} \int_0^\infty e^{-bz} z^{\nu/2+1} \frac{I_{\nu+1}^2(2b\sqrt{z\phi})}{I_\nu(2b\sqrt{z\phi})} dz \right\}$$

and, for large b (large sample size), $F(\lambda; \nu) \to \phi^{-1}$ *regardless the number of degrees of freedom* ν. Thus, Jeffrey's prior is consistent with the result from reference analysis. In fact, from the asymptotic behavior of the Bessel Function in the corresponding expressions of the pdf, one can already see that $F(\phi; \nu) \sim \phi^{-1}$. A cross check from a numeric integration is shown in Fig. 2.9 where, for $k = 3, 5, 7$ ($\nu = 1/2, 3/2, 5/2$), $F(\phi; \nu)$ is depicted as function of ϕ compared to $1/\phi$ in black for a sufficiently large

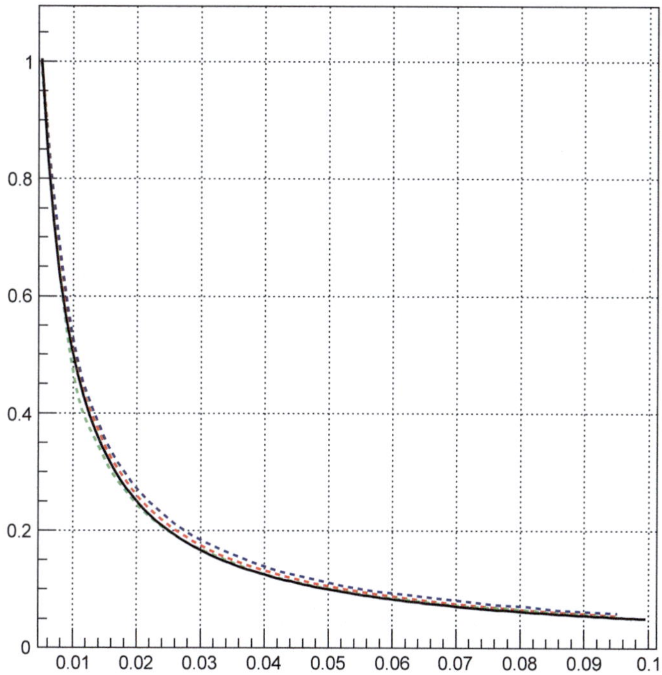

Fig. 2.9 Fisher's information (numeric integration) as function of ϕ for $k = 3, 5, 7$ (*discontinuous lines*) and $f(\phi) = \phi^{-1}$ (*continuous line*). All are scaled so that $F(\phi = 0.005, \nu) = 1$

value of b. Therefore we shall use $\pi(\phi) = \phi^{-1/2}$ for the cases of interest (dipole, quadrupole, ... any-pole).

The posterior densities are

- For $\phi = ||\boldsymbol{\mu}||^2$: $\quad p(\phi|\phi_m, \nu) = N\, e^{-b\phi}\, \phi^{-(\nu+1)/2}\, I_\nu(2b\sqrt{\phi_m}\sqrt{\phi})$ with

$$N = \frac{\Gamma(\nu+1)\, b^{1/2-\nu}\, \phi_m^{-\nu/2}}{\sqrt{\pi}\, M(1/2, \nu+1, b\phi_m)}$$

The Mellin Transform is

$$\mathcal{M}_\phi(s)_{\langle 1/2, \infty\rangle} = \frac{\Gamma(s-1/2)}{b^{s-1}\sqrt{\pi}}\, \frac{M(s-1/2, \nu+1, b\phi_m)}{M(1/2, \nu+1, b\phi_m)}$$

and therefore the moments

$$E[\phi^n] = M(n+1) = \frac{\Gamma(n+1/2)\, M(n+1/2, \nu+1, b\phi_m)}{\sqrt{\pi}\, b^n\, M(1/2, \nu+1, b\phi_m)}$$

In the limit $|b\phi_m|\to\infty$, $E[\phi^n]=\phi_m^n$.

• For $\rho=||\boldsymbol{\mu}||$: $p(\rho|\phi_m,\nu)=2\,N\,e^{-b\rho^2}\,\rho^{-\nu}\,I_\nu(2b\sqrt{\phi_m}\rho)$ and

$$\mathcal{M}_\rho(s)=\mathcal{M}_\phi(s/2+1/2)\;\longrightarrow\;E[\rho^n]=\frac{\Gamma(n/2+1/2)\,M(n/2+1/2,\nu+1,b\phi_m)}{\sqrt{\pi}\,b^{n/2}\,M(1/2,\nu+1,b\phi_m)}$$

In the particular case that $k=3$ (dipole; $\nu=1/2$), we have for $\delta=\sqrt{3/4\pi}\rho$ that the first two moments are:

$$E[\delta]=\frac{\text{erf}\,(z)}{a\delta_m M(1,3/2,-z^2)}\qquad\qquad E[\delta^2]=\frac{1}{a\,M(1,3/2,-z^2)}$$

with $z=2\delta_m\sqrt{b\pi/3}$ and, when $\delta_m\to 0$ we get

$$E[\delta]=\sqrt{\frac{2}{\pi a}}\simeq\frac{1.38}{\sqrt{n}}\qquad\qquad E[\delta^2]=\frac{1}{a}\qquad\qquad \sigma_\delta\simeq\frac{1.04}{\sqrt{n}}$$

and a one sided 95% upper credible region (see Sect. 2.11 for more details) of $\delta_{0.95}=\frac{3.38}{\sqrt{n}}$.

So far, the analysis has been done assuming that the variances σ_j^2 are of the same size (equal in fact) and the correlations are small. This is a very reasonable assumption but may not always be the case. The easiest way to proceed then is to perform a transformation of the parameters of interest ($\boldsymbol{\mu}$) to polar coordinates $\boldsymbol{\mu}(\rho,\Omega)$ and do a Monte Carlo sampling from the posterior:

$$p(\rho,\Omega|\mathbf{z},\boldsymbol{\Sigma}^{-1})\propto\left[\prod_{j=1}^{n}N(z_j|\mu_j(\rho,\Omega),\boldsymbol{\Sigma}^{-1})\right]\pi(\rho)d\rho\,dS^{n-1}$$

with a constant prior for δ or $\pi(\phi)\propto\phi^{-1/2}$ for ϕ.

References

1. G. D'Agostini, *Bayesian Reasoning in Data Analysis* (World Scientific Publishing Co, Singapore, 2003)
2. F. James, *Statistical Methods in Experimental Physics* (World Scientific Publishing Co, Singapore, 2006)
3. J.M. Bernardo, The concept of exchangeability and its applications. Far East J. Math. Sci. **4**, 111–121 (1996). www.uv.es/~bernardo/Exchangeability.pdf
4. J.M. Bernardo, A.F.M. Smith, *Bayesian Theory* (Wiley, New York, 1994)
5. R.E. Kass, L. Wasserman, The selection of prior distributions by formal Rules. J. Am. Stat. Assoc. V **91**(453), 1343–1370 (1996)

6. H. Jeffreys, *Theory of Probability* (Oxford University Press, Oxford, 1939)
7. E.T. Jaynes, *Prior Probabilities and Transformation Groups*, NSF G23778 (1964)
8. V.I. Bogachev, *Measure Theory* (Springer, Berlin, 2006)
9. M. Stone, Right haar measures for convergence in probability to invariant posterior distributions. Ann. Math. Stat. **36**, 440–453 (1965)
10. M. Stone, Necessary and sufficient conditions for convergence in probability to invariant posterior distributions. Ann. Math. Stat. **41**, 1349–1353 (1970)
11. H. Raiffa, R. Schlaifer, *Applied Statistical Decision Theory* (Harvard University Press, Cambridge, 1961)
12. S.R. Dalal, W.J. Hall, J.R. Stat, Soc. Ser. B **45**, 278–286 (1983)
13. B. Welch, H. Pears, J.R. Stat, Soc. B **25**, 318–329 (1963)
14. M. Gosh, R. Mukerjee, Biometrika **84**, 970–975 (1984)
15. G.S. Datta, M. Ghosh, Ann. Stat. **24**(1), 141–159 (1996)
16. G.S. Datta, R. Mukerjee, *Probability Matching Priors and Higher Order Asymptotics* (Springer, New York, 2004)
17. J.M. Bernardo, J.R. Stat, Soc. Ser. B **41**, 113–147 (1979)
18. J.O. Berger, J.M. Bernardo, D. Sun, Ann. Stat **37**(2), 905–938 (2009)
19. J.M. Bernardo, J.M. Ramón, The Statistician **47**, 1–35 (1998)
20. J.O. Berger, J.M. Bernardo, D. Sun, Objective priors for discrete parameter spaces. J. Am. Stat. Assoc. **107**(498), 636–648 (2012)
21. A. O'Hagan, J.R. Stat. Soc. **B57**, 99–138 (1995)
22. J.O. Berger, L.R. Pericchi, J. Am. Stat. Assoc. V **91**(433), 109–122 (1996)
23. R.E. Kass, A.E. Raftery, J. Am. Stat. Assoc. V **90**(430), 773–795 (1995)
24. G. Schwarz, Ann. Stat. **6**, 461–464 (1978)
25. Feldman G.J. and Cousins R.D. (1997); arXiV:physics/9711021v2
26. J.O. Berger, L.R. Pericchi, Ann. Stat. V **32**(3), 841–869 (2004)

Chapter 3
Monte Carlo Methods

Anyone who considers arithmetical methods of producing random digits is, of course, in a state of sin

J. Von Neumann

The Monte Carlo Method is a very useful and versatile numerical technique that allows to solve a large variety of problems difficult to tackle by other procedures. Even though the central idea is to simulate experiments on a computer and make inferences from the *"observed"* sample, it is applicable to problems that do not have an explicit random nature; it is enough if they have an adequate probabilistic approach. In fact, a frequent use of Monte Carlo techniques is the evaluation of definite integrals that at first sight have no statistical nature but can be interpreted as expected values under some distribution.

Detractors of the method used to argue that one uses Monte Carlo Methods because a manifest incapability to solve the problems by other *more academic* means. Well, consider a *"simple"* process in particle physics: $ee \rightarrow ee\mu\mu$. Just four particles in the final state; the differential cross section in terms of eight variables that are not independent due to kinematic constraints. To see what we expect for a particular experiment, it has to be integrated within the acceptance region with dead zones between subdetectors, different materials and resolutions that distort the momenta and energy, detection efficiencies,... Yes. Admittedly we are not able to get nice expressions. Nobody in fact and Monte Carlo comes to our help. Last, it may be a truism but worth to mention that Monte Carlo is not a magic black box and will not give the answer to our problem out of nothing. It will simply present the available information in a different and more suitable manner after more or less complicated calculations are performed but all the needed information has to be put in to start with in some way or another.

C. Maña, *Probability and Statistics for Particle Physics*,
UNITEXT for Physics, DOI 10.1007/978-3-319-55738-0_3

In this lecture we shall present and justify essentially all the procedures that are commonly used in particle physics and statistics leaving aside subjects like Markov Chains that deserve a whole lecture by themselves and for which only the relevant properties will be stated without demonstration. A general introduction to Monte Carlo techniques can be found in [1].

3.1 Pseudo-Random Sequences

Sequences of random numbers $\{x_1, x_2, \ldots, x_n\}$ are the basis of Monte Carlo simulations and, in principle, their production is equivalent to perform an experiment $e(n)$ sampling n times the random quantity $X \sim p(x|\boldsymbol{\theta})$. Several procedures have been developed for this purpose (real experiments, dedicated machines, digits of transcendental numbers,…) but, besides the lack of precise knowledge behind the generated sequences and the need of periodic checks, the complexity of the calculations we are interested in demands large sequences and fast generation procedures. We are then forced to devise simple and efficient arithmetical algorithms to be implemented in a computer. Obviously neither the sequences produced are random nor we can produce truly random sequences by arithmetical algorithms but we really do not need them. It is enough for them to *simulate* the relevant properties of truly random sequences and be such that if I give you one of these sequences and no additional information, you won't be able to tell after a bunch of tests [2] whether it is a truly random sequence or not (at least for the needs of the problem at hand). That's why they are called *pseudo-random* although, in what follows we shall call them *random*. The most popular (and essentially the best) algorithms are based on congruential relations (used for this purpose as far as in the 1940s) together with binary and/or shuffling operations with some free parameters that have to be fixed before the sequence is generated. They are fast, easy to implement on any computer and, with the adequate initial setting of the parameters, produce very long sequences with sufficiently good properties. And the easiest and fastest pseudo-random distribution to be generated on a computer is the *Discrete Uniform*.[1]

Thus, let's assume that we have a *good Discrete Uniform random number* generator[2] although, as Marsaglia said, "*A Random Number Generator is like sex: When it is good it is wonderful; when it is bad… it is still pretty good*". Each call in a computer algorithm will produce an output (x) that we shall represent as $x \Leftarrow Un(0, 1)$ and simulates a sampling of the random quantity $X \sim Un(x|0, 1)$. Certainly, we are not very much interested in the Uniform Distribution so the task is to obtain a sampling of densities $p(\boldsymbol{x}|\boldsymbol{\theta})$ other than Uniform from a *Pseudo-Uniform Random Number Generator* for which there are several procedures.

---

[1] See [3, 4] for a detailed review on random and quasi-random number generators.

[2] For the examples in this lecture I have used *RANMAR* [5] that can be found, for instance, at the CERN Computing Library.

Table 3.1 Estimation of π from a Binomial random process

Throws (N)	Accepted (n)	$\pi^\star$	$\sigma^\star$
100	83	3.3069	0.1500
1000	770	3.0789	0.0532
10000	7789	3.1156	0.0166
100000	78408	3.1363	0.0052
1000000	785241	3.1410	0.0016

Example 3.1 (*Estimate the value of* π) As a simple first example, let's see how we may estimate the value of π. Consider a circle of radius r inscribed in a square with sides of length $2r$. Imagine now that we throw random *points* evenly distributed inside the square and count how many have fallen inside the circle. It is clear that since the area of the square is $4r^2$ and the area enclosed by the circle is πr^2, the probability that a throw falls inside the circle is $\theta = \pi/4$.

If we repeat the experiment N times, the number n of throws falling inside the circle follows a Binomial law $Bi(n|N, p)$ and therefore, having observed n out of N trials we have that

$$p(\theta|n, N) \propto \theta^{n-1/2}(1-\theta)^{N-n-1/2}$$

Let's take $\pi^\star = 4E[\theta]$ as point estimator and $\sigma^\star = 4\sigma_\theta$ as a measure of the precision. The results obtained for samplings of different size are shown in Table 3.1.

It is interesting to see that the precision decreases with the sampling size as $1/\sqrt{N}$. This dependence is a general feature of Monte Carlo estimations *regardless* the number of dimensions of the problem.

A similar problem is that of Buffon's needle: *A needle of length l is thrown at random on a horizontal plane with stripes of width $d > l$. What is the probability that the needle intersects one of the lines between the stripes?* It is left as an exercise to shown that, as given already by Buffon in 1777, $P_{cut} = 2l/\pi d$. Laplace pointed out, in what may be the first use of the Monte Carlo method, that doing the experiment one may estimate the value of π "*... although with large error*".

3.2 Basic Algorithms

3.2.1 Inverse Transform

This is, at least formally, the easiest procedure. Suppose we want a sampling of the continuous one-dimensional random quantity $X \sim p(x)$[3] so

[3]Remember that if $\text{supp}(X) = \Omega \subseteq \mathcal{R}$, it is assumed that the density is $p(x)\mathbf{1}_\Omega(x)$.

$$P[X \in (-\infty, x]] = \int_{-\infty}^{x} p(x')dx' = \int_{-\infty}^{x} dF(x') = F(x)$$

Now, we define the new random quantity $U = F(X)$ with support in [0, 1]. How is it distributed? Well,

$$F_U(u) \equiv P[U \le u] = P[F(X) \le u)] = P[X \le F^{-1}(u)] = \int_{-\infty}^{F^{-1}(u|\cdot)} dF(x') = u$$

and therefore $U \sim Un(u|0, 1)$. The algorithm is then clear; at step i:

(i_1) $u_i \Leftarrow Un(u|0, 1)$
(i_2) $x_i = F^{-1}(u_i)$

After repeating the sequence n times we end up with a sampling $\{x_1, x_2, \ldots, x_n\}$ of $X \sim p(x)$.

Example 3.2 Let' see how we generate a sampling of the Laplace distribution $X \sim La(x|\alpha, \beta)$ with $\alpha \in \mathcal{R}$, $\beta \in (0, \infty)$ and density

$$p(x|\alpha, \beta) = \frac{1}{2\beta} e^{-|x-\alpha|/\beta} \mathbf{1}_{(-\infty,\infty)}(x)$$

The distribution function is

$$F(x) = \int_{-\infty}^{x} p(x'|\alpha, \beta)dx' = \begin{cases} \frac{1}{2}\exp\left(\frac{x-\alpha}{\beta}\right) & \text{if } x < \alpha \\ 1 - \frac{1}{2}\exp\left(-\frac{x-\alpha}{\beta}\right) & \text{if } u \ge \alpha \end{cases}$$

Then, if $u \Leftarrow Un(0, 1)$:

$$x = \begin{cases} \alpha + \beta \ln(2u) & \text{if } u < 1/2 \\ \alpha - \beta \ln(2(1-u)) & \text{if } u \ge 1/2 \end{cases}$$

The generalization of the *Inverse Transform* method to n-dimensional random quantities is trivial. We just have to consider the marginal and conditional distributions

$$F(x_1, x_2, \ldots, x_n) = F_n(x_n|x_{n-1}, \ldots, x_1) \cdots F_2(x_2|x_1) \cdots F_1(x_1)$$

or, for absolute continuous quantities, the probability densities

$$p(x_1, x_2, \ldots, x_n) = p_n(x_n|x_{n-1}, \ldots, x_1) \cdots p_2(x_2|x_1) \cdots p_1(x_1)$$

and then proceed sequentially; that is:

$(i_{2,1})$ $u_1 \Leftarrow Un(u|0,1)$ and $x_1 = F_1^{-1}(u_1)$;
$(i_{2,2})$ $u_2 \Leftarrow Un(u|0,1)$ and $x_2 = F_2^{-1}(u_2|x_1)$;
$(i_{2,1})$ $u_3 \Leftarrow Un(u|0,1)$ and $x_3 = F_3^{-1}(u_3|x_1, x_2)$;

$$\vdots$$

$(i_{2,n})$ $u_n \Leftarrow Un(u|0,1)$ and $x_n = F_n^{-1}(u_3|x_{n-1}, \ldots, x_1)$

If the random quantities are independent there is a unique decomposition

$$p(x_1, x_2, \ldots, x_n) = \prod_{i=1}^{n} p_i(x_i) \quad \text{and} \quad F(x_1, x_2, \ldots, x_n) = \prod_{i=1}^{n} F_i(x_i)$$

but, if this is not the case, note that there are $n!$ ways to do the decomposition and some may be easier to handle than others (see Example 3.3).

Example 3.3 Consider the probability density

$$p(x, y) = 2\, e^{-x/y}\, \mathbf{1}_{(0,\infty)}(x) \mathbf{1}_{(0,1]}(y)$$

We can express the Distribution Function as $F(x, y) = F(x|y)F(y)$ where:

$$p(y) = \int_0^{\infty} p(x, y)\, dx = 2\, y \qquad \longrightarrow \qquad F(y) = y^2$$

$$p(x|y) = \frac{p(x, y)}{p(y)} = \frac{1}{y}\, e^{-x/y} \qquad \longrightarrow \qquad F(x|y) = 1 - e^{-x/y}$$

Both $F(y)$ and $F(x|y)$ are easy to invert so:

(i_1) $u \Leftarrow Un(0,1)$ and get $y = u^{1/2}$
(i_2) $w \Leftarrow Un(0,1)$ and get $x = -y \ln w$

Repeating the algorithm n times, we get the sequence $\{(x_1, y_1), (x_2, y_2), \ldots\}$ that simulates a sampling from $p(x, y)$.

Obviously, we can also write $F(x, y) = F(y|x)F(x)$ and proceed in an analogous manner. However

$$p(x) = \int_0^1 p(x, y) dy = 2x \int_x^{\infty} e^{-u} u^{-2} du$$

is not so easy to sample.

Last, let's see how to use the *Inverse Transform* procedure for discrete random quantities. If X can take the values in $\Omega_X = \{x_0, x_1, x_2, \ldots\}$ with probabilities $P(X = x_k) = p_k$, the Distribution Function will be:

$$
\begin{aligned}
F_0 &= P(X \leq x_0) = p_0 \\
F_1 &= P(X \leq x_1) = p_0 + p_1 \\
F_2 &= P(X \leq x_2) = p_0 + p_1 + p_2 \\
&\dots
\end{aligned}
$$

Graphically, we can represent the sequence $\{0, F_0, F_1, F_2, \dots, 1\}$ as:

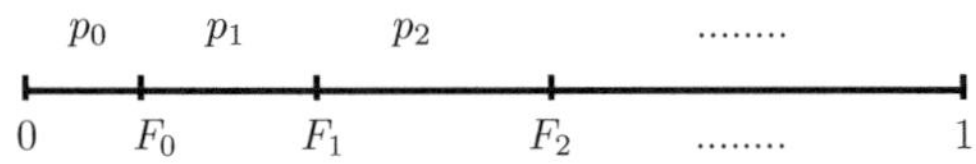

Then, it is clear that a random quantity u_i drawn from $U(x|0, 1)$ will determine a point in the interval $[0, 1]$ and will belong to the subinterval $[F_{k-1}, F_k]$ with probability $p_k = F_k - F_{k-1}$ so we can set up the following algorithm:

(i_1) Get $u_i \sim Un(u|0, 1)$;
(i_2) Find the value x_k such that $F_{k-1} < u_i \leq F_k$

The sequence $\{x_0, x_1, x_2, \dots\}$ so generated will be a sampling of the probability law $P(X = x_k) = p_k$. Even though discrete random quantities can be sampled in this way, some times there are specific properties based on which faster algorithms can be developed. That is the case for instance for the Poisson Distribution as the following example shows.

Example 3.4 (*Poisson Distribution*) $Po(k|\mu)$. From the recurrence relation

$$
p_k = e^{-\mu} \frac{\mu^k}{\Gamma(k+1)} = \frac{\mu}{k} p_{k-1}
$$

(i_1) $u_i \Leftarrow Un(0, 1)$
(i_2) Find the value $k = 0, 1, \dots$ such that $F_{k-1} < u_i \leq F_k$ and deliver $x_k = k$

For the Poisson Distribution, there is a faster procedure. Consider a sequence of n independent random quantities $\{X_1, X_2, \dots, X_n\}$, each distributed as $X_i \sim Un(x|0, 1)$, and introduce a new random quantity

$$
W_n = \prod_{k=1}^{n} X_k
$$

with $\text{supp}\{W_n\} = [0, 1]$. Then

$$
W_n \sim p(w_n|n) = \frac{(-\log w_n)^{n-1}}{\Gamma(n)} \qquad \longrightarrow \qquad P(W_n \leq a) = \frac{1}{\Gamma(n)} \int_{-\log a}^{\infty} e^{-t} t^{n-1} dt
$$

and if we take $a = e^{-\mu}$ we have, in terms of the Incomplete Gamma Function $P(a, x)$:

$$P(W_n \le e^{-\mu}) = 1 - P(n,\mu) = e^{-\mu}\sum_{k=0}^{n-1}\frac{\mu^k}{\Gamma(k+1)} = Po(X \le n-1|\mu)$$

Therefore,

(i_0) Set $w_p = 1$;
(i_1) $u_i \Leftarrow Un(0, 1)$ and set $w_p = w_p u$;
(i_2) Repeat step (i_1) while $w_p \le e^{-\mu}$, say k times, and deliver $x_k = k - 1$

Example 3.5 (*Binomial Distribution $Bn(k|N,\theta)$*) From the recurrence relation

$$p_k = \binom{N}{k}\,\theta^k\,(1-\theta)^{n-k} = \frac{\theta}{1-\theta}\,\frac{n-k+1}{k}\,p_{k-1}$$

with $p_0 = (1-\theta)^k$

(i_1) $u_i \Leftarrow Un(0, 1)$
(i_2) Find the value $k = 0, 1, \dots, N$ such that $F_{k-1} < u_i \le F_k$ and deliver $x_k = k$

Example 3.6 (*Simulation of the response of a Photomultiplier tube*) Photomultiplier tubes are widely used devices to detect electromagnetic radiation by means of the external photoelectric effect. A typical photomultiplier consists of a vacuum tube with an input window, a photocathode, a focusing and a series of amplifying electrodes (dynodes) and an electron collector (anode). Several materials are used for the input window (borosilicate glass, synthetic silica,...) which transmit radiation in different wavelength ranges and, due to absorptions (in particular in the UV range) and external reflexions, the transmittance of the window is never 100%. Most photocathodes are compound semiconductors consisting of alkali metals with a low work function. When the photons strike the photocathode the electrons in the valence band are excited and, if they get enough energy to overcome the vacuum level barrier, they are emitted into the vacuum tube as photoelectrons. The trajectory of the electrons inside the photomultiplier is determined basically by the applied voltage and the geometry of the focusing electrode and the first dynode. Usually, the photoelectron is driven towards the first dynode and originates an electron shower which is amplified in the following dynodes and collected at the anode. However, a fraction of the incoming photons pass through the photocathode and originates a smaller electron shower when it strikes the first dynode of the amplification chain.

To study the response of a photomultiplier tube, an experimental set-up has been made with a LED as photon source. We are interested in the response for isolated photons so we regulate the current and the frequency so as to have a low intensity source. Under this conditions, the number of photons that arrive at the window of the photomultiplier is well described by a Poisson law. When one of this photons strikes on the photocathode, an electron is ejected and driven towards the first dynode to start the electron shower. We shall assume that the number of electrons so produced follows also a Poisson law $n_{gen} \sim Po(n|\mu)$. The parameter μ accounts for the efficiency of this first process and depends on the characteristics of the photocathode, the applied voltage and the geometry of the focusing electrodes (essentially that of the first dynode). It has been estimated to be $\mu = 0.25$. Thus, we start our simulation with

(1) $n_{gen} \Leftarrow Po(n|\mu)$

electrons leaving the photocathode. They are driven towards the first dynode to start the electron shower but there is a chance that they miss the first and start the shower at the second. Again, the analysis of the experimental data suggest that this happens with probability $p_{d2} \simeq 0.2$. Thus, we have to decide how many of the n_{gen} electrons start the shower at the second dynode. A Binomial model is appropriate in this case:

(2) $n_{d2} \Leftarrow Bi(n_{d2}|n_{gen}, p_{d2})$ and therefore $n_{d1} = n_{gen} - n_{d2}$.

Obviously, we shall do this second step if $n_{gen} > 0$.

Now we come to the amplification stage. Our photomultiplier has 12 dynodes so let's see the response of each of them. For each electron that strikes upon dynode k $(k = 1, \ldots, 12)$, n_k electrons will be produced and directed towards the next element of the chain (dynode $k + 1$), the number of them again well described by a Poisson law $Po(n_k|\mu_k)$. If we denote by V the total voltage applied between the photocathode and the anode and by R_k the resistance previous to dynode k we have that the current intensity through the chain will be

$$I = \frac{V}{\sum_{i=1}^{13} R_i}$$

where we have considered also the additional resistance between the last dynode and the anode that collects the electron shower. Therefore, the parameters μ_k are determined by the relation

$$\mu_k = a\,(I\,R_k)^b$$

where a and b are characteristic parameters of the photomultiplier. In our case we have that $N = 12, a = 0.16459, b = 0.75$, a total applied voltage of 800 V and a resistance chain of $\{2.4, 2.4, 2.4, 1.0, 1.0, 1.0, 1.0, 1.0, 1.0, 1.0, 1.0, 1.2, 2.4\}$ Ohms. It is easy to see that if the response of dynode k to one electron is modeled as $Po(n_k|\mu_k)$, the response to n_i incoming electrons is described by $Po(n_k|n_i\mu_k)$. Thus, we simulate the amplification stage as:

(3.1) If $n_{d1} > 0$: do from k = 1 to 12:

$$\mu = \mu_k\, n_{d1} \quad \longrightarrow \quad n_{d1} \Leftarrow Po(n|\mu)$$

(3.2) If $n_{d2} > 0$: do from k = 2 to 12:

$$\mu = \mu_k\, n_{d2} \quad \longrightarrow \quad n_{d2} \Leftarrow Po(n|\mu)$$

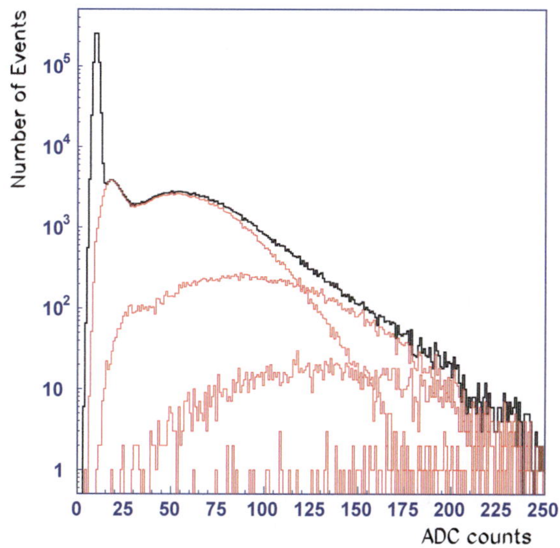

Fig. 3.1 Result of the simulation of the response of a photomultiplier tube. The histogram contains 10^6 events and shows the final ADC distribution detailing the contribution of the pedestal and the response to 1, 2 and 3 incoming photons

Once this is done, we have to convert the number of electrons at the anode in ADC counts. The electron charge is $Q_e = 1.602176\,10^{-19}$ C and in our set-up we have $f_{ADC} = 2.1\,10^{14}$ ADC counts per Coulomb so

$$ADC_{pm} = (n_{d1} + n_{d2})\,(Q_e\, f_{ADC})$$

Last, we have to consider the noise (*pedestal*). In our case, the number of pedestal ADC counts is well described by a mixture model with two Normal densities

$$p_{ped}(x|\cdot) = \alpha\, N_1(x|10.,\,1.) + (1-\alpha)\, N_1(x|10.,\,1.5)$$

with $\alpha = 0.8$. Thus, with probability α we obtain $ADC_{ped} \Leftarrow N_1(x|10,\,1.5)$, and with probability $1-\alpha$, $ADC_{ped} \Leftarrow N_1(x|10,\,1)$ so the total number of ADC counts will be

$$ADC_{tot} = ADC_{ped} + ADC_{pm}$$

Obviously, if in step 1) we get $n_{gen} = 0$, then $ADC_{tot} = ADC_{ped}$. Figure 3.1 shows the result of the simulation for a sampling size of 10^6 together with the main contributions (1, 2 or 3 initial photoelectrons) and the pedestal. From these results, the parameters of the device can be adjusted (voltage, resistance chain,...) to optimize the response for our specific requirements.

The *Inverse Transform* method is conceptually simple and easy to implement for discrete distributions and many continuous distributions of interest. Furthermore, is *efficient* in the sense that for each generated value u_i as $Un(x|0,1)$ we get a value x_i from $F(x)$. However, with the exception of easy distributions the inverse function $F^{-1}(x)$ has no simple expression in terms of elementary functions and may be difficult or time consuming to invert. This is, for instance, the case if you attempt to invert the Error Function for the Normal Distribution. Thus, apart from simple cases, the *Inverse Transform* method is used in combination with other procedures to be described next.

NOTE 5: Bootstrap. Given the iid sample $\{x_1, x_2, \dots, x_n\}$ of the random quantity $X \sim p(x|\theta)$ we know (Glivenko-Cantelly theorem; see lecture 1 (7.6)) that:

$$F_n(x) = 1/n \sum_{k=1}^{n} \mathbf{1}_{(-\infty,x]}(x_k) \qquad \stackrel{\text{unif.}}{\longrightarrow} \qquad F(x|\theta)$$

Essentially the idea behind the *bootstrap* is to sample from the empirical Distribution Function $F_n(x)$ that, as we have seen for discrete random quantities, is equivalent to draw samplings $\{x'_1, x'_2, \dots, x'_n\}$ of size n from the original sample with replacement. Obviously, increasing the number of resamplings does not provide more information than what is contained in the original data but, used with good sense, each *bootstrap* will lead to a posterior and can also be useful to give insight about the form of the underlying model $p(x|\theta)$ and the distribution of some statistics. We refer to [6] for further details.

3.2.2 *Acceptance-Rejection* (Hit-Miss; J. Von Neumann 1951)

The *Acceptance-Rejection* algorithm is easy to implement and allows to sample a large variety of n-dimensional probability densities with a less detailed knowledge of the function. But nothing is for free; these advantages are in detriment of the generation efficiency.

Let's start with the one dimensional case where $X \sim p(x|\theta)$ is a continuous random quantity with $\text{supp}(X) = [a, b]$ and $p_m = \max_x p(x|\theta)$. Consider now two independent random quantities $X_1 \sim Un(x_1|\alpha, \beta)$ and $X_2 \sim Un(x_2|0, \delta)$ where $[a, b] \subseteq \text{supp}(X_1) = [\alpha, \beta]$ and $[0, p_m] \subseteq \text{supp}(X_2) = [0, \delta]$. The covering does not necessarily have to be a rectangle in $\mathcal{R}^2$ (nor a hypercube $\mathcal{R}^{n+1}$) and, in fact, in some cases it may be interesting to consider other coverings to improve the efficiency but the generalization is obvious. Then

$$p(x_1, x_2|\cdot) = \frac{1}{\beta - \alpha}\,\frac{1}{\delta}$$

Now, let's find the distribution of X_1 conditioned to $X_2 \leq p(X_1|\theta)$:

$$P(X_1 \leq x | X_2 \leq p(x|\theta)) = \frac{P(X_1 \leq x, X_2 \leq p(x|\theta))}{P(X_2 \leq p(x|\theta))} = \frac{\int_\alpha^x dx_1 \int_0^{p(x_1|\theta)} p(x_1, x_2|\cdot)\, dx_2}{\int_\alpha^\beta dx_1 \int_0^{p(x_1|\theta)} p(x_1, x_2|\cdot)\, dx_2} =$$
$$= \frac{\int_\alpha^x p(x_1|\theta) \mathbf{1}_{[a,b]}(x_1)\, dx_1}{\int_\alpha^\beta p(x_1|\theta) \mathbf{1}_{[a,b]}(x_1)\, dx_1} = \int_a^x p(x_1|\theta)\, dx_1 = F(x|\theta)$$

so we set up the following algorithm:

(i_1) $u_i \Leftarrow Un(u|\alpha \leq a, \beta \geq b)$ and $w_i \Leftarrow Un(w|0, \delta \geq p_m)$;
(i_2) If $w_i \leq p(u_i|\theta)$ we *accept* x_i; otherwise we *reject* x_i and start again from (i_1)

Repeating the algorithm n times we get the sampling $\{x_1, x_2, \ldots, x_n\}$ from $p(x|\theta)$.

Besides its simplicity, the Acceptance-Rejection scheme does not require to have normalized densities for it is enough to know an upper bound and in some cases, for instance when the support of the random quantity X is determined by functional relations, it is easier to deal with a simpler covering of the support. However, the price to pay is a low *generation efficiency*:

$$\epsilon \stackrel{def.}{=} \frac{\text{accepted trials}}{\text{total trials}} = \frac{\text{area under } p(x|\theta)}{\text{area of the covering}} \leq 1$$

Note that the efficiency so defined refers only to the fraction of accepted trials and, obviously, the more adjusted the covering is the better but for the *Inverse Transform* $\epsilon = 1$ and it does not necessarily imply that it is more efficient attending to other considerations. It is interesting to observe that if we do not know the normalization factor of the density function, it can be estimated as

$$\int_X p(x|\theta)\, dx \simeq (\text{area of the covering})\, \epsilon$$

Let's see some examples before we proceed.

Example 3.7 Consider $X \sim Be(x|\alpha, \beta)$. In this case, what follows is just for pedagogical purposes since other procedures to be discussed later are more efficient. Anyway, the density is

$$p(x|\alpha, \beta) \propto x^{\alpha-1} (1-x)^{\beta-1}; \qquad x \in [0, 1]$$

Suppose that $\alpha, \beta > 1$ so the mode $x_o = (\alpha - 1)(\alpha + \beta - 2)^{-1}$ exists and is unique. Then

$$p_m \equiv \max_x\{p(x|\alpha, \beta)\} = p(x_0|\alpha, \beta) = \frac{(\alpha-1)^{\alpha-1} (\beta-1)^{\beta-1}}{(\alpha+\beta-2)^{\alpha+\beta-2}}$$

so let's take then the domain $[\alpha = 0, \beta = 1] \times [0, p_m]$ and

(i_1) Get $x_i \sim Un(x|0, 1)$ and $y_i \sim Un(y|0, p_m)$;
(i_2) If

(a) $y_i \leq p(x_i|\alpha, \beta)$ we deliver (*accept*) x_i
(r) $y_i > p(x_i|\alpha, \beta)$ we *reject* x_i and start again from (i_1)

Repeating the procedure n times, we get a sampling $\{x_1, x_2, \ldots, x_n\}$ from $Be(x|\alpha, \beta)$. In this case we know the normalization so the area under $p(x|\alpha, \beta)$ is $Be(x|\alpha, \beta)$ and the *generation efficiency* will be:

$$\epsilon = B(\alpha, \beta)\,\frac{(\alpha+\beta-2)^{\alpha+\beta-2}}{(\alpha-1)^{\alpha-1}\,(\beta-1)^{\beta-1}}$$

Example 3.8 Let's generate a sampling of the spatial distributions of a bounded electron in a Hydrogen atom. In particular, as an example, those for the principal quantum number $n = 3$. The wave-function is $\psi_{nlm}(r, \theta, \phi) = R_{nl}(r) Y_{lm}(\theta, \phi)$ with:

$$R_{30} \propto (1 - 2r/2 - 2r^2/27)e^{-r/3}; \qquad R_{31} \propto r(1 - r/6)e^{-r/3} \quad \text{and} \quad R_{32} \propto r^2 e^{-r/3}$$

the radial functions of the 3s, 3p and 3d levels and

$$|Y_{10}|^2 \propto \cos^2\theta; \quad |Y_{1\pm1}|^2 \propto \sin^2\theta; \quad |Y_{20}|^2 \propto (3\cos^2\theta - 1)^2; \quad |Y_{2\pm1}|^2 \propto \cos^2\theta \sin^2\theta;$$
$$|Y_{2\pm2}|^2 \propto \sin^4\theta$$

the angular dependence from the spherical harmonics. Since $d\mu = r^2 \sin\theta dr d\theta d\phi$, the probability density will be

$$p(r, \theta, \phi|n, l, m) = R^2_{nl}(r)\, |Y_{lm}|^2\, r^2 \sin\theta = p_r(r|n, l)\; p_\theta(\theta|l, m)\; p_\phi(\phi)$$

so we can sample independently r, θ and ϕ. It is left as an exercise to explicit a sampling procedure. Note however that, for the marginal radial density, the mode is at $r = 13$, 12 and 9 for $l = 0$, 1 and 2 and decreases exponentially so even if the support is $r \in [0, \infty)$ it will be a reasonable approximation to take a covering $r \in [0, r_{max})$ such that $P(r \geq r_{max})$ is small enough. After $n = 4000$ samplings for the quantum numbers $(n, l, m) = (3, 1, 0)$, $(3, 2, 0)$ and $(3, 2, \pm1)$, the projections on the planes π_{xy}, π_{xz} and π_{yz} are shown in Fig. 3.2.

The generalization of the *Acceptance-Rejection* method to sample a n-dimensional density $\boldsymbol{X} \sim p(\boldsymbol{x}|\boldsymbol{\theta})$ with $\dim(\boldsymbol{x}) = n$ is straight forward. Covering with an n+1 dimensional hypercube:

(i_1) Get a sampling $\{x_i^{1)}, x_i^{2)}, \ldots, x_i^{n)}; y_i\}$ where

$$\{x_i^{k)} \Leftarrow Un(x_i^{k)}|\alpha_k, \beta_k)\}_{k=1}^{n}; \quad y_i \Leftarrow Un(y|0, k) \quad \text{and} \quad k \geq \max_{\boldsymbol{x}} p(\boldsymbol{x}|\boldsymbol{\theta})$$

(i_2) Accept the n-tuple $\boldsymbol{x}_i = (x_i^{1)}, x_i^{2)}, \ldots, x_i^{n)})$ if $y_i \leq p(\boldsymbol{x}_i|\boldsymbol{\theta})$ or reject it otherwise.

3.2.2.1 Incorrect Estimation of $\max_x\{p(x|\cdot)\}$

Usually, we know the support $[\alpha, \beta]$ of the random quantity but the pdf is complicated enough to know the maximum. Then, we start the generation with our best guess for $\max_x p(x|\cdot)$, say k_1, and after having generated N_1 events (*generated*, not *accepted*) in $[\alpha, \beta] \times [0, k_1]$,... wham!, we generate a value x_m such that $p(x_m) > k_1$. Certainly,

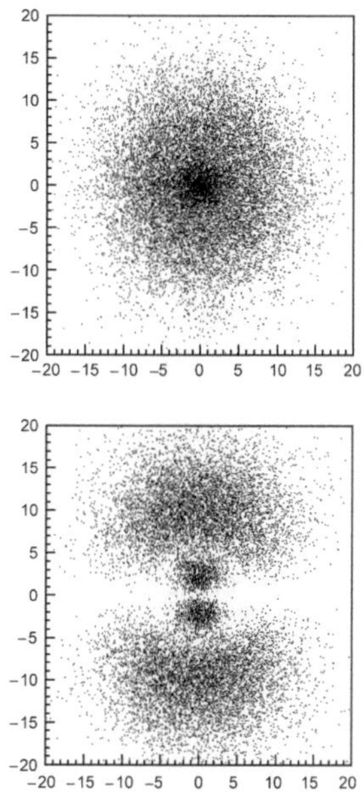

Fig. 3.2 Spatial probability distributions of an electron in a hydrogen atom corresponding to the quantum states $(n, l, m) = (3, 1, 0), (3, 2, 0)$ and $(3, 2, \pm 1)$ (columns 1, 2, 3) and projections (x, y) and $(x, z) = (y, z)$ (rows 1 and 2) (see Example 3.8)

our estimation of the maximum was not correct. A possible solution is to forget about what has been generated and start again with the new maximum $k_2 = p(x_m) > k_1$ but, obviously, this is not desirable among other things because we have no guarantee that this is not going to happen again. We better keep what has been done and proceed in the following manner:

(1) We have generated N_1 pairs (x_1, x_2) in $[\alpha, \beta] \times [0, k_1]$ and, in particular, X_2 uniformly in $[0, k_1]$. How many additional pairs N_a do we have to generate? Since the density of pairs is constant in both domains $[\alpha, \beta] \times [0, k_1]$ and $[\alpha, \beta] \times [0, k_2]$ we have that

$$\frac{N_1}{(\beta - \alpha)\, k_1} = \frac{N_1 + N_a}{(\beta - \alpha)\, k_2} \quad \longrightarrow \quad N_a = N_1 \left(\frac{k_2}{k_1} - 1 \right)$$

(2) How do we generate them? Obviously in the domain $[\alpha, \beta] \times [k_1, k_2]$ but from the truncated density

$$p_e(x|\cdot) = (p(x|\cdot) - k_1)\, \mathbf{1}_{p(x|\cdot) > k_1}(x)$$

(3) Once the N_a additional events have been generated (out of which some have been hopefully accepted) we continue with the usual procedure but on the domain $[\alpha, \beta]\times[0, k_2]$.

The whole process is repeated as many times as needed.

NOTE 6: Weighted events.

The *Acceptance-Rejection* algorithm just explained is equivalent to:

(i_1) Sample x_i from $Un(x|\alpha, \beta)$ and u_i from $Un(u|0, 1)$;
(i_2) Assign to each generated event x_i a *weight*: $w_i = p(x_i|\cdot)/p_m$; $0 \leq w_i \leq 1$ and accept the event if $u_i \leq w_i$ or reject it otherwise.

It is clear that:

- Events with a higher weight will have a higher chance to be accepted;
- After applying the *acceptance-rejection* criteria at step (i_2), all events will have a weight either 1 if it has been accepted or 0 if it was rejected.
- The generation efficiency will be

$$\epsilon = \frac{\text{accepted trials}}{\text{total trials}(N)} = \frac{1}{N}\sum_{i=1}^{N} w_i = \overline{w}$$

In some cases it is interesting to keep all the events, accepted or not.

Example 3.9 Let's obtain a sampling $\{\boldsymbol{x}_1, \boldsymbol{x}_2, \ldots\}$, $\dim(\boldsymbol{x}) = n$, of points inside a n-dimensional sphere centered at $\boldsymbol{x}^c$ and radius r. For a direct use of the *Acceptance-Rejection* algorithm we enclose the sphere in a n-dimensional hypercube

$$C_n = \prod_{i=1}^{n} [x_i^c - r, x_i^c + r]$$

and:

(1) $x_i \Leftarrow Un(x|x_i^c - r, x_i^c + r)$ for $i = 1, \ldots, n$
(2) Accept $\boldsymbol{x}_i$ if $\rho_i = ||\boldsymbol{x}_i - \boldsymbol{x}^c|| \leq r$ and reject otherwise.

The generation efficiency will be

$$\epsilon(n) = \frac{\text{volume of the sphere}}{\text{volume of the covering}} = \frac{2\,\pi^{n/2}}{n\,\Gamma(n/2)}\,\frac{1}{2^n}$$

Note that the sequence $\{\boldsymbol{x}_i/\rho_i\}_{i=1}^{n}$ will be a sampling of points uniformly distributed on the sphere of radius $r = 1$. This we can get also as:

(1) $z_i \Leftarrow N(z|0, 1)$ for $i = 1, \ldots, n$
(2) $\rho = ||\boldsymbol{z}_i||$ and $\boldsymbol{x}_i = \boldsymbol{z}_i/\rho$

Except for simple densities, the efficiency of the *Acceptance-Rejection* algorithm is not very high and decreases quickly with the number of dimensions. For instance, we have seen in the previous example that covering the n-dimensional sphere with a hypercube has a generation efficiency

$$\epsilon(n) = \frac{2\,\pi^{n/2}}{n\,\Gamma(n/2)}\,\frac{1}{2^n}$$

and $\lim_{n\to\infty}\epsilon(n) = 0$. Certainly, some times we can refine the covering since there is no need other than simplicity for a hypercube (see *Stratified Sampling*) but, in general, the origin of the problem will remain: when we generate points uniformly in whatever domain, we are sampling with constant density regions that have a very low probability content or even zero when they have null intersection with the support of the random quantity $\boldsymbol{X}$. This happens, for instance, when we want to sample from a differential cross-section that has very sharp peaks (sometimes of several orders of magnitude as in the case of bremsstrahlung). Then, the problem of having a low efficiency is not just the time expend in the generation but the accuracy and convergence of the evaluations. We need a more clever way to generate sequences and the *Importance Sampling* method comes to our help.

3.2.3 Importance Sampling

The *Importance Sampling* generalizes the *Acceptance-Rejection* method sampling the density function with higher frequency in regions of the domain where the probability of acceptance is larger (more *important*). Let's see the one-dimensional case since the extension to n-dimensions is straight forward.

Suppose that we want a sampling of $X \sim p(x)$ with support $\Omega_X \in [a, b]$ and $F(x)$ the corresponding distribution function. We can always express $p(x)$ as:

$$p(x) = c\,g(x)\,h(x)$$

where:

(1) $h(x)$ is a probability density function, i.e., non-negative and normalized in Ω_X;
(2) $g(x) \geq 0;\ \ \forall x \in \Omega_X$ and has a finite maximum $g_m = \max\{g(x);\, x \in \Omega_X\}$;
(3) $c > 0$ a constant normalization factor.

Now, consider a sampling $\{x_1, x_2, \ldots, x_n\}$ drawn from the density $h(x)$. If we apply the *Acceptance-Rejection* criteria with $g(x)$, how are the accepted values distributed? It is clear that, if $g_m = \max(g(x))$ and $Y \sim Un(y|0, g_m)$

$$P(X \leq x | Y \leq g(x)) = \frac{\int_a^x h(x)\,dx\,\int_0^{g(x)} dy}{\int_a^b h(x)\,dx\,\int_0^{g(x)} dy} = \frac{\int_a^x h(x)\,g(x)\,dx}{\int_a^b h(x)\,g(x)\,dx} = F(x)$$

and therefore, from a sampling of $h(x)$ we get a sampling of $p(x)$ applying the *Acceptance-Rejection* with the function $g(x)$. There are infinite options for $h(x)$. First, the simpler the better for then the Distribution Function can be easily inverted and the *Inverse Transform* applied efficiently. The Uniform Density $h(x) = Un(x|a, b)$ is the simplest one but then $g(x) = p(x)$ and this is just the *Acceptance-Rejection* over $p(x)$. The second consideration is that $h(x)$ be a fairly good approximation to $p(x)$ so that $g(x) = p(x)/h(x)$ is as smooth as possible and the *Acceptance-Rejection* efficient. Thus, if $h(x) > 0 \; \forall x \in [a, b]$:

$$p(x)\,dx = \frac{p(x)}{h(x)}\,h(x)\,dx = g(x)\,dH(x).$$

3.2.3.1 Stratified Sampling

The *Stratified Sampling* is a particular case of the *Importance Sampling* where the density $p(x);\ x \in \Omega_X$ is approximated by a simple function over Ω_X. Thus, in the one-dimensional case, if $\Omega = [a, b)$ and we take the partition (*stratification*)

$$\Omega = \cup_{i=1}^{n}\Omega_i = \cup_{i=1}^{n}[a_{i-1},\, a_i); \quad a_0 = a\,, \quad a_n = b$$

with measure $\lambda(\Omega_i) = (a_i - a_{i-1})$, we have

$$h(x) = \sum_{i=1}^{n} \frac{\mathbf{1}_{[a_{i-1},\, a_i)}(x)}{\lambda(\Omega_i)} \quad \longrightarrow \quad \int_{a_0}^{a_n} h(x)\,dx = 1$$

Denoting by $p_m(i) = \max_x\{p(x)|x \in \Omega_i\}$, we have that for the *Acceptance-Rejection* algorithm the volume of each sampling domain is $V_i = \lambda(\Omega_i)\, p_m(i)$. In consequence, for a partition of size n, if $Z \in \{1, 2, \ldots, n\}$ and define

$$P(Z = k) = \frac{V_k}{\sum_{i=1}^{n} V_i}; \qquad F(k) = P(Z \le k) = \sum_{j=1}^{k} P(Z = j)$$

we get a sampling of $p(x)$ from the following algorithm:

(i_1) $u_i \Leftarrow Un(u|0, 1)$ and select the partition $k = \text{Int}[\min\{F_i \mid F_i > n{\cdot}u_i\}]$;
(i_2) $x_i \Leftarrow Un(x|a_{k-1}, a_k)$, $y_i \Leftarrow Un(y|0, p_m(k))$ and accept x_i if $y_i \le p(x_i)$ (reject otherwise).

3.2.4 Decomposition of the Probability Density

Some times it is possible to express in a simple manner the density function as a linear combination of densities; that is

$$p(x) = \sum_{j=1}^{k} a_j \, p_j(x); \quad a_j \; > \; 0 \; \forall j = 1, 2, \ldots, k$$

that are easier to sample. Since normalization imposes that

$$\int_{-\infty}^{\infty} p(x)\, dx = \sum_{j=1}^{k} a_j \int_{-\infty}^{\infty} p_j(x)\, dx = \sum_{j=1}^{k} a_m = 1$$

we can sample from $p(x)$ selecting, at each step i, one of the k densities $p_i(x)$ with probability $p_i = a_i$ from which we shall obtain x_i and therefore sampling with higher frequency from those densities that have a higher relative weight. Thus:

(i_1) Select which density $p_i(x)$ are we going to sample at step (i_2) with probability $p_i = a_i$;

(i_2) Get x_i from $p_i(x)$ selected at (i_1).

It may happen that some densities $p_j(x)$ can not be easily integrated so we do not know a priory the relative weights. If this is the case, we can sample from $f_j(x) \propto p_j(x)$ and estimate with the generated events from $f_i(x)$ the corresponding normalizations I_i with, for instance, from the sample mean

$$I_i = \frac{1}{n} \sum_{k=1}^{n} f_i(x_k)$$

Then, since $p_i(x) = f_i(x)/I_i$ we have that

$$p(x|\cdot) = \sum_{i=1}^{K} a_i \, f_i(x|\cdot) = \sum_{i=1}^{K} a_i \, I_i \, \frac{f_i(x)}{I_i} = \sum_{i=1}^{K} a_i \, I_i \, p_i(x)$$

so each generated event from $f_i(x)$ has a weight $w_i = a_i I_i$.

Example 3.10 Suppose we want to sample from the density

$$p(x) = \frac{3}{8}\,(1 + x^2); \quad x \in [-1, 1]$$

Then, we can take:

$$\left. \begin{array}{l} p_1(x) \propto 1 \\ p_2(x) \propto x^2 \end{array} \right\} \quad \longrightarrow \quad \text{normalization} \quad \longrightarrow \quad \left\{ \begin{array}{l} p_1(x) = 1/2 \\ p_2(x) = 3\,x^2/2 \end{array} \right.$$

so:

$$p(x) = \frac{3}{4}\, p_1(x) + \frac{1}{4}\, p_2(x)$$

Then:

(i_1) Get u_i and w_i as $Un(u|0,1)$;
(i_2) Get x_i as:

if $u_i \leq 3/4$ then $x_i = 2\, w_i - 1$
if $u_i > 3/4$ then $x_i = (2\, w_i - 1)^{1/3}$

In this case, 75% of the times we sample from the trivial density $Un(x|-1,1)$.

3.3 Everything at Work

3.3.1 *The Compton Scattering*

When problems start to get complicated, we have to combine several of the aforementioned methods; in this case *Importance Sampling*, *Acceptance-Rejection* and *Decomposition* of the probability density.

Compton Scattering is one of the main processes that occur in the interaction of photons with matter. When a photon interacts with one of the atomic electrons with an energy greater than the binding energy of the electron, is suffers an inelastic scattering resulting in a photon of less energy and different direction than the incoming one and an ejected free electron from the atom. If we make the simplifying assumptions that the atomic electron initially at rest and neglect the binding energy we have that if the incoming photon has an energy E_γ its energy after the interaction ($E_\gamma^{'}$) is:

$$\epsilon = \frac{E_\gamma^{'}}{E_\gamma} = \frac{1}{1 + a\,(1 - \cos\theta)}$$

where $\theta \in [0, \pi]$ is the angle between the momentum of the outgoing photon and the incoming one and $a = E_\gamma / m_e$. It is clear that if the dispersed photon goes in the forward direction (that is, with $\theta = 0$), it will have the maximum possible energy ($\epsilon = 1$) and when it goes backwards (that is, $\theta = \pi$) the smallest possible energy ($\epsilon = (1 + 2a)^{-1}$). Being a two body final state, given the energy (or the angle) of the outgoing photon the rest of the kinematic quantities are determined uniquely:

$$E_e^{'} = E_\gamma \left(1 + \frac{1}{a} - \epsilon\right) \qquad \text{and} \qquad \tan\theta_e = \frac{\cot\theta/2}{1 + a}$$

The cross-section for the Compton Scattering can be calculated perturbatively in Relativistic Quantum Mechanics resulting in the Klein-Nishina expression:

$$\frac{d\,\sigma_0}{d\,x} = \frac{3\,\sigma_T}{8}\,f(x)$$

where $x = \cos(\theta)$, $\sigma_T = 0.665\,\text{barn} = 0.665 \cdot 10^{-24}\,\text{cm}^2$ is the *Thomson cross-section* and

$$f(x) = \frac{1}{[1 + a\,(1-x)]^2}\left(1 + x^2 + \frac{a^2\,(1-x)^2}{1 + a\,(1-x)}\right)$$

has all the angular dependence. Due to the azimuthal symmetry, there is no explicit dependence with $\phi \in [0, 2\pi]$ and has been integrated out. Last, integrating this expression for $x \in [-1, 1]$ we have the total cross-section of the process:

$$\sigma_0(E_\gamma) = \frac{\sigma_T}{4}\left[\left(\frac{1+a}{a^2}\right)\left(\frac{2\,(1+a)}{1+2\,a} - \frac{\ln\,(1+2a)}{a}\right) + \frac{\ln\,(1+2a)}{2a} - \frac{1+3a}{(1+2\,a)^2}\right]$$

For a material with Z electrons, the atomic cross-section can be approximated by $\sigma = Z\,\sigma_0\,\text{cm}^2/\text{atom}$.

Let's see how to simulate this process sampling the angular distribution $p(x) \sim f(x)$. Figure 3.3 (left) shows this function for incoming photon energies of 10, 100 and 1000 MeV. It is clear that it is peaked at x values close to 1 and gets sharper with the incoming energy; that is, when the angle between the incoming and outgoing photon momentum becomes smaller. In consequence, for high energy photons the *Acceptance-Rejection* algorithm becomes very inefficient. Let's then define the functions

$$f_n(x) = \frac{1}{[1 + a\,(1-x)]^n}$$

and express $f(x)$ as $f(x) = (f_1(x) + f_2(x) + f_3(x)) \cdot g(x)$ where

$$g(x) = 1 - (2 - x^2)\,\frac{f_1(x)}{1 + f_1(x) + f_2(x)}$$

The functions $f_n(x)$ are easy enough to use the *Inverse Transform* method and apply afterward the *Acceptance-Rejection* on $g(x) > 0\ \forall x \in [-1, 1]$. The shape of this function is shown in Fig. 3.3 (right) for different values of the incoming photon energy and clearly is much more smooth than $f(x)$ so the *Acceptance-Rejection* will be significantly more efficient. Normalizing properly the densities

$$p_i(x) = \frac{1}{w_i}\,f_i(x) \qquad \text{such that} \qquad \int_{-1}^{1} p_i(x)\,dx = 1;\ \ i = 1, 2, 3$$

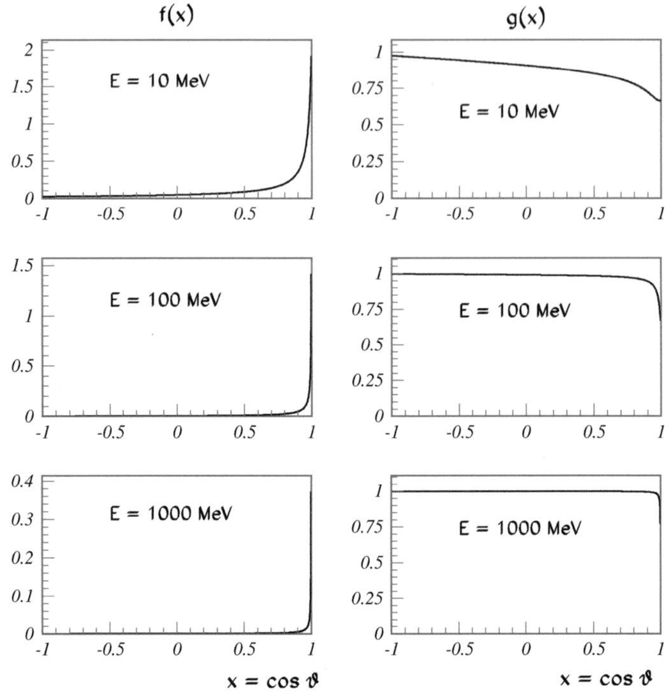

Fig. 3.3 Functions $f(x)$ (*left*) and $g(x)$ (*right*) for different values of the incoming photon energy

we have that, with $b = 1 + 2a$:

$$w_1 = \frac{1}{a}\ln b \quad w_2 = \frac{2}{a^2\,(b^2-1)} \quad w_3 = \frac{2\,b}{a^3\,(b^2-1)^2}$$

and therefore

$$\begin{aligned} f(x) &= (f_1(x) + f_2(x) + f_3(x)) \cdot g(x) \;=\; (w_1\,p_1(x) + w_2\,p_2(x) + w_3\,p_3(x)) \cdot g(x) \\ &= \;w_t\;(\alpha_1\,p_1(x) + \alpha_2\,p_2(x) + \alpha_3\,p_3(x)) \cdot g(x) \end{aligned}$$

where $w_t = w_1 + w_2 + w_3$,

$$\alpha_i = \frac{w_i}{w_t} > 0;\; i = 1, 2, 3 \qquad \text{and} \qquad \sum_{i=1}^{i=3} \alpha_i = 1$$

Thus, we set up the following algorithm:

(1) Generate $u \Leftarrow Un(u|0,1)$,

(1.1) if $u \leq \alpha_1$ we sample $x_g \sim p_1(x)$;
(1.2) if $\alpha_1 < u \leq \alpha_1 + \alpha_2$ we sample $x_g \sim p_2(x)$ and
(1.3) if $\alpha_1 + \alpha_2 < u$ we sample $x_g \sim p_2(x)$;

(2) Generate $w \Leftarrow Un(w|0, g_M)$ where

$$g_M \equiv \max[g(x)] = g(x=-1) = 1 - \frac{b}{1+b+b^2}$$

If $w \leq g(x_g)$ we accept x_g; otherwise we go back to step (1).

Let's see now to sample from the densities $p_i(x)$. If $u \Leftarrow Un(u|0,1)$ and

$$F_i(x) = \int_{-1}^{x} p_i(s)\,ds \quad i = 1, 2, 3$$

then:

- $x \sim p_1(x)$: $F_1(x) = 1 - \dfrac{\ln(1+a(1-x))}{\ln(b)} \longrightarrow x_g = \dfrac{1+a-b^u}{a}$
- $x \sim p_2(x)$: $F_2(x) = \dfrac{b^2-1}{2\,(b-x)} - \dfrac{1}{2\,a} \longrightarrow x_g = b - \dfrac{a\,(b^2-1)}{1+2au}$
- $x \sim p_3(x)$: $F_3(x) = \dfrac{1}{4a(1+a)}\left(\dfrac{(b+1)^2}{(b-x)^2} - 1\right) \longrightarrow x_g = b - \dfrac{b+1}{[1+4a(1+a)u]^{1/2}}$

Once we have x_g we can deduce the remaining quantities of interest from the kinematic relations. In particular, the energy of the outcoming photon will be

$$\epsilon_g = \frac{E_g^{'}}{E} = \frac{1}{1+a(1-x_g)}$$

Last, we sample the azimuthal outgoing photon angle as $\phi \Leftarrow Un(u|0, 2\pi)$.

Even though in this example we are going to simulate only the Compton effect, there are other processes by which the photon interacts with matter. At low energies (essentially ionization energies: $\leq E_\gamma \leq 100\,\mathrm{KeV}$) the dominant interaction is the photoelectric effect

$$\gamma + \text{atom} \longrightarrow \text{atom}^+ + e^-;$$

at intermediate energies ($E_\gamma \sim 1 - 10\,\mathrm{MeV}$) the Compton effect

$$\gamma + \text{atom} \longrightarrow \gamma + e^- + \text{atom}^+$$

and at high energies ($E_\gamma \geq 100\,\text{MeV}$) the dominant one is pair production

$$\gamma + \text{nucleus} \longrightarrow e^+ + e^- + \text{nucleus}$$

To first approximation, the contribution of other processes is negligible. Then, at each step in the evolution of the photon along the material we have to decide first which interaction is going to occur next. The cross section is a measure of the interaction probability expressed in cm^2 so, since the total interaction cross section will be in this case:

$$\sigma_t = \sigma_{\text{phot.}} + \sigma_{\text{Compt.}} + \sigma_{\text{pair}}$$

we decide upon the process i that is going to happen next with probability $p_i = \sigma_i/\sigma_t$; that is, $u \Leftarrow Un(0, 1)$ and

(1) if $u \leq p_{\text{phot.}}$ we simulate the photoelectric interaction;
(2) if $p_{\text{phot.}} < u \leq (p_{\text{phot.}} + p_{\text{Compt.}})$: we simulate the Compton effect and otherwise
(3) we simulate the pair production

Once we have decided which interaction is going to happen next, we have to decide where. The probability that the photon interacts after traversing a distance x (cm) in the material is given by

$$F_{int} = 1 - e^{-x/\lambda}$$

where λ is the *mean free path*. Being A the atomic mass number of the material, N_A the Avogadro's number, ρ the density of the material in g/cm^3, and σ the cross-section of the process under discussion, we have that

$$\lambda = \frac{A}{\rho\, N_A\, \sigma} \quad [\text{cm}]$$

Thus, if $u \Leftarrow Un(0, 1)$, the next interaction is going to happen at $x = -\lambda \ln u$ along the direction of the photon momentum.

As an example, we are going to simulate what happens when a beam of photons of energy $E_\gamma = 1\,\text{MeV}$ (X rays) incide normally on the side of a rectangular block of carbon ($Z = 6$, $A = 12.01$, $\rho = 2.26$) of $10 \times 10\,\text{cm}^2$ surface y 20 cm depth. Behind the block, we have hidden an iron coin ($Z = 26$, $A = 55.85$, $\rho = 7.87$) centered on the surface and in contact with it of 2 cm radius and 1 cm thickness. Last, at 0.5 cm from the coin there is a photographic film that collects the incident photons.

The beam of photons is wider than the block of carbon so some of them will go right the way without interacting and will burn the film. We have assumed for simplicity that when the photon energy is below 0.01 MeV, the photoelectric effect is dominant and the ejected electron will be absorbed in the material. The photon will then be lost and we shall start with the next one. Last, an irrelevant technical issue:

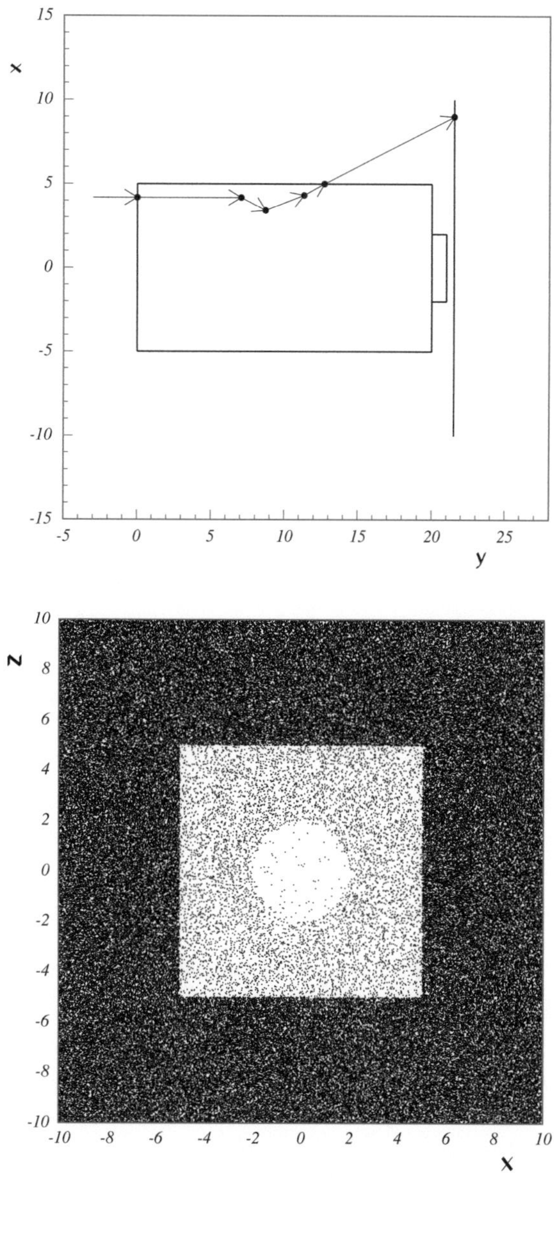

Fig. 3.4 The *upper figure* shows an sketch of the experimental set-up and the trajectory of one of the simulated photons until it is detected on the screen. The *lower figure* shows the density of photons collected at the screen for a initially generated sample 10^5 events

the angular variables of the photon after the interaction are referred to the direction of the incident photon so in each case we have to do the appropriate rotation.

Figure 3.4 (up) shows the sketch of the experimental set-up and the trajectory of one of the traced photons of the beam collected by the film. The *radiography* obtained after tracing 100,000 photons is shown in Fig. 3.4 (down). The black zone corresponds to photons that either go straight to the screen or at some point leave the block before getting to the end. The mid zone are those photons that cross the carbon block and the central circle, with less interactions, those that cross the carbon block and afterward the iron coin.

3.3.2 An Incoming Flux of Particles

Suppose we have a detector and we want to simulate a flux of isotropically distributed incoming particles. It is obvious that generating them one by one in the space and tracing them backwards is extremely inefficient. Consider a large cubic volume V that encloses the detector, both centered in the reference frame S_0. At time t_0, we have for particles uniformly distributed inside this volume that:

$$p(\boldsymbol{r}_0)\, d\mu_0 = \frac{1}{V}\, dx_0\, dy_0\, dz_0$$

Assume now that the velocities are isotropically distributed; that is:

$$p(\boldsymbol{v})\, d\mu_v = \frac{1}{4\pi}\, \sin\theta\, d\theta\, d\phi\, f(v)\, dv$$

with $f(v)$ properly normalized. Under independence of positions and velocities at t_0, we have that:

$$p(\boldsymbol{r}_0, \boldsymbol{v})\, d\mu_0\, d\mu_v = \frac{1}{V}\, dx_0\, dy_0\, dz_0\, \frac{1}{4\pi}\, \sin\theta\, d\theta\, d\phi\, f(v)\, dv$$

Given a square of surface $S = (2l)^2$, parallel to the (x, y) plane, centered at $(0, 0, z_c)$ and well inside the volume V, we want to find the probability and distribution of particles that, coming from the top, cross the surface S in unit time.

For a particle having a coordinate z_0 at $t_0 = 0$, we have that $z(t) = z_0 + v_z t$. The surface S is parallel to the plane (x, y) at $z = z_c$ so particles will cross this plane at time $t_c = (z_c - z_0)/v_z$ from above iff:

(0) $z_0 \geq z_c$; obvious for otherwise they are below the plane S at $t_0 = 0$;
(1) $\theta \in [\pi/2, \pi)$; also obvious because if they are above S at $t_0 = 0$ and cut the plane at some $t > 0$, the only way is that $v_z = v\cos\theta < 0 \rightarrow \cos\theta < 0 \rightarrow \theta \in [\pi/2, \pi)$.

But to cross the squared surface S of side $2l$ we also need that

(2) $-l \leq x(t_c) = x_0 + v_x\, t_c \leq l$ and $-l \leq y(t_c) = y_0 + v_y\, t_c \leq l$

Last, we want particles crossing in unit time; that is $t_c \in [0, 1]$ so $0 \leq t_c = (z_c - z_0)/v_z \leq 1$ and therefore

(3) $z_0 \in [z_c, z_c - v\cos\theta]$

Then, the desired subspace with conditions (1), (2), (3) is

$$\Omega_c = \{\theta \in [\pi/2, \pi);\ z_0 \in [z_c, z_c - v\cos\theta];\ x_0 \in [-l - v_x t_c, l - v_x t_c];\ y_0 \in [-l - v_y t_c, l - v_y t_c]\}$$

After integration:

$$\int_{z_c}^{z_c - v\cos\theta} dz_0 \int_{-l-v_x t_c}^{l-v_x t_c} dx_0 \int_{-l-v_y t_c}^{l-v_y t_c} dy_0 = -(2l)^2\, v\, \cos\theta$$

Thus, we have that for the particles crossing the surface $S = (2l)^2$ from above in unit time

$$p(\theta, \phi, v)\, d\theta d\phi dv = -\frac{(2l)^2}{V}\frac{1}{4\pi}\sin\theta\,\cos\theta\, d\theta\, d\phi\, f(v)\, v\, dv$$

with $\theta \in [\pi/2, \pi)$ and $\phi \in [0, 2\pi)$. If we define the *average velocity*

$$E[v] = \int_{\Omega_v} v\, f(v)\, dv$$

the probability to have a cut per unit time is

$$P_{cut}(t_c \leq 1) = \int_{\Omega_c \times \Omega_v} p(\theta, \phi, v)\, d\theta d\phi dv = \frac{S\, E[v]}{4\, V}$$

and the pdf for the angular distribution of velocities (direction of crossing particles) is

$$p(\theta, \phi)\, d\theta d\phi = -\frac{1}{\pi}\sin\theta\,\cos\theta\, d\theta\, d\phi = \frac{1}{2\pi}\, d(\cos^2\theta)\, d\phi$$

If we have a density of n particles per unit volume, the expected number of crossings per unit time due to the $n_V = n\, V$ particles in the volume is

$$n_c = n_V\, P_{cut}(t_c \leq 1) = \frac{n\, E[v]}{4}\, S$$

so the *flux*, number of particles crossing the surface from one side per unit time and unit surface is

$$\Phi_c^0 = \frac{n_c}{S} = \frac{n\, E[v]}{4}$$

Note that the requirement that the particles cross the square surface S in a finite time ($t_c \in [0, 1]$) modifies the angular distribution of the direction of particles. Instead of

$$p_1(\theta, \phi) \propto \sin\theta; \theta \in [0, \pi); \quad \phi \in [0, 2\pi)$$

we have

$$p_2(\theta, \phi) \propto -\sin\theta \cos\theta; \theta \in [\pi/2, \pi); \quad \phi \in [0, 2\pi)$$

The first one spans a solid angle of

$$\int_0^{\pi} d\theta \int_0^{2\pi} d\phi p_1(\theta, \phi) = 4\pi$$

while for the second one we have that

$$\int_0^{\pi/2} d\theta \int_0^{2\pi} d\phi p_2(\theta, \phi) = \pi$$

that is; one fourth the solid angle spanned by the sphere. Therefore, the flux expressed as number of particles crossing from one side of the square surface S per unit time and solid angle is

$$\Phi_c = \frac{\Phi_c^0}{\pi} = \frac{n_c}{\pi S} = \frac{n\, E[v]}{4\,\pi}$$

Thus, if we generate a total of $n_T = 6n_c$ particles on the of the surface of a cube, each face of area S, with the angular distribution

$$p(\theta, \phi)\, d\theta d\phi = \frac{1}{2\pi} d(\cos^2\theta)\, d\phi$$

for each surface with $\theta \in [\pi/2, \pi)$ and $\phi \in [0, 2\pi)$ defined with $\vec{k}$ normal to the surface, the equivalent *generated flux* per unit time, unit surface and solid angle is $\Phi_T = n_T/6\pi S$ and corresponds to a density of $n = 2n_T/3SE[v]$ particles per unit volume.

NOTE 7: Sampling some continuous distributions of interest

These are some procedures to sample from continuous distributions of interest. There are several algorithms for each case with efficiency depending on the parameters but those outlined here have in general high efficiency. In all cases, it is assumed that $u \Leftarrow Un(0, 1)$.

• **Beta**: $Be(x|\alpha,\beta)$; $\alpha,\beta\in(0,\infty)$:

$$p(x|\cdot)=\frac{1}{B(\alpha,\beta)}x^{\alpha-1}(1-x)^{\beta-1}\mathbf{1}_{(0,1)}(x)\quad\longrightarrow\quad x=\frac{x_1}{x_1+x_2}$$

where $x_1\Leftarrow Ga(x|1/2,\alpha)$ and $x_2\Leftarrow Ga(x|1/2,\beta)$

• **Cauchy**: $Ca(x|\alpha,\beta)$; $\alpha\in\mathcal{R}$; $\beta\in(0,\infty)$:

$$p(x|\cdot)=\frac{\beta/\pi}{\left(1+\beta^2(x-\alpha)^2\right)}\mathbf{1}_{(-\infty,\infty)}(x)\quad\longrightarrow\quad x=\alpha+\beta^{-1}\tan(\pi(u-1/2))$$

• **Chi-squared**: For $\chi^2(x|\nu)$ see $Ga(x|1/2,\nu/2)$.

• **Dirichlet** $Di(\boldsymbol{x}|\boldsymbol{\alpha})$; $\dim(\boldsymbol{x},\boldsymbol{\alpha})=$ n, $\alpha_j\in(0,\infty)$, $x_j\in(0,1)$ and $\sum_{j=1}^n x_j=1$

$$p(\boldsymbol{x}|\boldsymbol{\alpha})=\frac{\Gamma(\alpha_1+\cdots+\alpha_n)}{\Gamma(\alpha_1)\cdots\Gamma(\alpha_n)}\prod_{j=1}^{n}x_j^{\alpha_j-1}\mathbf{1}_{(0,1)}(x_j)\quad\longrightarrow\quad\{x_j=z_j/z_0\}_{j=1}^n$$

where $z_j\Leftarrow Ga(z|1,\alpha_j)$ and $z_0=\sum_{j=1}^n z_j$.

Generalized Dirichlet $GDi(\boldsymbol{x}|\boldsymbol{\alpha},\boldsymbol{\beta})$; $\dim(\boldsymbol{\beta})=$ n, $\beta_j\in(0,\infty)$, $\sum_{j=1}^{n-1}x_j<1$

$$p(x_1,\ldots,x_{n-1}|\boldsymbol{\alpha},\boldsymbol{\beta})=\prod_{i=1}^{n-1}\frac{\Gamma(\alpha_i+\beta_i)}{\Gamma(\alpha_i)\Gamma(\beta_i)}x_i^{\alpha_i-1}\left(1-\sum_{k=1}^{i}x_k\right)^{\gamma_i}$$

with

$$\gamma_i=\begin{cases}\beta_i-\alpha_{i+1}-\beta_{i+1} & \text{for } i=1,2,\ldots,n-2\\ \beta_{n-1}-1 & \text{for } i=n-1\end{cases}$$

When $\beta_i=\alpha_{i+1}+\beta_{i+1}$ reduces to the usual Dirichlet. If $z_k\Leftarrow \text{Be}(z|\alpha_k,\beta_k)$ then $x_k=z_k(1-\sum_{j=1}^{k-1}x_j)$ for $k=1,\ldots,n-1$ and $x_n=1-\sum_{i=1}^{n-1}x_i$.

• **Exponential**: $Ex(x|\alpha)$; $\alpha\in(0,\infty)$:

$$p(x|\cdot)=\alpha\exp\{-\alpha x\}\mathbf{1}_{[0,\infty)}(x)\quad\longrightarrow\quad x=-\alpha^{-1}\ln u$$

• **Gamma Distribution** $Ga(x|\alpha,\beta)$; $\alpha,\beta\in(0,\infty)$.
The probability density is

$$p(x|\alpha,\beta)=\frac{\alpha^{\beta}}{\Gamma(\beta)}e^{-\alpha x}x^{\beta-1}\mathbf{1}_{(0,\infty)}(x)$$

Note that $Z = \alpha X \sim Ga(z|1, \beta)$ so let's see how to simulate a sampling of $Ga(z|1, \beta)$ and, if $\alpha \neq 1$, take $x = z/\alpha$. Depending on the value of the parameter β we have that

- ▷ $\beta = 1$: This is the Exponential distribution $Ex(x|1)$ already discussed;
- ▷ $\beta = m \in \mathcal{N}$: As we know, the sum $X_s = X_1 + \cdots + X_n$ of n independent random quantities $X_i \sim Ga(x_i|\alpha, \beta_i)$; $i = 1, \ldots, n$ is a random quantity distributed as $Ga(|x_s|\alpha, \beta_1 + \cdots + \beta_n)$. Thus, if we have m independent samplings $x_i \Leftarrow Ga(x|1, 1) = Ex(x|1)$, that is

  $$x_1 = -\ln u_1, \ldots, x_m = -\ln u_m$$

  with $u_i \Leftarrow Un(0, 1)$, then

  $$x_s = x_1 + x_2 \cdots + x_m = -\ln \prod_{i=1}^{m} u_i$$

  will be a sampling of $Ga(x_s|1, \beta = m)$.
- ▷ $\beta > 1 \in \mathcal{R}$: Defining $m = [\beta]$ we have that $\beta = m + \delta$ with $\delta \in [0, 1)$. Then, if $u_i \Leftarrow Un(0, 1)$; $i = 1, \ldots, m$ and $w \Leftarrow Ga(w|1, \delta)$,

  $$z = -\ln \prod_{i=1}^{m} u_i + w$$

  will be a sampling from $Ga(x|1, \beta)$. The problem is reduced to get a sampling $w \Leftarrow Ga(w|1, \delta)$ with $\delta \in (0, 1)$.
- ▷ $0 < \beta < 1$: In this case, for small values of x the density is dominated by $p(x) \sim x^{\beta-1}$ and for large values by $p(x) \sim e^{-x}$. Let's then take the approximant

  $$g(x) = x^{\beta-1}\, \mathbf{1}_{(0,1)}(x) + e^{-x}\, \mathbf{1}_{[1,\infty)}(x)$$

  Defining

  $$\begin{aligned} p_1(x) &= \beta x^{\beta-1} \mathbf{1}_{(0,1)}(x) \quad \longrightarrow \quad F_1(x) = x^{\beta} \\ p_2(x) &= e^{-(x-1)} \mathbf{1}_{[1,\infty)}(x) \quad \longrightarrow \quad F_2(x) = 1 - e^{-(x-1)} \end{aligned}$$

  $w_1 = e/(e + \beta)$ and $w_2 = \beta/(e + \beta)$ we have that

  $$g(x) = w_1\, p_1(x) + w_2\, p_2(x)$$

and therefore:

(1) $u_i \Leftarrow Un(0, 1)$; $i = 1, 2, 3$
(2) If $u_1 \leq w_1$, set $x = u_2^{1/\beta}$ and accept x if $u_3 \leq e^{-x}$; otherwise go to (1);
If $u_1 > w_1$, set $x = 1 - \ln u_2$ and accept x if $u_3 \leq x^{\beta-1}$; otherwise go to (1);

The sequence of accepted values will simulate a sampling from $Ga(x|1, \beta)$. It is easy to see that the generation efficiency is

$$\epsilon(\beta) = \frac{e}{e+\beta}\,\Gamma(\beta+1)$$

and $\epsilon_{min}(\beta \simeq 0.8) \simeq 0.72$.

• **Laplace**: $La(x|\alpha, \beta)$; $\alpha \in \mathcal{R}$, $\beta \in (0, \infty)$:

$$p(x|\alpha, \beta) = \frac{1}{2\beta}\,e^{-|x-\alpha|/\beta}\mathbf{1}_{(-\infty,\infty)}(x) \quad \longrightarrow \quad x = \begin{cases} \alpha + \beta\ln(2u) & \text{if} \quad u < 1/2 \\ \alpha - \beta\ln(2(1-u)) & \text{if} \quad u \geq 1/2 \end{cases}$$

• **Logistic**: $Lg(x|\alpha, \beta)$; $\alpha \in \mathcal{R}$; $\beta \in (0, \infty)$:

$$p(x|\cdot) = \beta\frac{\exp\{-\beta(x-\alpha)\}}{(1+\exp\{-\beta(x-\alpha)\})^2}\mathbf{1}_{(-\infty,\infty)}(x) \quad \longrightarrow \quad x = \alpha + \beta^{-1}\ln\left(\frac{u}{1-u}\right)$$

• **Normal Distribution** $N(\mathbf{x}|\boldsymbol{\mu}, \mathbf{V})$.

There are several procedures to generate samples from a Normal Distribution. Let's start with the one-dimensional case $X \sim N(x|\mu, \sigma)$ considering two independent standardized random quantities $X_i \sim N(x_i|0, 1)$; $i = 1, 2$ [7] with joint density

$$p(x_1, x_2) = \frac{1}{2\pi}\,e^{-(x_1^2+x_2^2)/2}$$

After the transformation $X_1 = R\cos\Theta$ and $X_2 = R\sin\Theta$ with $R \in [0, \infty)$; $\Theta \in [0, 2\pi)$ we have

$$p(r, \theta) = \frac{1}{2\pi}\,e^{-r^2/2}\,r$$

Clearly, both quantities R and Θ are independent and their distribution functions

$$F_r(r) = 1 - e^{-r^2/2} \quad \text{and} \quad F_\theta(\theta) = \frac{\theta}{2\pi}$$

are easy to invert so, using then the *Inverse Transform* algorithm:

(1) $u_1 \Leftarrow Un(0, 1)$ and $u_2 \Leftarrow Un(0, 1)$;
(2) $r = \sqrt{-2\ln u_1}$ and $\theta = 2\pi u_2$;
(3) $x_1 = r\cos\theta$ and $x_2 = r\sin\theta$.

Thus, we get two independent samplings x_1 and x_2 from $N(x|0, 1)$ and

(4) $z_1 = \mu_1 + \sigma_1 x_1$ and $z_2 = \mu_2 + \sigma_2 x_2$

will be two independent samplings from $N(x|\mu_1, \sigma_1)$ and $N(x|\mu_2, \sigma_2)$.

For the n-dimensional case, $\boldsymbol{X} \sim N(\boldsymbol{x}|\boldsymbol{\mu}, \boldsymbol{V})$ and $\boldsymbol{V}$ the covariance matrix, we proceed from the conditional densities

$$p(\boldsymbol{x}|\boldsymbol{\mu}, \boldsymbol{V}) = p(x_n|x_{n-1}, x_{n-2} \ldots, x_1; \cdot)\, p(x_{n-1}|x_{n-2} \ldots, x_1; \cdot) \cdots p(x_1|\cdot)$$

For high dimensions this is a bit laborious and it is easier if we do first a bit of algebra. We know from *Cholesky's Factorization Theorem* that if $\mathbf{V} \in \mathcal{R}^{n \times n}$ is a symmetric positive defined matrix there is a unique lower triangular matrix $\mathbf{C}$, with positive diagonal elements, such that $\mathbf{V} = \mathbf{C}\mathbf{C}^T$. Let then $\boldsymbol{Y}$ be an n-dimensional random quantity distributed as $N(\mathbf{y}|\mathbf{0}, \mathbf{I})$ and define a new random quantity

$$\boldsymbol{X} = \boldsymbol{\mu} + \mathbf{C}\,\boldsymbol{Y}$$

Then $\mathbf{V}^{-1} = [\mathbf{C}^{-1}]^T\, \mathbf{C}^{-1}$ and

$$\boldsymbol{Y}^T\boldsymbol{Y} = (\boldsymbol{X} - \boldsymbol{\mu})^T\, [\mathbf{C}^{-1}]^T\, [\mathbf{C}^{-1}]\, (\boldsymbol{X} - \boldsymbol{\mu}) = (\boldsymbol{X} - \boldsymbol{\mu})^T\, [\mathbf{V}^{-1}]\, (\boldsymbol{X} - \boldsymbol{\mu})$$

After some algebra, the elements if the matrix $\mathbf{C}$ can be easily obtained as

$$\begin{aligned}
\mathbf{C}_{i1} &= \frac{\mathbf{V}_{i1}}{\sqrt{\mathbf{V}_{11}}} && 1 \le i \le n \\
\mathbf{C}_{ij} &= \frac{\mathbf{V}_{ij} - \sum_{k=1}^{j-1} C_{ik} C_{jk}}{C_{jj}} && 1 < j < i \le n \\
\mathbf{C}_{ii} &= \left(\mathbf{V}_{ii} - \sum_{k=1}^{i-1} C_{ik}^2 \right)^{1/2} && 1 < i \le n
\end{aligned}$$

and, being lower triangular, $C_{ij} = 0\, \forall j > i$. Thus, we have the following algorithm:

(1) Get the matrix $\mathbf{C}$ from the covariance matrix $\mathbf{V}$;
(2) Get n independent samplings $z_i \Leftarrow N(0, 1)$ with $i = 1, \ldots, n$;
(3) Get $x_i = \mu_i + \sum_{j=1}^{n} \mathbf{C}_{ij} z_j$

In particular, for a two-dimensional random quantity we have that

$$\mathbf{V} = \begin{pmatrix} \sigma_1^2 & \rho\sigma_1\sigma_2 \\ \rho\sigma_1\sigma_2 & \sigma_2^2 \end{pmatrix}$$

and therefore:

$$\begin{aligned}
\mathbf{C}_{11} &= \frac{\mathbf{V}_{11}}{\sqrt{\mathbf{V}_{11}}} = \sigma_1; \quad \mathbf{C}_{12} = 0 \\
\mathbf{C}_{21} &= \frac{\mathbf{V}_{21}}{\sqrt{\mathbf{V}_{11}}} = \rho\,\sigma_2; \quad \mathbf{C}_{22} = (\mathbf{V}_{22} - \mathbf{C}_{21}^2)^{1/2} = \sigma_2\,\sqrt{1 - \rho^2}
\end{aligned}$$

so:

$$\mathbf{C} = \begin{pmatrix} \sigma_1 & 0 \\ \rho\sigma_2 & \sigma_2\sqrt{1 - \rho^2} \end{pmatrix}$$

Then, if $z_{1,2} \Leftarrow N(z|0,1)$ we have that:

$$\begin{pmatrix} x_1 \\ x_2 \end{pmatrix} = \begin{pmatrix} \mu_1 \\ \mu_2 \end{pmatrix} + \begin{pmatrix} z_1\,\sigma_1 \\ \sigma_2(z_1\,\rho + z_2\,\sqrt{1-\rho^2}) \end{pmatrix}$$

- **Pareto**: $Pa(x|\alpha,\beta)$; $\alpha,\beta \in (0,\infty)$:

$$p(x|\cdot) = \alpha\beta^{\alpha}x^{-(\alpha+1)}\mathbf{1}_{(\beta,\infty)}(x) \quad\longrightarrow\quad x = \beta\,u^{-1/\alpha}$$

- **Snedecor**: $Sn(x|\alpha,\beta)$; $\alpha,\beta \in (0,\infty)$:

$$p(x|\cdot) \propto x^{\alpha/2-1}\,(\beta+\alpha x)^{-(\alpha+\beta)/2}\mathbf{1}_{(0,\infty)}(x) \quad\longrightarrow\quad x = \frac{x_1/\alpha}{x_2/\beta}$$

where $x_1 \Leftarrow Ga(x|1/2,\alpha/2)$ and $x_2 \Leftarrow Ga(x|1/2,\beta/2)$.

- **Student** $St(x|\nu)$; $\nu \in (0,\infty)$:

$$p(x|\cdot) \propto \frac{1}{\left(1+x^2/\nu\right)^{(\nu+1)/2}}\mathbf{1}_{(-\infty,\infty)}(x) \quad\longrightarrow\quad x = \sqrt{\nu(u_1^{-2/\nu}-1)}\,\sin(2\pi u_2)$$

where $u_{1,2} \Leftarrow Un(0,1)$.

- **Uniform** $Un(x|a,b)$; $a < b \in \mathcal{R}$

$$p(x|\cdot) = (b-a)^{-1}\mathbf{1}_{[a,b]}(x) \quad\longrightarrow\quad x = (b-1) + a\,u$$

- **Weibull**: $We(x|\alpha,\beta)$; $\alpha,\beta \in (0,\infty)$:

$$p(x|\cdot = \alpha\beta^{\alpha}x^{\alpha-1}\exp\{-(x/\beta)^{\alpha}\}\mathbf{1}_{(0,\infty)}(x) \quad\longrightarrow\quad x = \beta\,(-\ln u)^{1/\alpha}\,.$$

3.4 Markov Chain Monte Carlo

With the methods we have used up to now we can simulate samples from distributions that are more or less easy to handle. Markov Chain Monte Carlo allows to sample from more complicated distributions. The basic idea is to consider each sampling as a state of a system that evolves in consecutive steps of a Markov Chain converging (asymptotically) to the desired distribution. In the simplest version were introduced by Metropolis in the 1950 s and were generalized by Hastings in the 1970s.

Let's start for simplicity with a discrete distribution. Suppose that we want a sampling of size n from the distribution

$$P(X = k) = \pi_k \quad \text{with} \qquad k = 1, 2, \ldots, N$$

that is, from the *probability vector*

$$\boldsymbol{\pi} = (\pi_1, \pi_2, \ldots, \pi_N); \quad \pi_i \in [0, 1]\, \forall i = 1, \ldots, N \;\; \text{and} \;\; \sum_{i=1}^{N} \pi_i = 1$$

and assume that it is difficult to generate a sample from this distribution by other procedures. Then, we may start from a sample of size n generated from a simpler distribution; for instance, a Discrete Uniform with

$$P_0(X = k) = \frac{1}{N}\,; \qquad \forall k$$

and from the sample obtained $\{n_1, n_2, \ldots, n_N\}$, where $n = \sum_{i=1}^{N} n_i$, we form the *initial sample probability vector*

$$\boldsymbol{\pi}^{(0)} = (\pi_1^{(0)}, \pi_2^{(0)}, \ldots, \pi_N^{(0)}) = (n_1/n, n_2/n, \ldots, n_N/n)$$

Once we have the n events distributed in the N classes of the sample space $\Omega = \{1, 2, \ldots, N\}$ we just have to redistribute them according to some criteria in different steps so that eventually we have a sample of size n drawn from the desired distribution $P(X = k) = \pi_k$.

We can consider the process of redistribution as an evolving system such that, if at step i the system is described by the probability vector $\boldsymbol{\pi}^{(i)}$, the new state at step $i + 1$, described by $\boldsymbol{\pi}^{(i+1)}$, depends only on the present state of the system (i) and not on the previous ones; that is, as a Markov Chain. Thus, we start from the state $\boldsymbol{\pi}^{(0)}$ and the aim is to find a *Transition Matrix* $\mathbf{P}$, of dimension $N \times N$, such that $\boldsymbol{\pi}^{(i+1)} = \boldsymbol{\pi}^{(i)}\mathbf{P}$ and allows us to reach the desired state $\boldsymbol{\pi}$. The matrix $\mathbf{P}$ is

$$\mathbf{P} = \begin{pmatrix} P_{11} & P_{12} & \cdots & P_{1N} \\ P_{21} & P_{22} & \cdots & P_{2N} \\ \vdots & \vdots & \ldots & \vdots \\ P_{N1} & P_{N2} & \cdots & P_{NN} \end{pmatrix}$$

where each element $(\mathbf{P})_{ij} = P(i \rightarrow j) \in [0, 1]$ represents the probability for an event in class i to move to class j in one step. Clearly, at any step in the evolution the probability that an event in class i goes to any other class $j = 1, \ldots, N$ is 1 so

$$\sum_{j=1}^{N} (\mathbf{P})_{ij} = \sum_{j=1}^{N} P(i \rightarrow j) = 1$$

and therefore is a *Probability Matrix*. If the Markov Chain is:

(1) *irreducible*; that is, all the states of the system communicate among themselves;
(2) *ergodic*; that is, the states are:

  (2.1) *recurrent*: being at one state we shall return to it at some point in the evolution with probability 1;
  (2.2) *positive*: we shall return to it in a finite number of steps in the evolution;
  (2.3) *aperiodic*: the system is not trapped in cycles;

then there is a *stationary distribution* $\boldsymbol{\pi}$ such that:

(1) $\boldsymbol{\pi} = \boldsymbol{\pi}\,\mathbf{P}$;
(2) Starting at any arbitrary state $\boldsymbol{\pi}^{(0)}$ of the system, the sequence

$$\begin{aligned}
&\boldsymbol{\pi}^{(0)}\\
&\boldsymbol{\pi}^{(1)} = \boldsymbol{\pi}^{(0)}\,\mathbf{P}\\
&\boldsymbol{\pi}^{(2)} = \boldsymbol{\pi}^{(1)}\,\mathbf{P} = \boldsymbol{\pi}^{(0)}\,\mathbf{P}^2\\
&\vdots\\
&\boldsymbol{\pi}^{(n)} = \boldsymbol{\pi}^{(0)}\,\mathbf{P}^n\\
&\vdots
\end{aligned}$$

converges asymptotically to the *fix vector* $\boldsymbol{\pi}$;

(3)

$$\lim_{n\to\infty}\mathbf{P}^n = \begin{pmatrix} \pi_1 & \pi_2 & \cdots & \pi_N \\ \pi_1 & \pi_2 & \cdots & \pi_N \\ \vdots & \vdots & \cdots & \vdots \\ \pi_1 & \pi_2 & \cdots & \pi_N \end{pmatrix}$$

There are infinite ways to choose the transition matrix $\mathbf{P}$. A *sufficient* (although not necessary) condition for this matrix to describe a Markov Chain with fixed vector $\boldsymbol{\pi}$ is that the *Detailed Balance* condition is satisfied (i.e., a reversible evolution); that is

$$\pi_i\,(\mathbf{P})_{ij} = \pi_j\,(\mathbf{P})_{ji} \iff \pi_i\,P(i\to j) = \pi_j\,P(j\to i)$$

It is clear that if this condition is satisfied, then $\boldsymbol{\pi}$ is a fixed vector since:

$$\boldsymbol{\pi}\,\mathbf{P} = \left(\sum_{i=1}^{N}\pi_i\,(\mathbf{P})_{i1}, \sum_{i=1}^{N}\pi_i\,(\mathbf{P})_{i2}, \ldots \sum_{i=1}^{N}\pi_i\,(\mathbf{P})_{iN}\right) = \boldsymbol{\pi}$$

due to the fact that

$$\sum_{i=1}^{N} \pi_i \, (\mathbf{P})_{ik} = \sum_{i=1}^{N} \pi_k \, (\mathbf{P})_{ki} = \pi_k \quad \text{for} \qquad k = 1, 2, \ldots, N$$

Imposing the *Detailed Balance* condition, we have freedom to choose the elements $(\mathbf{P})_{ij}$. We can obviously take $(\mathbf{P})_{ij} = \pi_j$ so that it is satisfied trivially $(\pi_i \pi_j = \pi_j \pi_i)$ but this means that being at class i we shall select the new possible class j with probability $P(i \to j) = \pi_j$ and, therefore, to sample directly the desired distribution that, in principle, we do not know how to do. The basic idea of Markov Chain Monte Carlo simulation is to take

$$(\mathbf{P})_{ij} = q(j|i) \cdot a_{ij}$$

where

$q(j|i)$: is a probability law to select the possible new class $j = 1, \ldots, N$ for an event that is actually in class i;

a_{ij}: is the probability to accept the proposed new class j for an event that is at i taken such that the *Detailed Balance* condition is satisfied for the desired distribution $\boldsymbol{\pi}$.

Thus, at each step in the evolution, for an event that is in class i we propose a new class j to go according to the probability $q(j|i)$ and accept the transition with probability a_{ij}. Otherwise, we reject the transition and leave the event in the class where it was. The Metropolis-Hastings [8] algorithm consists in taking the acceptance function

$$a_{ij} = \min \left\{ 1, \; \frac{\pi_j \, q(i|j)}{\pi_i \, q(j|i)} \right\}$$

It is clear that this election of a_{ij} satisfies the *Detailed Balance* condition. Indeed, if $\pi_i q(j|i) > \pi_j q(i|j)$ we have that:

$$a_{ij} = \min \left\{ 1, \; \frac{\pi_j \, q(i|j)}{\pi_i \, q(j|i)} \right\} = \frac{\pi_j \, q(i|j)}{\pi_i \, q(j|i)} \quad \text{and} \quad a_{ji} = \min \left\{ 1, \; \frac{\pi_i \, q(j|i)}{\pi_j \, q(i|j)} \right\} = 1$$

and therefore:

$$\begin{aligned} \pi_i \, (\mathbf{P})_{ij} &= \pi_i \, q(j|i) \, a_{ij} = \pi_i \, q(j|i) \, \frac{\pi_j \cdot q(i|j)}{\pi_i \cdot q(j|i)} = \\ &= \pi_j \, q(i|j) = \pi_j \, q(i|j) \, a_{ji} = \pi_j \, (\mathbf{P})_{ji} \end{aligned}$$

The same holds if $\pi_i q(j|i) < \pi_j q(i|j)$ and is trivial if both sides are equal. Clearly, if $q(i|j) = \pi_i$ then $a_{ij} = 1$ so the closer $q(i|j)$ is to the desired distribution the better.

A particularly simple case is to choose a symmetric probability $q(j|i) = q(i|j)$ [9]

$$a_{ij} = \min\left\{1, \frac{\pi_j}{\pi_i}\right\}$$

In both cases, it is clear that since the acceptance of the proposed class depends upon the ratio π_j/π_i, the normalization of the desired probability is not important.

The previous expressions are directly applicable in the case we want to sample an absolute continuous random quantity $X \sim \pi(x)$. If reversibility holds, $p(x'|x)\pi(x) = p(x|x')\pi(x')$ and therefore

$$\int_X p(x'|x)\,dx' = 1 \quad\longrightarrow\quad \int_X p(x'|x)\pi(x)\,dx = \pi(x')\int_X p(x|x')\,dx = \pi(x')$$

The transition kernel is expressed as

$$p(x'|x) \equiv p(x \rightarrow x') = q(x'|x) \cdot a(x \rightarrow x')$$

and the acceptance probability given by

$$a(x \rightarrow x') = \min\left\{1, \frac{\pi(x')\,q(x|x')}{\pi(x)\,q(x'|x)}\right\}$$

Let's see one example.

Example 3.11 (*The Binomial Distribution*) Suppose we want a sampling of size n of the random quantity $X \sim Bi(x|N, \theta)$ Since $x = 0, 1, \ldots, N$ we have $i = 1, 2, \ldots, N + 1$ classes and the desired probability vector, of dimension $N + 1$, is

$$\boldsymbol{\pi} = (p_0,\, p_1, \ldots, p_N) \quad \text{where} \quad p_k = P(X = k|\cdot) = \binom{N}{k}\,\theta^k\,(1-\theta)^{N-k}$$

Let's take for this example $N = 10$ (that is, 11 classes), $\theta = 0.45$ and $n = 100{,}000$. We start from a sampling of size n from a uniform distribution (Fig. 3.5(1)). At each step of the evolution we swap over the n generated events. For and event that is in bin j we choose a new possible bin to go $j = 1, \ldots, 10$ with uniform probability $q(j|i)$. Suppose that we look at an event in bin $i = 7$ and choose j with equal probability among the 10 possible bins. If, for instance, $j = 2$, then we accept the move with probability

$$a_{72} = a(7 \rightarrow 2)\min\left(1, \frac{\pi_2 = p_2}{\pi_7 = p_7}\right) = 0.026$$

if, on the other hand, we have $j = 6$,

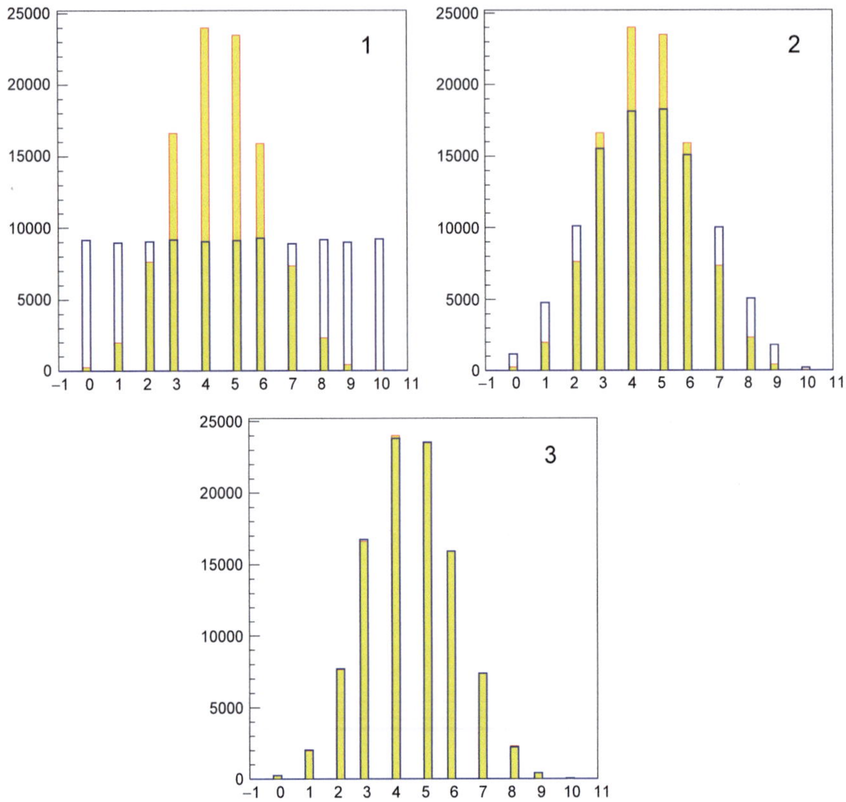

Fig. 3.5 Distributions at steps 0, 2 and 100 (**1**, **2**, **3**; *blue*) of the Markov Chain with the desired Binomial distribution superimposed in *yellow*

$$a_{76} = a(7 \to 6) \min\left(1, \frac{\pi_6 = p_6}{\pi_7 = p_7}\right) = 1.$$

so we make the move of the event. After two swaps over all the sample we have the distribution shown in Fig. 3.5(2) and after 100 swaps that shown in Fig. 3.5(3), both compared to the desired distribution:

$$\begin{aligned}
\pi^{0)} &= (0.091, 0.090, 0.090, 0.092, 0.091, 0.091, 0.093, 0.089, 0.092, 0.090, 0.092) \\
\pi^{2)} &= (0.012, 0.048, 0.101, 0.155, 0.181, 0.182, 0.151, 0.100, 0.050, 0.018, 0.002) \\
\pi^{100)} &= (0.002, 0.020, 0.077, 0.167, 0.238, 0.235, 0.159, 0.074, 0.022, 0.004, 0.000) \\
\pi &= (0.000, 0.021, 0.076, 0.166, 0.238, 0.234, 0.160, 0.075, 0.023, 0.004, 0.000)
\end{aligned}$$

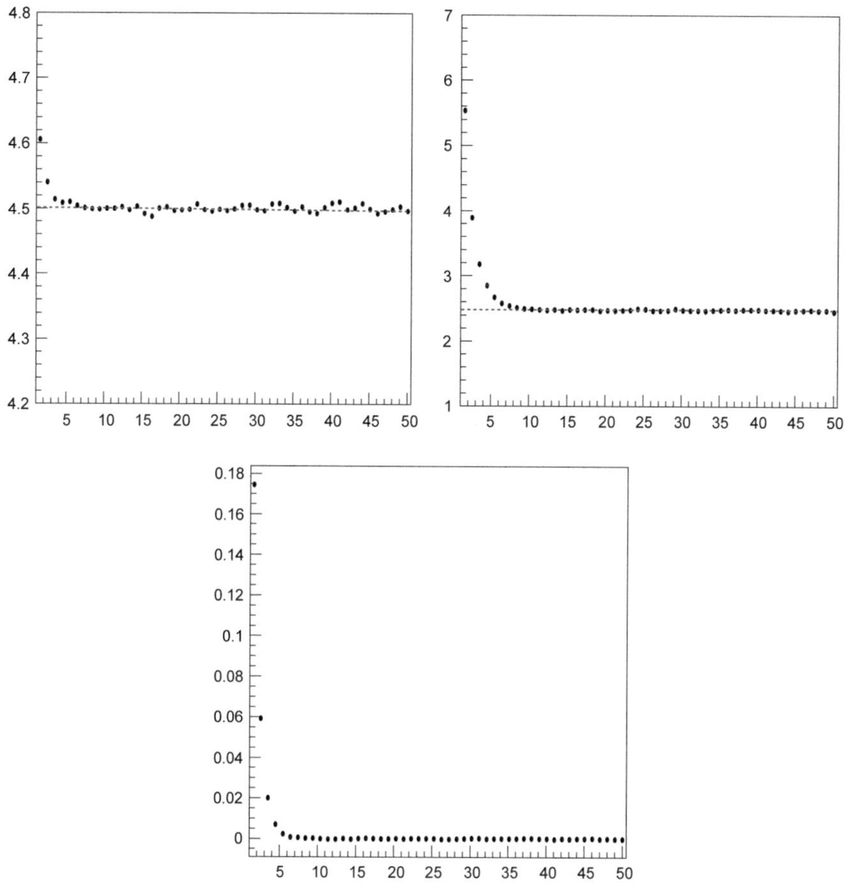

Fig. 3.6 Distributions of the mean value, variance and logarithmic discrepancy vs the number of steps. For the first two, the *red line* indicates what is expected for the Binomial distribution

The evolution of the moments, in this case the mean value and the variance with the number of steps is shown in Fig. 3.6 together with the Kullback-Leibler logarithmic discrepancy between each state and the new one defined as

$$\delta_{KL}\{\boldsymbol{\pi}|\boldsymbol{\pi}^{n)}\} = \sum_{k=1}^{10} \pi_k^{n)} \ln \frac{\pi_k^{n)}}{\pi_k}$$

As the previous example shows, we have to let the system evolve some steps (i.e., some initial sweeps for *"burn-out" or "thermalization"*) to reach the stable running conditions and get close to the stationary distribution after starting from an arbitrary state. Once this is achieved, each step in the evolution will be a sampling from the desired distribution so we do not have necessarily to generate a sample of the desired size to start with. In fact, we usually don't do that; we choose one admissible state and let the system evolve. Thus, for instance if we want a sample of $X \sim p(x|\cdot)$ with $x \in \Omega_X$, we may start with a value $x_0 \in \Omega_X$. At a given step i the system will be in the state $\{x\}$ and at the step $i+1$ the system will be in a new state $\{x'\}$ if we accept the change $x \to x'$ or in the state $\{x\}$ if we do not accept it. After thermalization, each trial will simulate a sampling of $X \sim p(x|\cdot)$. Obviously, the sequence of states of the system is not independent so, if correlations are important for the evaluation of the quantities of interest, it is a common practice to reduce them by taking for the evaluations one out of few steps.

As for the *thermalization* steps, there is no universal criteria to tell whether stable conditions have been achieved. One may look, for instance, at the evolution of the discrepancy between the desired probability distribution and the probability vector of the state of the system and at the moments of the distribution evaluated with a fraction of the last steps. More details about that are given in [10]. It is interesting also to look at the acceptance rate; i.e. the number of accepted new values over the number of trials. If the rate is low, the proposed new values are rejected with high probability (are far away from the more likely ones) and therefore the chain will mix slowly. On the contrary, a high rate indicates that the steps are short, successive samplings move slowly around the space and therefore the convergence is slow. In both cases we should think about tuning the parameters of the generation.

Example 3.12 (*The Beta distribution*) Let's simulate a sample of size 10^7 from a Beta distribution $Be(x|4,2)$; that is:

$$\pi(x) \propto x^3\,(1-x) \quad \text{with} \quad x \in [0,1]$$

In this case, we start from the admissible state $\{x = 0.3\}$ and select a new possible state x' from the density $q(x'|x) = 2x'$; not symmetric and independent of x. Thus we generate a new possible state as

$$F_q(x') = \int_0^{x'} q(s|x)ds = \int_0^{x'} 2s ds = x'^2 \quad \longrightarrow \quad x' = u^{1/2} \quad \text{with} \quad u \Leftarrow Un(0,1)$$

The acceptance function will then be

$$a(x \to x') = \min\left\{1,\ \frac{\pi(x')\cdot q(x|x')}{\pi(x)\cdot q(x'|x)}\right\} = \min\left\{1,\ \frac{x'^2(1-x')}{x^2(1-x)}\right\}$$

depending on which we set at the state $i+1$ the system in x or x'. After evolving the system for thermalization, the distribution is shown in Fig. 3.7 where we have taken one x value out of 5 consecutive ones. The red line shows the desired distribution $Be(x|4,2)$.

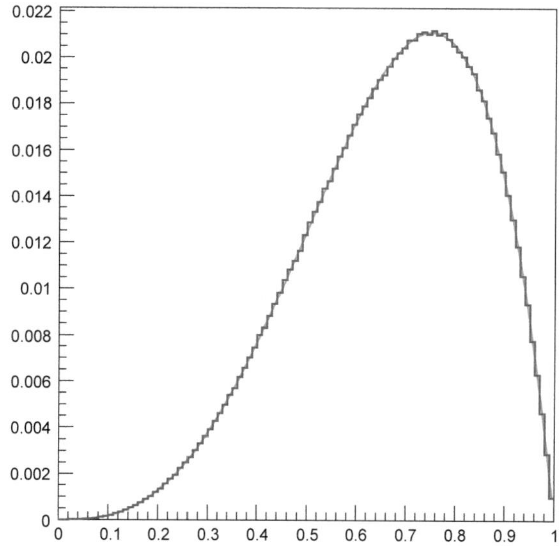

Fig. 3.7 Sampling of the Beta distribution $Be(x|4, 2)$ (*blue*) of the Example 3.12 compared to the desired distribution (continuous line) (color figure online)

Example 3.13 (*Path Integrals in Quantum Mechanics*)

In Feynman's formulation of non-relativistic Quantum Mechanics, the probability amplitude to find a particle in x_f at time t_f when at t_i was at x_i is given by

$$K(x_f, t_f|x_i, t_i) = \int_{paths} e^{i/\hbar\, S[x(t)]}\, D[x(t)]$$

where the integral is performed over all possible trajectories $x(t)$ that connect the initial state $x_i = x(t_i)$ with the final state $x_f = x(t_f)$, $S[x(t)]$ is the classic action functional

$$S[x(t)] = \int_{t_i}^{t_f} L(\dot{x}, x, t)\, dt$$

that corresponds to each trajectory and $L(\dot{x}, x, t)$ is the Lagrangian of the particle. All trajectories contribute to the amplitude with the same weight but different phase. In principle, small differences in the trajectories cause big changes in the action compared to $\hbar$ and, due to the oscillatory nature of the phase, their contributions cancel. However, the action does not change, to first order, for the trajectories in a neighborhood of the one for which the action is extremal and, since they have similar phases (compared to $\hbar$) their contributions will not cancel. The set of trajectories around the extremal one that produce changes in the action of the order of $\hbar$ define the limits of classical mechanics and allow to recover is laws expressed as the *Extremal Action Principle*.

The transition amplitude (*propagator*) allows to get the wave-function $\Psi(x_f, t_f)$ from $\Psi(x_i, t_i)$ as:

$$\Psi(x_f, t_f) = \int K(x_f, t_f|x_i, t_i)\,\Psi(x_i, t_i)\,dx_i \qquad \text{for} \qquad t_f > t_i$$

In non-relativistic Quantum Mechanics there are no trajectories evolving backwards in time so in the definition of the propagator a Heaviside step function $\theta(t_f - t_i)$ is implicit. Is this clear from this equation that $K(x_f, t|x_i, t) = \delta(x_f - x_i)$.

For a local Lagrangian (additive actions), it holds that:

$$K(x_f, t_f|x_i, t_i) = \int K(x_f, t_f|x, t)\,K(x, t|x_i, t_i)\,dx$$

analogous expression to the Chapman-Kolmogorov equations that are satisfied by the conditional probabilities of a Markov process. If the Lagrangian is not local, the evolution of the system will depend on the intermediate states and this equation will not be true. On the other hand, if the Classical Lagrangian has no explicit time dependence the propagator admits an expansion (*Feynman-Kac Expansion Theorem*) in terms of a compete set of eigenfunctions $\{\phi_n\}$ of the Hamiltonian as:

$$K(x_f, t_f|x_i, t_i) = \sum_n e^{-i/\hbar\,E_n\,(t_f - t_i)}\,\phi_n(x_f)\,\phi_n^\star(x_i)$$

where the sum is understood as a sum for discrete eigenvalues and as an integral for continuous eigenvalues. Last, remember that expected value of an operator $A(x)$ is given by:

$$\langle A \rangle = \int A[x(t)]\,e^{i/\hbar\,S[x(t)]}\,D[x(t)] \Big/ \int e^{i/\hbar\,S[x(t)]}\,D[x(t)]$$

Let's see how to do the integral over paths to get the propagator in a one-dimensional problem. For a particle that follows a trajectory $x(t)$ between $x_i = x(t_i)$ and $x_f = x(t_f)$ under the action of a potential $V(x(t))$, the Lagrangian is:

$$L(\dot{x}, x, t) = \frac{1}{2}\,m\,\dot{x}(t)^2 - V(x(t))$$

and the corresponding action:

$$S[x(t)] = \int_{t_i}^{t_f} \left(\frac{1}{2}\,m\,\dot{x}(t)^2 - V(x(t))\right) dt$$

so we have for the propagator:

$$\begin{aligned} K(x_f, t_f|x_i, t_i) &= \int_{Tr} e^{i/\hbar\,S[x(t)]}\,D[x(t)] = \\ &= \int_{Tr} \exp\left\{\frac{i}{\hbar}\int_{t_i}^{t_f}\left(\frac{1}{2}\,m\,\dot{x}(t)^2 - V(x(t))\right) dt\right\}\,D[x(t)] \end{aligned}$$

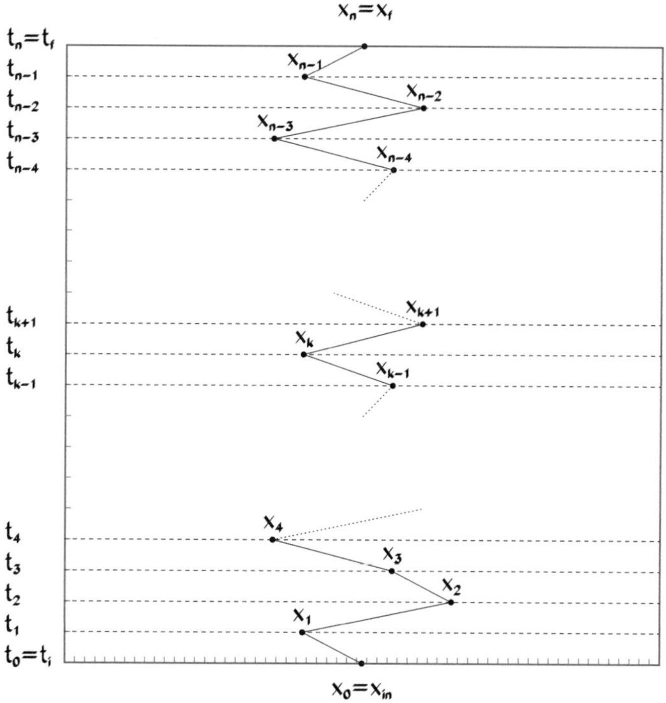

Fig. 3.8 Trajectory in a space with discretized time

where the integral is performed over the set Tr of all possible trajectories that start at $x_i = x(t_i)$ and end at $x_f = x(t_f)$. Following Feynman, a way to perform this integrals is to make a partition of the interval (t_i, t_f) in N subintervals of equal length ϵ (Fig. 3.8); that is, with

$$\epsilon = \frac{t_f - t_i}{N} \qquad \text{so that} \qquad t_j - t_{j-1} = \epsilon; \quad j = 1, 2, \dots, N$$

Thus, if we identify $t_0 = t_i$ and $t_N = t_f$ we have that

$$[t_i, t_f) = \cup_{j=0}^{N-1} [t_j, t_{j+1})$$

On each interval $[t_j, t_{j+1})$, the possible trajectories $x(t)$ are approximated by straight segments so they are defined by the sequence

$$\{x_0 = x_i = x(t_i),\ x_1 = x(t_1),\ x_2 = x(t_2), \dots\ x_{N-1} = x(t_{N-1}),\ x_N = x_f = x(t_f)\}$$

Obviously, the trajectories so defined are continuous but not differentiable so we have to redefine the velocity. An appropriate prescription is to substitute $\dot{x}(t_j)$ by

$$\dot{x}(t_j) \longrightarrow \frac{x_j - x_{j-1}}{\epsilon}$$

so the action is finally expressed as:

$$S_N[x(t)] = \epsilon \sum_{j=1}^{N} \left[\frac{1}{2} m \left(\frac{x_j - x_{j-1}}{\epsilon} \right)^2 - V(x_j) \right]$$

Last, the integral over all possible trajectories that start at x_0 and end at x_N is translated in this space with discretized time axis as an integral over the quantities $x_1, x_2, \ldots, x_{N-1}$ so the differential measure for the trajectories $D[x(t)]$ is substituted by

$$D[x(t)] \longrightarrow A_N \prod_{j=1}^{j=N-1} dx_j$$

with A_N a normalization factor. The propagator is finally expressed as:

$$K_N(x_f, t_f | x_i, t_i) =$$

$$= A_N \int_{x_1} dx_1 \cdots \int_{x_{N-1}} dx_{N-1} \exp \left\{ \frac{i}{\hbar} \sum_{j=1}^{N} \left[\frac{1}{2} m \left(\frac{x_j - x_{j-1}}{\epsilon} \right)^2 - V(x_j) \right] \cdot \epsilon \right\}$$

After doing the integrals, taking the limit $\epsilon \rightarrow 0$ (or $N \rightarrow \infty$ since the product $N\epsilon = (t_f - t_i)$ is fixed) we get the expression of the propagator. Last, note that the interpretation of the integral over trajectories as the limit of a multiple Riemann integral is valid only in Cartesian coordinates.

To derive the propagator from path integrals is a complex problem and there are few potentials that can be treated exactly (the "*simple*" Coulomb potential for instance was solved in 1982). The Monte Carlo method allows to attack satisfactorily this type of problems but before we have first to convert the complex integral in a positive real function. Since the propagator is an analytic function of time, it can be extended to the whole complex plane of t and then perform a rotation of the time axis (*Wick's rotation*) integrating along

$$\tau = e^{i\pi/2}\, t = i\, t; \text{ that is, } \quad t \longrightarrow -i\, \tau$$

Taking as prescription the analytical extension over the imaginary time axis, the oscillatory exponentials are converted to decreasing exponentials, the results are consistent with those derived by other formulations (Schrodinger or Heisenberg for instance) and it is manifest the analogy with the partition function of Statistical Mechanics. Then, the action is expressed as:

$$S[x(t)] \longrightarrow i \int_{\tau_i}^{\tau_f} \left(\frac{1}{2} m\, \dot{x}(t)^2 + V(x(t)) \right) dt$$

Note that the integration limits are real as corresponds to integrate along the imaginary axis and not to just a simple change of variables. After partitioning the time interval, the propagator is expressed as:

$$K_N(x_f, t_f | x_i, t_i) = A_N \int_{x_1} dx_1 \cdots \int_{x_{N-1}} dx_{N-1} \, \exp\left\{-\frac{1}{\hbar} S_N(x_0, x_1, \ldots, x_N)\right\}$$

where

$$S_N(x_0, x_1, \ldots, x_N) = \sum_{j=1}^{N} \left[\frac{1}{2} m \left(\frac{x_j - x_{j-1}}{\epsilon}\right)^2 + V(x_j)\right] \cdot \epsilon$$

and the expected value of an operator $A(x)$ will be given by:

$$\langle A \rangle = \frac{\int \prod_{j=1}^{j=N-1} dx_j \, A(x_0, x_1, \ldots, x_N) \, \exp\left\{-\frac{1}{\hbar} S_N(x_0, x_1, \ldots, x_N)\right\}}{\int \prod_{j=1}^{j=N-1} dx_j \, \exp\left\{-\frac{1}{\hbar} S_N(x_0, x_1, \ldots, x_N)\right\}}$$

Our goal is to generate N_{gen} trajectories with the Metropolis criteria according to

$$p(x_0, x_1, \ldots, x_N) \propto \exp\left\{-\frac{1}{\hbar} S_N(x_0, x_1, \ldots, x_N)\right\}$$

Then, over these trajectories we shall evaluate the expected value of the operators of interest $A(x)$

$$\langle A \rangle = \frac{1}{N_{gen}} \sum_{k=1}^{N_{gen}} A(x_0, x_1^{(k)}, \ldots, x_{N-1}^{(k)}, x_N)$$

Last, note that if we take $(\tau_f, x_f) = (\tau, x)$ and $(\tau_i, x_i) = (0, x)$ in the Feynman-Kac expansion we have that

$$K(x, \tau | x, 0) = \sum_n e^{-1/\hbar \, E_n \, \tau} \, \phi_n(x) \, \phi_n^\star(x)$$

and therefore, for sufficiently large times

$$K(x, \tau | x, 0) \approx e^{-1/\hbar \, E_0 \, \tau} \, \phi_0(x) \, \phi_0^\star(x) + \cdots$$

so basically only the fundamental state will contribute.

Well, now we have everything we need. Let's apply all that first to an harmonic potential

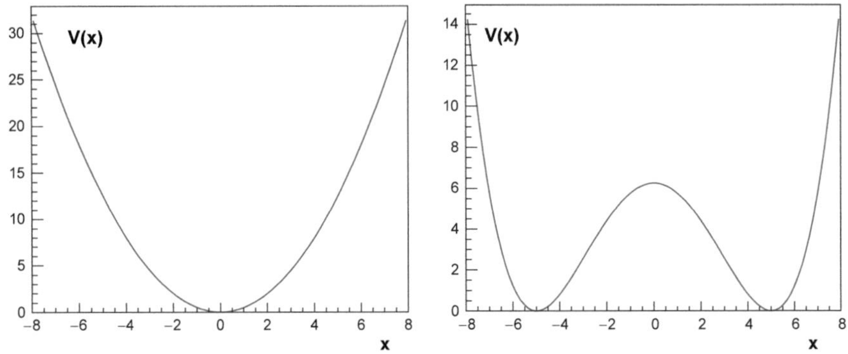

Fig. 3.9 Potential wells studied in the Example 3.13

$$V(x) = \frac{1}{2}\,k\,x^2$$

so the discretized action will be:

$$S_N(x_0, x_1, \ldots, x_N) = \sum_{j=1}^{N} \left[\frac{1}{2}\, m \left(\frac{x_j - x_{j-1}}{\epsilon} \right)^2 + \frac{1}{2}\, k\, x_j^2 \right] \cdot \epsilon$$

To estimate the energy of the fundamental state we use the *Virial Theorem*. Since:

$$\langle T \rangle_\Psi = \frac{1}{2}\, \langle \vec{x} \cdot \vec{\nabla} V(\vec{x}) \rangle_\Psi$$

we have that $\langle T \rangle_\Psi = \langle V \rangle_\Psi$ and therefore

$$\langle E \rangle_\Psi \;=\; \langle T \rangle_\Psi + \langle V \rangle_\Psi = k\,\langle x^2 \rangle_\Psi$$

In this example we shall take $m = k = 1$ (Fig. 3.9).

We start with an initial trajectory from $x_0 = x(t_i) = 0$ to $x_f = x(t_f) = 0$ and the intermediate values $x_1, x_2, \ldots, x_{N-1}$ drawn from $Un(-10., 10.)$, sufficiently large in this case since their support is $(-\infty, \infty)$. For the parameters of the grid we took $\epsilon = 0.25$ and $N = 2000$. The parameter ϵ has to be small enough so that the results we obtain are close to what we should have for a continuous time and N sufficiently large so that $\tau = N\epsilon$ is large enough to isolate the contribution of the fundamental state. With this election we have that $\tau = 2000{\cdot}0.25 = 500$. Obviously, we have to check the stability of the result varying both parameters. Once the grid is fixed, we sweep over all the points $x_1, x_2, \ldots, x_{N-1}$ of the trajectory and for each x_j, $j = 1, \ldots, N-1$ we propose a new candidate $x_j^{'}$ with support Δ. Then, taking $\hbar = 1$ we have that:

$$P(x_j \longrightarrow x_j^{'}) = \exp\left\{-S_N(x_0, x_1, \ldots x_j^{'}, \ldots, x_N)\right\}$$

and

$$P(x_j \longrightarrow x_j) = \exp\left\{-S_N(x_0, x_1, \ldots x_j, \ldots, x_N)\right\}$$

so the acceptance function will be:

$$a(x_j \longrightarrow x_j^{'}) = \min\left\{1, \frac{P(x_j \longrightarrow x_j^{'})}{P(x_j \longrightarrow x_j)}\right\}$$

Obviously we do not have to evaluate the sum over all the nodes because when dealing with node j, only the intervals (x_{j-1}, x_j) and (x_j, x_{j+1}) contribute to the sum. Thus, at node j we have to evaluate

$$a(x_j \longrightarrow x_j^{'}) = \min\left\{1, \exp\left\{-S_N(x_{j-1}, x_j^{'}, x_{j+1}) + S_N(x_{j-1}, x_j, x_{j+1})\right\}\right\}$$

Last, the trajectories obtained with the Metropolis algorithm will follow eventually the desired distribution $p(x_0, x_1, \ldots, x_N)$ in the asymptotic limit. To have a reasonable approximation to that we shall not use the first N_{term} trajectories (*thermalization*). In this case we have taken $N_{term} = 1000$ and again, we should check the stability of the result. After this, we have generated $N_{gen} = 3000$ and, to reduce correlations, we took one out of three for the evaluations; that is $N_{used} = 1000$ trajectories, each one determined by $N = 2000$ nodes. The distribution of the accepted values x_j will be an approximation to the probability to find the particle at position x for the fundamental state; that is, $|\Psi_0(x)|^2$. Figure 3.10 shows the results of the simulation compared to

$$|\Psi_0(x)|^2 \propto e^{-x^2}$$

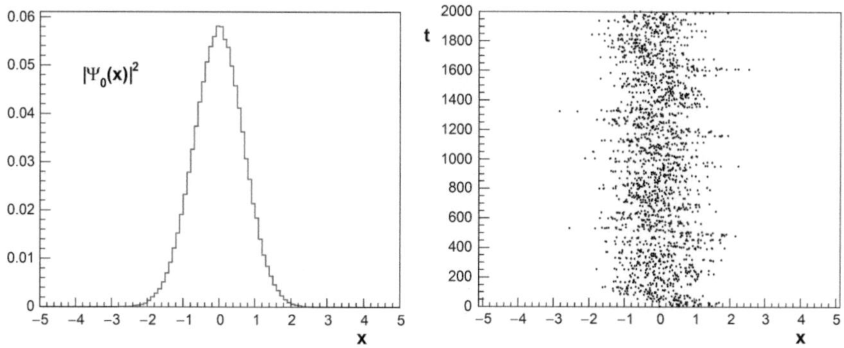

Fig. 3.10 Squared norm of the fundamental state wave-function for the harmonic potential and one of the simulated trajectories

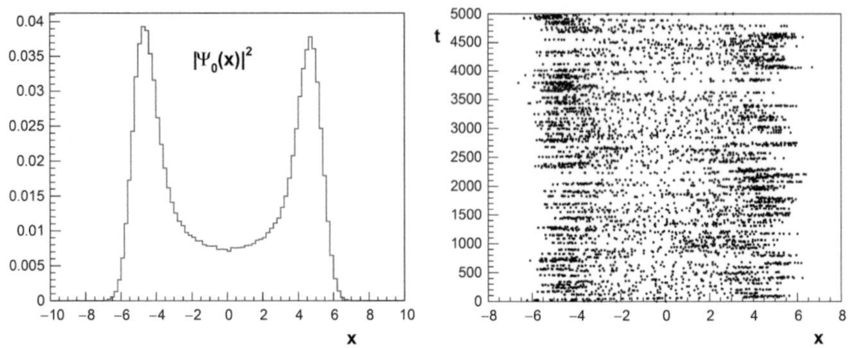

Fig. 3.11 Squared norm of the fundamental state wave-function for the quadratic potential and one of the simulated trajectories

together with one of the many trajectories generated. The sampling average $\langle x^2 \rangle = 0.486$ is a good approximation to the energy of the fundamental state $E_0 = 0.5$.

As a second example, we have considered the potential well (Fig. 3.9)

$$V(x) = \frac{a^2}{4} \left[(x/a)^2 - 1 \right]^2$$

and, again from the Virial Theorem:

$$\langle E \rangle_\Psi = \frac{3}{4a^2} \langle x^4 \rangle_\Psi - \langle x^2 \rangle_\Psi + \frac{a^2}{4}$$

We took $a = 5$, a grid of $N = 9000$ nodes and $\epsilon = 0.25$ (so $\tau = 2250$), and as before $N_{term} = 1000$, $N_{gen} = 3000$ and $N_{used} = 1000$. From the generated trajectories we have the sample moments $\langle x^2 \rangle = 16.4264$ and $\langle x^4 \rangle = 361.4756$ so the estimated fundamental state energy is $\langle E_0 \rangle = 0.668$ to be compared with the exact result $E_0 = 0.697$. The norm of the wave-function for the fundamental state is shown in Fig. 3.11 together with one of the simulated trajectories exhibiting the tunneling between the two wells.

3.4.1 Sampling from Conditionals and Gibbs Sampling

In many cases, the distribution of an n-dimensional random quantity is either not known explicitly or difficult to sample directly but sampling the conditionals is easier. In fact, sometimes it may help to introduce an additional random quantity an consider the conditional densities (see the Example 3.14). Consider then the n-dimensional random quantity $\mathbf{X} = (X_1, \ldots, X_n)$ with density $p(x_1, \ldots, x_n)$, the usually simpler conditional densities

$$
\begin{aligned}
&p(x_1|x_2, x_3 \ldots, x_n) \\
&p(x_2|x_1, x_3 \ldots, x_n) \\
&\quad\vdots \\
&p(x_n|x_1, x_2, \ldots, x_{n-1})
\end{aligned}
$$

and an arbitrary initial value $\boldsymbol{x}^{0)} = \{x_1^{0)}, x_2^{0)}, \ldots, x_n^{0)}\} \in \Omega_{\mathbf{X}}$. If we take the approximating density $q(x_1, x_2, \ldots, x_n)$ and the conditional densities

$$
\begin{aligned}
&q(x_1|x_2, x_3, \ldots, x_n) \\
&q(x_2|x_1, x_3, \ldots, x_n) \\
&\quad\vdots \\
&q(x_n|x_1, x_2, \ldots, x_{n-1})
\end{aligned}
$$

we generate for x_1 a proposed new value $x_1^{1)}$ from $q(x_1|x_2^{0)}, x_3^{0)}, \ldots, x_n^{0)})$ and accept the change with probability

$$
\begin{aligned}
a(x_1^{0)} \rightarrow x_1^{1)}) &= \min\left\{1, \frac{p(x_1^{1)}, x_2^{0)}, x_3^{0)}, \ldots, x_n^{0)})q(x_1^{0)}, x_2^{0)}, x_3^{0)}, \ldots, x_n^{0)})}{p(x_1^{0)}, x_2^{0)}, x_3^{0)}, \ldots, x_n^{0)})q(x_1^{1)}, x_2^{0)}, x_3^{0)}, \ldots, x_n^{0)})}\right\} = \\
&= \min\left\{1, \frac{p(x_1^{1)}|x_2^{0)}, x_3^{0)}, \ldots, x_n^{0)})q(x_1^{0)}|x_2^{0)}, x_3^{0)}, \ldots, x_n^{0)})}{p(x_1^{0)}|x_2^{0)}, x_3^{0)}, \ldots, x_n^{0)})q(x_1^{1)}|x_2^{0)}, x_3^{0)}, \ldots, x_n^{0)})}\right\}
\end{aligned}
$$

After this step, let's denote the value of x_1 by x'_1 (that is, $x'_1 = x_1^{1)}$ or $x'_1 = x_1^{0)}$ if it was not accepted). Then, we proceed with x_2. We generate a proposed new value $x_2^{1)}$ from $q(x_2|x'_1, x_3^{0)}, \ldots, x_n^{0)})$ and accept the change with probability

$$
\begin{aligned}
a(x_2^{0)} \rightarrow x_2^{1)}) &= \min\left\{1, \frac{p(x'_1, x_2^{1)}, x_3^{0)}, \ldots, x_n^{0)})q(x'_1, x_2^{0)}, x_3^{0)}, \ldots, x_n^{0)})}{p(x'_1, x_2^{0)}, x_3^{0)}, \ldots, x_n^{0)})q(x'_1, x_2^{1)}, x_3^{0)}, \ldots, x_n^{0)})}\right\} = \\
&= \min\left\{1, \frac{p(x_2^{1)}|x'_1, x_3^{0)}, \ldots, x_n^{0)})q(x_2^{0)}|x'_1, x_3^{0)}, \ldots, x_n^{0)})}{p(x_2^{0)}|x'_1, x_3^{0)}, \ldots, x_n^{0)})q(x_2^{1)}|x'_1 x_3^{0)}, \ldots, x_n^{0)})}\right\}
\end{aligned}
$$

After we run over all the variables, we are in a new state $\{x'_1, x'_2, \ldots, x'_n\}$ and repeat the whole procedure until we consider that stability has been reached so that we are sufficiently close to sample the desired density. The same procedure can be applied if we consider more convenient to express the density

$$
p(x_1, x_2, x_3 \ldots, x_n) = p(x_n|x_{n-1}, \ldots, x_2 \ldots, x_1) \cdots p(x_2|x_1)\, p(x_1)
$$

Obviously, we need only one one admissible starting value $x_1^{0)}$.

Gibbs sampling is a particular case of this approach and consists on sampling sequentially all the random quantities directly from the conditional densities; that is:

$$q(x_i|x_1, \ldots, x_{i-1}, x_{i+1}, \ldots, x_n) = p(x_i|x_1, \ldots, x_{i-1}, x_{i+1}, \ldots, x_n)$$

so the acceptance factor $a(x \to x') = 1$. This is particularly useful for Bayesian inference since, in more than one dimension, densities are usually specified in conditional form after the ordering of parameters.

Example 3.14 Sometimes it may be convenient to introduce additional random quantities in the problem to ease the treatment. Look for instance at the Student's distribution $X \sim St(x|\nu)$ with

$$p(x|\nu) \propto \left(1 + x^2/\nu\right)^{-(\nu+1)/2}$$

Since

$$\int_0^{\infty} e^{-au} u^{b-1} du = \Gamma(b)\, a^{-b}$$

we can introduce an additional random quantity $U \sim Ga(u|a, b)$ in the problem with $a = 1 + x^2/\nu$ and $b = (\nu + 1)/2$ so that

$$p(x, u|\nu) \propto e^{-au} u^{b-1} \quad \text{and} \quad p(x) \propto \int_0^{\infty} p(x, u|\nu) du \propto a^{-b} = \left(1 + x^2/\nu\right)^{-(\nu+1)/2}$$

The conditional densities are

$$p(x|u, \nu) = \frac{p(x, u|\nu)}{p(u|\nu)} = N(x|0, \sigma); \qquad \sigma^2 = \nu(2u)^{-1}$$
$$p(u|x, \nu) = \frac{p(x, u|\nu)}{p(x|\nu)} = Ga(u|a, b); \qquad a = 1 + x^2/\nu\,, \; b = (\nu + 1)/2$$

so, if we start with an arbitrary initial value $x \in \mathcal{R}$, we can

(1) Sample $U|X$: $u \Leftarrow Ga(u|a, b)$ with $a = 1 + x^2/\nu$ and $b = (\nu + 1)/2$
(2) Sample $X|U$: $x \Leftarrow N(x|0, \sigma)$ with $\sigma^2 = \nu(2u)^{-1}$

and repeat the procedure so that, after equilibrium, $X \sim St(x|\nu)$. We can obviously start from $u \in \mathcal{R}$ and reverse the steps (1) and (2). Thus, instead of sampling from the Student's distribution we may sample the conditional densities: Normal and a Gamma distributions. Following this approach for $\nu = 2$ and 10^3 thermalization sweeps (far beyond the needs), the results of 10^6 draws are shown in Fig. 3.12 together with the Student's distribution $St(x|2)$.

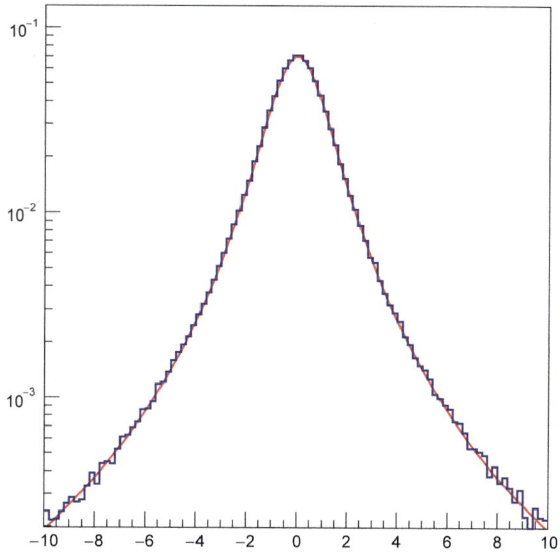

Fig. 3.12 Sampling of the Student's Distribution $St(x|2)$ (*blue*) compared to the desired distribution (*red*)

Example 3.15 We have $j = 1, \ldots, J$ groups of observations each with a sample of size n_j; that is $\boldsymbol{x}_j = \{x_{1j}, x_{2j}, \ldots, , x_{n_j j}\}$. Within each of the J groups, observations are considered an exchangeable sequence and assumed to be drawn from a distribution $x_{i,j} \sim N(x|\mu_j, \sigma^2)$ where $i = 1, \ldots, n_j$. Then:

$$p(\boldsymbol{x}_j|\mu_j, \sigma) = \prod_{i=1}^{n_j} N(x_{ij}|\mu_j, \sigma^2) \propto \sigma^{-n_j} \exp\left\{-\sum_{i=1}^{n_j} \frac{(x_{ij}-\mu_j)^2}{2\,\sigma^2}\right\}$$

Then, for the J groups we have the parameters $\boldsymbol{\mu} = \{\mu_1, \ldots \mu_J\}$ that, in turn, are also considered as an exchangeable sequence drawn from a parent distribution $\mu_j \sim N(\mu_j|\mu, \sigma_\mu^2)$. We reparameterize the model in terms of $\eta = \sigma_\mu^{-2}$ and $\phi = \sigma^{-2}$ and consider conjugated priors for the parameters considered independent; that is

$$\pi(\mu, \eta, \phi) = N(\mu|\mu_0, \sigma_0^2)\; Ga(\eta|c, d)\; Ga(\phi|a, b)$$

Introducing the sample means $\overline{x}_j = n_j^{-1} \sum_{i=1}^{n_j} x_{ij}$, $\overline{x} = J^{-1} \sum_{j=1}^{J} \overline{x}_j$ and defining $\overline{\mu} = J^{-1} \sum_{j=1}^{J} \mu_j$ and defining After some simple algebra it is easy to see that the marginal densities are:

$$\begin{aligned}
\mu_j &\sim N\left(\frac{n_j\sigma_\mu^2\overline{x}_j+\mu\sigma^2}{n_j\sigma_\mu^2+\sigma^2},\frac{\sigma_\mu^2\sigma^2}{n_j\sigma_\mu^2+\sigma^2}\right)\\
\mu &\sim N\left(\frac{\sigma_\mu^2\mu_0+\sigma_0^2J\overline{\mu}}{\sigma_\mu^2+J\sigma_0^2},\frac{\sigma_\mu^2\sigma_0^2}{\sigma_\mu^2+J\sigma_0^2}\right)\\
\eta=\sigma_\mu^{-2} &\sim Ga\left(\frac{1}{2}\sum_{j=1}^{J}(\mu_j-\mu)^2+c,\ \frac{J}{2}+d\right)\\
\phi=\sigma^{-2} &\sim Ga\left(\frac{1}{2}\sum_{j=1}^{J}\sum_{i=1}^{n_j}(x_{ij}-\mu_j)^2+a,\ \frac{1}{2}\sum_{j=1}^{J}n_j+b\right)
\end{aligned}$$

Thus, we set initially the parameters $\{\mu_0,\sigma_0,a,b,c,d\}$ and then, at each step

1. Get $\{\mu_1,\ldots,\mu_J\}$ each as $\mu_j\sim N(\cdot,\cdot)$
2. Get $\mu\sim N(\cdot,\cdot)$
3. Get $\sigma_\mu=\eta^{-1/2}$ with $\eta\sim Ga(\cdot,\cdot)$
4. Get $\sigma=\phi^{-1/2}$ with $\phi\sim Ga(\cdot,\cdot)$

and repeat the sequence until equilibrium is reached and samplings for evaluations can be done.

3.5 Evaluation of Definite Integrals

A frequent use of Monte Carlo sampling is the evaluation of definite integrals. Certainly, there are many numerical methods for this purpose and for low dimensions they usually give a better precision when fairly compared. In those cases one rarely uses Monte Carlo… although sometimes the domain of integration has a very complicated expression and the Monte Carlo implementation is far easier. However, as we have seen the uncertainty of Monte Carlo estimations decreases with the sampling size N as $1/\sqrt{N}$ regardless the number of dimensions so, at some point, it becomes superior. And, besides that, it is fairly easy to estimate the accuracy of the evaluation. Let's see in this section the main ideas.

Suppose we have the n-dimensional definite integral

$$I=\int_\Omega f(x_1,x_2,\ldots,x_n)\,dx_1\,dx_2\,\ldots\,dx_n$$

where $(x_1,x_2,\ldots,x_n)\in\Omega$ and $f(x_1,x_2,\ldots,x_n)$ is a Riemann integrable function. If we consider a random quantity $\boldsymbol{X}=(X_1,X_2,\ldots,X_n)$ with distribution function $P(\boldsymbol{x})$ and support in Ω, the mathematical expectation of $\boldsymbol{Y}=g(\boldsymbol{X})\equiv f(\boldsymbol{X})/p(\boldsymbol{X})$ is given by

$$E[Y] = \int_{\Omega} g(\boldsymbol{x})\, dP(\boldsymbol{x}) = \int_{\Omega} \frac{f(\boldsymbol{x})}{p(\boldsymbol{x})}\, dP(\boldsymbol{x}) = \int_{\Omega} f(\boldsymbol{x})\, d\boldsymbol{x} = I$$

Thus, if we have a sampling $\{\boldsymbol{x}_1, \boldsymbol{x}_2, \ldots, \boldsymbol{x}_N\}$, of size N, of the random quantity $\boldsymbol{X}$ under $P(\boldsymbol{x})$ we know, by the Law of Large Numbers that, as $N \to \infty$, the sample means

$$I_N^{(1)} = \frac{1}{N} \sum_{i=1}^{N} g(\boldsymbol{x}_i) \quad \text{and} \quad I_N^{(2)} = \frac{1}{N} \sum_{i=1}^{N} g^2(\boldsymbol{x}_i)$$

converge respectively to $E[Y]$ (and therefore to I) and to $E[Y^2]$ (as for the rest, all needed conditions for existence are assumed to hold). Furthermore, if we define

$$SI^2 = \frac{1}{N} \left(I_K^{(2)} - (I_K^{(1)})^2 \right)$$

we know by the Central Limit Theorem that the random quantity

$$Z = \frac{I_K^{(1)} - I}{SI}$$

is, in the limit $N \to \infty$, distributed as $N(x|0, 1)$. Thus, Monte Carlo integration provides a simple way to estimate the integral I and a quantify the accuracy. Depending on the problem at hand, you can envisage several tricks to further improve the accuracy. For instance, if $g(\boldsymbol{x})$ is a function *"close"* to $f(\boldsymbol{x})$ with the same support and known integral I_g one can write

$$I = \int_{\Omega} f(\boldsymbol{x})\, d\boldsymbol{x} = \int_{\Omega} (f(\boldsymbol{x}) - g(\boldsymbol{x}))\, d\boldsymbol{x} + I_g$$

and in consequence estimate the value of the integral as

$$\tilde{I} = \frac{1}{N} \sum_{i=1}^{N} (f(\boldsymbol{x}_i) - g(\boldsymbol{x}_i)) + I_g$$

reducing the uncertainty.

References

1. F. James, Monte Carlo theory and practice. Rep. Prog. Phys. **43**, 1145–1189 (1980)
2. D.E. Knuth, *The Art of Computer Programming*, vol. 2 (Addison-Wesley, Menlo Park, 1981)
3. F. James, J. Hoogland, R. Kleiss, Comput. Phys. Commun. **2–3**, 180–220 (1999)
4. P. L'Ecuyer, *Handbook of Simulations, Chap. 4* (Wiley, New York, 1998)

5. G. Marsaglia, A. Zaman, *Toward a Universal Random Number Generator*, Florida State University Report FSU-SCRI-87-50 (1987)
6. D.B. Rubin, Ann. Stat. **9**, 130–134 (1981)
7. G.E.P. Box, M.E. Müller, A note on the generation of random normal deviates. Ann. Math. Stat. **29**(2), 610–611 (1958)
8. W.K. Hastings, Biometrika **57**, 97–109 (1970)
9. N. Metropolis, A.W. Rosenbluth, M.W. Rosenbluth, A.H. Teller, E. Teller, J. Chem. Phys. **21**, 1087–1092 (1953)
10. A.B. Gelman, J.S. Carlin, H.S. Stern, D.B. Rubin, *Bayesian Data Analysis* (Chapman & Hall, London, 1995)

Chapter 4
Information Theory

Sir, the reason is very plain; knowledge is of two kinds. We know the subject ourselves, or we know where we can find information upon it

S. Johnson

The ultimate goal of doing experiments and make observations is to learn about the way nature behaves and, eventually, unveil the mathematical laws governing the Universe and predict yet-unobserved phenomena. In less pedantic words, to get *information* about the natural world. *Information* plays a relevant role in a large number is disciplines (physics, mathematics, biology, image processing,…) and, in particular, it is an important concept in Bayesian Inference. It is useful for instance to quantify the similarities or differences between distributions and to evaluate the different ways we have to analyse the observed data because, in principle, not all of them provide the same amount of information on the same questions. The first steep will be to quantify the amount of information that we get from a particular observation.

4.1 Quantification of Information

When we observe the result of an experiment, we get some amount of information about the underlying random process. How can we quantify the information we have received? Let's start with a discrete random quantity X that can take the values $\{x_1, x_2, \ldots, x_k, \ldots\}$ with probabilities $p_i = p(x_i)$; $i = 1, 2, \ldots$ It is reasonable to assume that the information we get when we observe the event $X = x_i$ will depend on its probability of occurrence $p_i = p(x_i)$; that is, $I(x_i) = g(p_i)$. Now, if I tell you that I have seen a lion in a photo-safari in Kenya, you will not be surprised. In fact, it is quite

C. Maña, *Probability and Statistics for Particle Physics*,
UNITEXT for Physics, DOI 10.1007/978-3-319-55738-0_4

natural, a very likely observation, and I hardly give you any valuable information. However, if I tell you that I have seen a lone lion walking along Westminster Bridge, you will be quite surprised. This is not expected; a very unlikely observation worth of further investigation. I give you a lot of information. *Surprise is Information*. Thus, it is also sensible to assume that if the probability for an event to occur is large we receive a small amount of information and, conversely, if the probability is very small we receive a large amount of information. In fact, if the event is a sure event, $p(x_i) = 1$ and its occurrence will provide no information at all. Therefore, we start assuming two reasonable hypothesis:

H$_1$: $I(x_i) = f(1/p_i)$ with $f(x)$ a non-negative increasing function;
H$_2$: $f(1) = 0$.

Now, imagine that we repeat the experiment n-times under the same conditions and obtain the sequence of *independent* events $\{x_1, x_2, \ldots, x_n\}$. We shall assume that the information provided by this n-tuple of observed results is equal to the sum of the information provided by each observation separately; that is:

$$I(x_1, x_2, \ldots, x_n) = \sum_{=1}^{n} I(x_i)$$

Being all independent, we have that $p(x_1, x_2, \ldots, x_n) = \prod_{i=1}^{n} p(x_i)$ and, in consequence:

H$_3$: $f\left(\frac{1}{p_1 \cdots p_n}\right) = \sum_{k=1}^{n} f\left(\frac{1}{p_k}\right)$

Those are the three hypothesis we shall make. Since $p_i \in [0, 1]$, we have that $w_i = 1/p_i \in [1, \infty)$ and therefore we are looking for a function $f(w)$ such that:

(1) $f : w \in [1, \infty) \longrightarrow [0, \infty)$ and increasing;
(2) $f(1) = 0$;
(3) $f(w_1 \cdot w_2 \cdots w_n) = f(w_1) + \cdots + f(w_n)$

The third condition implies that $f(w^n) = nf(w)$ so, taking derivatives with respect to w:

$$w^n \frac{\partial f(w^n)}{\partial w^n} = w \frac{\partial f(w)}{\partial w}$$

and this has to hold for all $n \in \mathcal{N}$ and $w \in [1, \infty)$; hence, we can write

$$w \frac{\partial f(w)}{\partial w} = c \qquad \longrightarrow \qquad f(w) = c \log w$$

with c a positive constant and $f(w)$ an increasing function since $w \geq 1$. Taking $c = 1$, we define:

- The **amount of information** we receive about the random process **after** we have observed the occurrence of the event $X = x_i$ is:

$$I(x_i) = \log \frac{1}{p(x_i)} = -\log p(x_i)$$

This is the expression Claude Shannon derived in 1948 [1] as a quantification of information in the context of *Communication Theory*. The integration constant c determines the base of the logarithms and therefore the units of information. In particular:

$$I(x_i) = -\ln p(x_i) \quad \text{"}nats\text{"} = -\lg_2 p(x_i) \quad \text{"}bits\text{"} = -\lg_{10} p(x_i) \quad \text{"}hartleys\text{"}$$

and therefore: $1 nat = \lg_2 e \;\; bits \; (\simeq 1.44) = \lg_{10} e \;\; hartleys \; (\simeq 0.43)$. In general, the units will be irrelevant for us and we shall work with natural logarithms.

4.2 Expected Information and Entropy

The amount of information we receive **after** we have observed the event $X = x_i$, $I(x_i) = -\log\, p(x_i)$, depends **only** on the probability of this particular event but **before** we do the experiment we do not know which result we shall get; we know only that each of the possible outcomes $\{x_1, x_2, \ldots, x_k, \ldots\}$ has a probability $p(x_i)$ to occur. Then, we define

- The **amount of information we expect** to get from the realization of the random experiment is:

$$I(X) = \sum_i p(x_i)\, I(x_i) = -\sum_i p(x_i) \ln p(x_i)$$

with the prescription $\lim_{x \to 0^+} x \ln x = 0$. It is clear that the **expected information** $I(X)$ does not depend on any particular result but on the probability distribution associated to the random process.

We can look at this expression from another point of view. If from the experiment we are going to do we expect to get the amount of information $I(X)$, before we do the experiment we have a **lack of information** $I(X)$ relative to what we shall have after the experiment is done. Interpreted in this way, the quantity $I(X)$ is called **entropy** ($H(X)$) and quantifies the amount of ignorance about the random process that we expect to reduce after the observation.

Example 4.1 Consider a binary random process described by a random quantity X that can take the values $\{0,\; 1\}$ with probabilities p and $1 - p$ respectively. Then, if we observe the result $X = 0$ we get $I(X = 0) = -\log p$ units of information and, if we observe the event $X = 1$, we get $I(X = 1) = -\log(1 - p)$. The information we expect to get from the realization of the experiment $e(1)$ is:

$$I(X) = -p \log p - (1 - p) \log(1 - p)$$

that, in the case the two results are equally likely ($p = 1/2$) and we take logarithms in base 2, becomes:

$$I = \frac{1}{2} \log_2 2 + \frac{1}{2} \log_2 2 = \log_2 2 = 1 \text{ bit}$$

Example 4.2 Consider a discrete random quantity with support on the finite set $\{x_1, x_2, \ldots, x_n\}$ and probabilities $p_i = p(x_i)$ and consider an experiment that consist in one observation of X. What is the distribution for which the lack of information is maximal? Defining

$$\phi(\mathbf{p}, \lambda) = -\sum_{i=1}^{n} p_i \ln p_i + \lambda \left(\sum_{i=1}^{n} p_i - 1 \right)$$

we have that

$$\left.\begin{array}{lll} \dfrac{\partial \phi(\mathbf{p}, \lambda)}{\partial p_i} = -\ln p_i - 1 + \lambda = 0 & \longrightarrow & p_i = e^{\lambda - 1} \\ \dfrac{\partial \phi(\mathbf{p}, \lambda)}{\partial \lambda} = \sum_{i=1}^{n} p_i - 1 = 0 & \longrightarrow & e^{\lambda - 1} = \frac{1}{n} \end{array}\right\} \longrightarrow p(x_i) = \frac{1}{n}; \qquad i = 1, \ldots, n$$

Therefore the entropy, the lack of information, is maximal for the Discrete Uniform Distribution and its value will be

$$H_M(X) = \sum_{i=1}^{n} \frac{1}{n} \ln n = \ln n$$

Thus, for any random quantity with finite support of dimension n, $0 \leq H(X) \leq \ln n$ and the maximum amount of information we expect to get from one observation is $I(X) = \ln n$.

If the sample space is countable, it is obvious that the distribution that maximizes the entropy can not be the Discrete Uniform. Suppose that we know the expected values of k functions $\{f_j(x)\}_{j=1}^{k}$; that is:

$$\mu_j = \sum_{i=1}^{\infty} p(x_i) f_j(x_i) \qquad \text{with} \qquad j = 1, 2, \ldots, k$$

and let's define

$$\phi(\mathbf{p}, \lambda, \boldsymbol{\lambda}) = -\sum_{i=1}^{\infty} p_i \ln p_i + \lambda \left(\sum_{i=1}^{\infty} p_i - 1 \right) + \sum_{j=1}^{k} \lambda_j \left(\mu_j - \sum_{i=1}^{\infty} p_i f_j(x_i) \right)$$

with $\boldsymbol{\lambda} = \{\lambda_1, \ldots, \lambda_k\}$. Then:

$$\frac{\partial \phi(\mathbf{p}, \lambda, \boldsymbol{\lambda})}{\partial p_i} = -\ln p_i - 1 + \lambda - \sum_{j=1}^{k} \lambda_j f_j(x_i) = 0$$

$$\longrightarrow \quad p_i = \exp\{\lambda - 1\} \exp\left\{ -\sum_{j=1}^{k} \lambda_j f_j(x_i) \right\}$$

The derivative with respect to λ gives the normalization condition

$$\exp(\lambda - 1) \sum_{i=1}^{\infty} \exp\left(-\sum_{j=1}^{k} \lambda_j f_j(x_i) \right) = 1$$

so if we define

$$Z(\boldsymbol{\lambda}) = \left[\sum_{i=1}^{\infty} \exp\left(-\sum_{j=1}^{k} \lambda_j f_j(x_i) \right) \right]^{-1}$$

we have that

$$p_i = Z(\boldsymbol{\lambda}) \exp\left(-\sum_{j=1}^{k} \lambda_j f_j(x_i) \right)$$

The remaining Lagrange multipliers $\boldsymbol{\lambda}$ are determined by the conditions

$$\mu_j = \sum_{i=1}^{\infty} p(x_i) f_j(x_i) \qquad \text{with} \qquad j = 1, 2, \ldots, k$$

Suppose for instance that X may take the values $\{0, 1, 2, \ldots\}$ and the have only one condition:

$$\mu = E[X] = \sum_{n=0}^{\infty} p_n n$$

that is, $f(x_n) = n$. Then $p_n = Z(\lambda_1)\exp\{-\lambda_1 n\}$ and therefore

$$\mu = \sum_{n=0}^{\infty} p_n\, n = Z(\lambda_1) \sum_{n=0}^{\infty} n \exp\{-\lambda_1 n\} \quad \longrightarrow \quad Z(\lambda_1) = \mu\, e^{\lambda_1}\,(1 - e^{-\lambda_1})^2$$

Imposing finally that $\sum_{n=0}^{\infty} p_n = 1$ we get $\lambda_1 = \ln\left(\frac{1}{\mu} + 1\right)$ and in consequence

$$p_n = \frac{\mu^n}{(1+\mu)^{1+n}} \qquad \text{with} \qquad n = 0, 1, 2, \ldots \qquad \text{and} \quad \mu > 0$$

is the distribution for which the lack of information is maximal. In this case:

$$H_M(X) = -\sum_{i=1}^{\infty} p(x_i) \ln p(x_i) = \ln\left[\frac{(1+\mu)^{1+\mu}}{\mu^{\mu}}\right]$$

is the maximum entropy for any random quantity with countable support and known mean value $\mu = E[X] > 0$.

4.3 Conditional and Mutual Information

Consider two discrete random quantities, X and Y, with supports $\Omega_x = \{x_1, x_2, \ldots\}$ and $\Omega_y = \{y_1, y_2, \ldots\}$ and joint probability

$$P(X = x_i, Y = y_j) = p(x_i, y_j) = p(x_i|y_j)\, p(y_j) = p(y_j|x_i)\, p(x_i)$$

The Expected Information about (X, Y) from an observation will be

$$\begin{aligned} I(X, Y) &= -\sum_{\Omega_x}\sum_{\Omega_y} p(x_i, y_j) \ln p(x_i, y_j) \\ &= -\sum_{\Omega_x}\sum_{\Omega_y} p(x_i, y_j) \ln p(x_i|y_j) - \sum_{\Omega_y} p(y_j) \ln p(y_j) \end{aligned}$$

If we define[1]

[1]The non-negativity of this and the following expressions of Information can be easily derived from the Jensen's inequality for convex functions: Given the probability space $(\mathcal{R}, \mathcal{B}, \mu)$, a μ-integrable function X and a convex function ϕ over the range of X, then $\phi(\int_{\mathcal{R}} X\, d\mu) \leq \int_{\mathcal{R}} \phi(X)\, d\mu$ provided the last integral exist; that is, $\phi(E[X]) \leq E[\phi(X)]$. Observe that if ϕ is a concave function, then $-\phi$ is convex so the inequality sign is reversed and that if ϕ is twice continuously differentiable on $[a, b]$, it is convex on that interval iff $\phi''(x) \geq 0$ for all $x \in [a, b]$. Frequent and useful convex functions are $\phi(x) = \exp(x)$ and $\phi(x) = -\log x$.

$$I(X|Y) = -\sum_{\Omega_x} \sum_{\Omega_y} p(x_i, y_j) \ln p(x_i|y_j) \geq 0$$

we can write:

$$I(X, Y) = I(X|Y) + I(Y) = I(Y|X) + I(X) = I(Y, X)$$

Now, $I(Y)$ is the amount of information we expect to get about Y and, if (X, Y) are not independent, the knowledge of Y gives some information on X so the remaining information we expect to get about X is not $I(X)$ but the smaller quantity $I(X|Y) < I(X)$ because we already know something about it. In entropy language, $H(X|Y)$ is the amount of ignorance about X that remains after Y is known. It is clear that if X and Y are independent, $I(X|Y) = I(X)$ so the knowledge of Y doesn't say anything about X and therefore the remaining information we expect to get about X is $I(X)$. The interesting question is: How much information on X is contained in Y? (or, entropy wise, By how much the ignorance about X will be reduced if we observe first the quantity Y?). Well, if observing X we expect to get $I(X)$ and after observing Y we expect to get $I(X|Y)$, the amount of information that the knowledge of Y provides on X is the **Mutual Information**:

$$I(X : Y) = I(X) - I(X|Y)$$

Again, if they are independent $I(X|Y) = I(X)$ so $I(X : Y) = 0$. From the previous properties, it is clear that

$$I(X : Y) = I(X) - I(X|Y) = I(Y) - I(Y|X) = I(X) + I(Y) - I(X, Y) = I(Y : X)$$

so it is symmetric: *the observation of Y provides as much information about X as the observation of X about Y*. Expliciting the terms:

$$\begin{aligned} I(X : Y) &= I(X) + I(Y) - I(X, Y) = \\ &= \sum_{\Omega_x} \sum_{\Omega_y} p(x_i, y_j) \ln p(x_i, y_j) - \sum_{\Omega_x} p(x_i) \ln p(x_i) - \sum_{\Omega_y} p(y_j) \ln p(y_i) \end{aligned}$$

and, in consequence:

- The **Mutual Information**, the amount of information we expect to get about X (or Y) from the knowledge of Y (or X), is given by

$$I(X : Y) = I(Y : X) = \sum_{\Omega_x} \sum_{\Omega_y} p(x_i, y_j) \ln \left(\frac{p(x_i, y_j)}{p(x_i)\, p(y_j)} \right)$$

Again from the Jensen's inequality for convex functions $I(X:Y) \geq 0$ with the equality satisfied if and only if X and Y are independent random quantities so the *Mutual Information* is therefore a measure of the statistical dependence between them (see Note 3).

4.4 Generalization for Absolute Continuous Random Quantities

Up to now, we have been dealing with discrete random quantities. Consider now a continuous random quantity $X \sim p(x)$ with support on the compact set $\Omega_x = [a, b]$ and let's get a discrete approximation. Given a partition

$$\Omega_X = \bigcup_{n=1}^{N} \Delta_n; \qquad \Delta_n = [a + (n-1)\epsilon,\ a + n\epsilon]$$

with large N and $\epsilon = (b-a)/N > 0$, if $p(x)$ is continuous on Δ_n we can use the Mean Value Theorem and write

$$P(X \in \Delta_n) = \int_{\Delta_n} p(x)\,dx = p(x'_n)\,\epsilon$$

with x'_n an interior point of Δ_n. Then, we define the discrete approximation X_D of X that takes values $\{x'_1, x'_2, \ldots, x'_N\}$ with probabilities $p'_k = p(x'_k)\epsilon$ such that $\sum_{k=1}^{N} p'_k = 1$ and write

$$I_D(X_D) = -\sum_{k=1}^{N} [p(x'_k)\,\epsilon] \log [p(x'_i)\,\epsilon] = -\sum_{k=1}^{N} p'_k \log p(x'_k) - \log \epsilon \sum_{k=1}^{N} p'_k$$

so, in limit $\epsilon \rightarrow 0^+$, X_D tends to X in distribution and

$$I_D(X_D) \longrightarrow -\int_a^b p(x) \log p(x)\,dx - \log \epsilon$$

One may be tempted to regularize this expression and define

$$I(X) \stackrel{def.}{=} \lim_{\epsilon \longrightarrow 0} [I_D(X_D) + \log \epsilon] = -\int_{\Omega_X} p(x) \log p(x)\,dx$$

... but this doesn't work. This *"naive"* generalisation is not the limit of the information for a discrete quantity and is not an appropriate measure of information, among other things, because $I(X)$ so defined may be negative. For example, if $X \sim Un(x|0, b)$, have that $I(X) = \ln b$ and for $0 < b < 1$ this is negative.[2]

However, this reasoning is meaningful for the *Mutual Information* because, being the argument of the logarithm a ratio of probability densities, the limit is well defined. Thus, we can state that

- Given two continuous random quantities X and Y with probability density $p(x, y)$, the amount of information on X that is contained in Y is the **Mutual Information**:

$$I(X:Y) = \int_{\Omega_x} dx \int_{\Omega_y} dy\, p(x, y) \ln\left(\frac{p(x, y)}{p(x)\, p(y)}\right)$$
$$= \int_{\Omega_x} dx \int_{\Omega_y} dy\, p(x, y) \ln\left(\frac{p(x|y)}{p(x)}\right).$$

4.5 Kullback–Leibler Discrepancy and Fisher's Matrix

Consider the random quantity $X \sim p(x|\boldsymbol{\theta})$, the sequence of observations $\boldsymbol{x} = \{x_1, x_2, \ldots, x_n\}$ and the prior density $\pi(\boldsymbol{\theta})$ that describes the knowledge we have on $\boldsymbol{\theta}$ **before** we do the experiment. Then $p(\boldsymbol{x}, \boldsymbol{\theta}) = p(\boldsymbol{\theta}|\boldsymbol{x})p(\boldsymbol{x})$ and therefore, the Mutual Information

$$I(X:\theta) = \int_{\Omega_X} d\boldsymbol{x} \int_{\Omega_\theta} d\boldsymbol{\theta}\, p(\boldsymbol{x}, \boldsymbol{\theta}) \ln\left(\frac{p(\boldsymbol{x}, \boldsymbol{\theta})}{p(\boldsymbol{x})\pi(\boldsymbol{\theta})}\right)$$
$$= \int_{\Omega_X} d\boldsymbol{x}\, p(\boldsymbol{x}) \int_{\Omega_\theta} d\boldsymbol{\theta}\, p(\boldsymbol{\theta}|\boldsymbol{x}) \ln\left(\frac{p(\boldsymbol{\theta}|\boldsymbol{x})}{\pi(\boldsymbol{\theta})}\right)$$

quantifies (Lindley; 1956) the *Information* we expect to get from the experiment $e(n)$ on the parameter $\boldsymbol{\theta} \in \Omega_\theta$ when the *prior* knowledge is represented by $\pi(\boldsymbol{\theta})$. Therefore, we have that:

- The amount of Information provided by the data sample $\boldsymbol{x} = \{x_1, x_2, \ldots, x_n\}$ on the parameter $\boldsymbol{\theta} \in \Omega_\theta$ with respect to the prior density $\pi(\boldsymbol{\theta})$ is:

$$I(\boldsymbol{x}|\pi(\boldsymbol{\theta})) = \int_{\Omega_\theta} p(\boldsymbol{\theta}|\boldsymbol{x}) \ln\left(\frac{p(\boldsymbol{\theta}|\boldsymbol{x})}{\pi(\boldsymbol{\theta})}\right) d\boldsymbol{\theta}$$

  that is; given the experimental sample $\boldsymbol{x}$, $I(\boldsymbol{x}|\theta)$ is the information we need to actualize the *prior* knowledge $\pi(\boldsymbol{\theta})$ and substitute it by $p(\boldsymbol{\theta}|\boldsymbol{x})$;

[2]Despite of that, the *"Differential Entropy"* $h(p) = -\int_{\Omega_X} p(x) \log p(x)\, dx$ is a useful quantity in a different context. It is left as an exercise to show that among all continuous distributions with support on $[a, b]$, then the Uniform distribution $Un(x|a, b)$ is the one that maximizes the Differential Entropy, among those with support on $[0, \infty)$ and specified first order moment is the Exponential $Ex(x|\mu)$ and, if the second order moment is also constrained, we get the Normal density $N(x|\mu, \sigma)$.

- The amount of Expected Information from the experiment $e(n)$ on the parameter $\boldsymbol{\theta} \in \Omega_\theta$ with respect to the knowledge contained in $\pi(\boldsymbol{\theta})$ will be

$$I(e|\pi(\boldsymbol{\theta})) = \int_{\Omega_X} p(\boldsymbol{x})\, I(\boldsymbol{x}|\pi(\boldsymbol{\theta}))\, d\boldsymbol{x} = \int_{\Omega_X} p(\boldsymbol{x})\, d\boldsymbol{x} \int_{\Omega_\theta} p(\boldsymbol{\theta}|\boldsymbol{x}) \ln\left(\frac{p(\boldsymbol{\theta}|\boldsymbol{x})}{\pi(\boldsymbol{\theta})}\right) d\boldsymbol{\theta} =$$
$$= \int_{\Omega_X}\int_{\Omega_\theta} p(\boldsymbol{x},\boldsymbol{\theta}) \ln\left(\frac{p(\boldsymbol{x},\boldsymbol{\theta})}{\pi(\boldsymbol{\theta})\, p(\boldsymbol{x})}\right) d\boldsymbol{x}\, d\boldsymbol{\theta} = I(X:\boldsymbol{\theta})$$

As a general criteria to compare a probability density $p(\boldsymbol{x})$ with an approximation $\pi(\boldsymbol{x})$, Solomon Kullback [2] y Richard Leibler introduced in 1951 the so called *Kullback–Leibler Discrepancy* as:

$$D_{KL}[p(.)\|\pi(.)] = \int_{\Omega_X} d\boldsymbol{x}\; p(\boldsymbol{x}) \ln\left(\frac{p(\boldsymbol{x})}{\pi(\boldsymbol{x})}\right)$$

It is easy to check that:

(i) $D_{KL}[p(.)\|\pi(.)] \geq 0$ with the equality iff $p(\boldsymbol{x}) = \pi(\boldsymbol{x})$ almost everywhere;
(ii) $D_{KL}[p(.)\|\pi(.)]$ is a convex function with respect to the pair $(p(\boldsymbol{x}), \pi(\boldsymbol{x}))$.

The Kullback–Leibler discrepancy $D_{KL}[p(.)\|\pi(.)]$ is not a metric distance because it is not symmetric; that is, $D_{KL}[p(.)\|\pi(.)] \neq D_{KL}[\pi(.)\|p(.)]$ and therefore does not satisfy either the triangular inequality. However, it can be symmetrized as $J[p(.);\pi(.)] = D_{KL}[p(.)\|\pi(.)] + D_{KL}[\pi(.)\|p(.)]$.

From the previous expressions, we have that

$$I(e|\pi(\boldsymbol{\theta})) = D_{KL}[p(\boldsymbol{x},\boldsymbol{\theta})\|\pi(\boldsymbol{\theta})p(\boldsymbol{x})] \qquad \text{and} \qquad I(\boldsymbol{x}|\pi(\boldsymbol{\theta}) = D_{KL}[p(\boldsymbol{\theta}|\boldsymbol{x})\|\pi(\boldsymbol{\theta})]$$

In the context of Bayesian Inference, the Kullback–Leibler discrepancy is a natural measure of information and clearly evidences that amount of knowledge on the parameters $\boldsymbol{\theta}$ that we get from the experiment and is contained in the posterior density, is relative to our prior knowledge. This is an important relation for the *Reference Analysis* (Sect. 6.7) developed in [3, 4].

4.5.1 Fisher's Matrix

What is the capacity of the experiment to distinguish two infinitesimally close values $\boldsymbol{\theta}_0$ and $\boldsymbol{\theta}_1 = \boldsymbol{\theta}_0 + \Delta\boldsymbol{\theta}_0$ of the parameters? or, in other words, How much information has to be provided by the experiment so that we can discern between two infinitesimally close values of $\boldsymbol{\theta}$? Let's analyze the local behaviour of

$$I(\boldsymbol{\theta}_0 : \boldsymbol{\theta}_1 = \boldsymbol{\theta}_0 + \Delta\boldsymbol{\theta}_0) = \int_{\Omega_X} p(x|\boldsymbol{\theta}_0) \ln\left(\frac{p(x|\boldsymbol{\theta}_0)}{p(x|\boldsymbol{\theta}_1)}\right) dx$$

If $\dim\{\boldsymbol{\theta}\} = n$, after a Taylor expansion of $\ln\, p(x|\boldsymbol{\theta}_1)$ around $\boldsymbol{\theta}_0$ and considering that

$$\int_{\Omega_X} p(x|\theta)\, dx = 1 \longrightarrow \int_{\Omega_X} p(x|\theta)\, \frac{\partial}{\partial\theta} \ln p(x|\theta)\, dx = \frac{\partial}{\partial\theta} \int_{\Omega_X} p(x|\theta)\, dx = 0$$

we get

$$I(\boldsymbol{\theta}_0 : \boldsymbol{\theta}_0 + \Delta\boldsymbol{\theta}) \simeq -\frac{1}{2!} \sum_{i=1}^{n}\sum_{j=1}^{n} \Delta\theta_i\, \Delta\theta_j \left[\int_{\Omega_X} p(\boldsymbol{x}|\boldsymbol{\theta})\, \frac{\partial^2 \ln p(\boldsymbol{x}|\boldsymbol{\theta})}{\partial\theta_i \partial\theta_j}\, d\boldsymbol{x}\right]_{\theta_0} + \cdots$$

Thus, if we define the

- **Fisher's matrix**

$$\boldsymbol{I}_{ij}(\boldsymbol{\theta}) = -\int_{\Omega_X} p(\boldsymbol{x}|\boldsymbol{\theta})\, \frac{\partial^2 \ln p(\boldsymbol{x}|\boldsymbol{\theta})}{\partial\theta_i \partial\theta_j}\, d\boldsymbol{x} = E_X\left[-\frac{\partial^2 \ln p(\boldsymbol{x}|\boldsymbol{\theta})}{\partial\theta_i \partial\theta_j}\right]$$

we can write

$$I(\boldsymbol{\theta}_0 : \boldsymbol{\theta}_0 + \Delta\boldsymbol{\theta}) \simeq \frac{1}{2} \sum_{i=1}^{n}\sum_{j=1}^{n} \Delta\theta_i\, \boldsymbol{I}_{ij}(\boldsymbol{\theta}_0)\Delta\theta_j + \cdots$$

The *Fisher's matrix* is a non-negative symmetric matrix that plays a very important role in statistical inference... provided it exists. This is the case for regular distributions where:

(1) $\text{supp}_x\{p(x|\boldsymbol{\theta})\}$ does not depend on $\boldsymbol{\theta}$;
(2) $p(x|\boldsymbol{\theta}) \in C_k(\boldsymbol{\theta})$ for $k \geq 2$ and
(3) The integrand is well-behaved so $\frac{\partial}{\partial\theta} \int_{\Omega_X} (\bullet) d\boldsymbol{x} = \int_{\Omega_X} \frac{\partial(\bullet)}{\partial\theta} d\boldsymbol{x}$

Interchanging the derivatives with respect to the parameters θ_i and the integrals over Ω_X it is easy to obtain the equivalent expressions:

$$\boldsymbol{I}_{ij}(\boldsymbol{\theta}) = E_X\left[-\frac{\partial^2 \ln p(\boldsymbol{x}|\boldsymbol{\theta})}{\partial\theta_i \partial\theta_j}\right] = E_X\left[\frac{\partial \ln p(\boldsymbol{x}|\boldsymbol{\theta})}{\partial\theta_i}\, \frac{\partial \ln p(\boldsymbol{x}|\boldsymbol{\theta})}{\partial\theta_j}\right]$$

$$\mathbf{I}_{ii}(\boldsymbol{\theta}) = E_X\left[-\frac{\partial^2 \ln p(\boldsymbol{x}|\boldsymbol{\theta})}{\partial\theta_i^2}\right] = E_X\left[\left(\frac{\partial \ln p(\boldsymbol{x}|\boldsymbol{\theta})}{\partial\theta_i}\right)^2\right]$$

If $\boldsymbol{X} = (X_1, \ldots, X_n)$ is a n-dimensional random quantity and $\{X_i\}_{i=1}^n$ are independent, then $I_X(\boldsymbol{\theta}) = \sum_i I_{X_i}(\boldsymbol{\theta})$. Obviously, if they are iid then $I_{X_i}(\boldsymbol{\theta}) = I_X(\boldsymbol{\theta})$ for all i and $I_X(\boldsymbol{\theta}) = nI_X(\boldsymbol{\theta})$.

4.5.2 Asymptotic Behaviour of the Likelihood Function

Suppose that the experiment $e(n)$ provides an independent and exchangeable sequence of observations $\{x_1, x_2, \ldots, x_n\}$ from the model $p(x|\boldsymbol{\theta})$ with $\dim(\boldsymbol{\theta}) = d$. The information that the experiment provides about $\boldsymbol{\theta}$ is contained in the likelihood function and, being non-negative function, consider for simplicity its logarithm:

$$w(\boldsymbol{\theta}|\cdot) = \sum_{i=1}^{n} \log\, p(x_i|\boldsymbol{\theta})$$

and the Taylor expansion around the maximum $\widehat{\boldsymbol{\theta}}$:

$$w(\boldsymbol{\theta}|\cdot) = w(\widehat{\boldsymbol{\theta}}|\cdot) - \frac{n}{2}\sum_{k=1}^{d}\sum_{m=1}^{d}\left[\frac{1}{n}\sum_{i=1}^{n}\left(-\frac{\partial^2 \log p(x_i|\boldsymbol{\theta})}{\partial\theta_k\partial\theta_m}\right)_{\widehat{\theta}}\right](\theta_k - \widehat{\theta}_k)(\theta_m - \widehat{\theta}_m) + \cdots$$

where the second term has been multiplied and divided by n. Under sufficiently regular conditions we have that $\widehat{\boldsymbol{\theta}}$ converges in probability to the true value $\boldsymbol{\theta}_0$ so we can neglect higher order terms and, by the Law of Large Numbers, approximate

$$\lim_{n\to\infty}\left[\frac{1}{n}\sum\nolimits_{i=1}^{n}\left(-\frac{\partial^2 \log p(x_i|\boldsymbol{\theta})}{\partial\theta_k\partial\theta_m}\right)\right]_{\widehat{\theta}} \simeq E_X\left[-\frac{\partial^2 \log p(x|\boldsymbol{\theta})}{\partial\theta_k\partial\theta_m}\right]_{\widehat{\theta}} \simeq \boldsymbol{I}_{km}(\widehat{\boldsymbol{\theta}})$$

Therefore

$$w(\boldsymbol{\theta}|\cdot) = w(\widehat{\boldsymbol{\theta}}|\cdot) - \frac{1}{2}\sum_{k=1}^{d}\sum_{m=1}^{d}(\theta_k - \widehat{\theta}_k)\left[n\boldsymbol{I}_{km}(\widehat{\boldsymbol{\theta}})\right](\theta_m - \widehat{\theta}_m) + \cdots$$

and, under regularity conditions, we can approximate the likelihood function by a Normal density with mean $\widehat{\boldsymbol{\theta}}$ and covariance matrix $\boldsymbol{\Sigma}^{-1} = n\boldsymbol{I}(\widehat{\boldsymbol{\theta}})$. In fact, for a Normal density

$$p(\boldsymbol{x}|\boldsymbol{\mu}, \boldsymbol{\sigma}) = \frac{1}{(2\pi)^{n/2}\det^{1/2}[\mathbf{V}]}\exp\{-\frac{1}{2}(\boldsymbol{x}-\boldsymbol{\mu})^T\mathbf{V}^{-1}(\boldsymbol{x}-\boldsymbol{\mu})\}$$

we have for the parameters $\boldsymbol{\mu}$ that

$$\frac{\partial \ln p(\boldsymbol{x}|\boldsymbol{\mu}, \boldsymbol{\sigma})}{\partial\mu_i} = -\sum\nolimits_{k=1}^{n}[\mathbf{V}^{-1}]_{ik}\,(x_k - \mu_k)$$

and therefore

$$\begin{aligned} \boldsymbol{I}_{ij}(\boldsymbol{\mu}) &= \sum\nolimits_{k=1}^{n} \sum\nolimits_{p=1}^{n} [\mathbf{V}^{-1}]_{ik} [\mathbf{V}^{-1}]_{jp} \, E_X \left[(x_k - \mu_k)(x_p - \mu_p) \right] = \\ &= \sum\nolimits_{k=1}^{n} \sum\nolimits_{p=1}^{n} [\mathbf{V}^{-1}]_{ik} [\mathbf{V}^{-1}]_{jp} \, [\mathbf{V}]_{kp} = [\mathbf{V}^{-1}]_{ij} \end{aligned}$$

that is; the Fisher's Matrix $\boldsymbol{I}_{ij}(\boldsymbol{\mu})$ is the inverse of the Covariance Matrix.

Example 4.3 The quantity $\mathbf{I}_{ij}(\boldsymbol{\theta})$ as an intrinsic measure of the information that an experiment will provide about the parameters $\boldsymbol{\theta}$ was introduced by R.A. Fisher around 1920 in the context of Experimental Design. It depends only on the conditional density $p(\text{data}|\text{parameters})$ and therefore is not a quantification relative to the knowledge we have on the parameters before the experiment is done but it is very useful and has many interesting applications; for instance, to compare different procedures to analyze the data. Let's see as an example the charge asymmetry of the angular distribution of the process $e^+e^- \longrightarrow \mu^+\mu^-$. If we denote by θ the angle between the incoming electron and the outgoing μ^- and by $x = \cos\theta \in [-1, 1]$, the angular distribution can be expressed in first order of electroweak perturbations as

$$p(x|a) = \frac{3}{8}\,(1 + x^2) \, + a\,x$$

where a, the asymmetry coefficient, is bounded by $|a| \leq a_m = 3/4$ since $p(x|a) \geq 0$.

Now, suppose that an experiment $e(n)$ provides n independent observations $\boldsymbol{x} = \{x_1, x_2, \ldots, x_n\}$ so we have the joint density

$$p(\boldsymbol{x}|A) = \prod_{i=1}^{n} \left(\frac{3}{8}\,(1 + x_i^2) + a\,x_i \right)$$

In this case, there are no minimal sufficient statistics but, instead of working with the whole sample, we may simplify the analysis and classify the events in two categories: *Forward* (F) if $\theta = \in [0, \pi/2] \rightarrow x \in [0, 1])$ or *Backward* (B) if $\theta \in [\pi/2, \pi] \longrightarrow x \in [-1, 0]$. Then

$$p_F = \int_0^1 p(x|a)\,dx = \frac{1}{2}(1 + a) \qquad \text{and} \qquad p_B = \int_{-1}^0 p(x|a)\,dx = \frac{1}{2}(1 - a) = 1 - p_F$$

and the model we shall use for inferences on a is the simpler Binomial model

$$P(n_F|n, a) = \frac{1}{2^n}\binom{n}{n_F}\,(1 + a)^{n_F}\,(1 - a)^{n - n_F}; \qquad n_F = 0,\ 1, \ldots, n$$

For this second analysis ($A2$) we have that

$$I_{A2}(a) = \sum_{n_F=0}^{n} P(n_F|n,a) \left(\frac{\partial \ln P(n_F|n,a)}{\partial a} \right)^2 = \frac{n}{1-a^2}$$

In contrast, for the first analysis (A_1) we have that

$$I_{A1}(a) = n \int_{-1}^{1} \frac{x^2}{p(x|a)} \, dx = n \, \frac{8}{3} \left(2 + \alpha \ln \frac{1-\alpha}{1+\alpha} + \frac{\pi}{2} \frac{2\,\alpha^2 - 1}{\sqrt{1-\alpha^2}} \right)$$

where $\alpha = a/a_m \in [-1, 1]$. For anys $a \in [-3/4, 3/4]$, it holds that $I_{A1}(a) > I_{A2}(a)$ so it is preferable to do the first analysis.

4.6 Some Properties of Information

- **If we do n independent samplings from the same distribution, Shall we get n times the Information on the parameters of interest that we get from one observation?**

Consider the random quantity $X \sim p(x|\theta)$, the sample $\boldsymbol{x} = \{x_1, \ldots, x_n\}$ and the Mutual Information

$$I(X_1, \ldots, X_n : \theta) = \int_{\Omega_\theta} d\theta \, \pi(\theta) \int_{\Omega_X} p(\boldsymbol{x}|\theta) \, \ln \left(\frac{p(\boldsymbol{x}|\theta)}{p(\boldsymbol{x})} \right) d\boldsymbol{x}$$

If the n observations are independent, $p(\boldsymbol{x}|\theta) = p(x_1|\theta) \cdots p(x_n|\theta)$ and therefore

$$\begin{aligned} \int_{\Omega_X} d\boldsymbol{x} \, p(\boldsymbol{x}|\theta) \, \ln \, p(\boldsymbol{x}|\theta) &= \sum_{i=1}^{n} \int_{\Omega_X} dx_i \, p(x_i|\theta) \ln p(x_i|\theta) \\ &= n \int_{\Omega_X} dx \, p(x|\theta) \ln p(x|\theta) \end{aligned}$$

On the other hand:

$$\int_{\Omega_\theta} d\theta \, \pi(\theta) \int_{\Omega_X} d\boldsymbol{x} \, p(\boldsymbol{x}|\theta) \ln p(\boldsymbol{x}) = \int_{\Omega_X} d\boldsymbol{x} \, p(\boldsymbol{x}) \ln p(\boldsymbol{x})$$

so:

$$\begin{aligned} I(X_1, \ldots, X_n : \theta) &= n \int_{\Omega_\theta} d\theta \, \pi(\theta) \int_{\Omega_X} dx \, p(x|\theta) \ln p(x|\theta) - \int_{\Omega_X} d\boldsymbol{x} \, p(\boldsymbol{x}) \, \ln \, p(\boldsymbol{x}) = \\ &= n \, I(X : \theta) - D_{KL}[p(\boldsymbol{x}) \| p(x_1) \cdots p(x_n)] \end{aligned}$$

and $D_{KL}[p(\boldsymbol{x})\|p(x_1)\cdots p(x_n)] = 0$ if, and only if, $p(\boldsymbol{x}) = p(x_1)\cdots p(x_n)$. However, since

$$p(\boldsymbol{x}) = \int_{\Omega_\theta} d\theta \, p(x_1|\theta)\dots p(x_n|\theta)\,\pi(\theta)$$

the random quantities $X_1, X_2, \dots, X_n$ are correlated through θ and therefore are not independent. Thus, since $D_{KL}[p(\boldsymbol{x})\|p(x_1)\cdots p(x_n)] > 0$ the information provided by the experiment $e(n)$ is less than n times the one provided by $e(1)$. This is reasonable because, as the number of samplings grows, the knowledge about the parameter θ increases, the prior distribution $\pi(\theta)$ is actualized by $p(\theta|x_1,\dots,x_n)$ and further independent realizations of the same experiment (*under the same conditions*) will provide less information.

This is not the case for Fisher's criteria of Information. In fact, if the observations are independent, it is trivial to see that for $\boldsymbol{x} = \{x_1, x_2, \dots, x_n\}$

$$\begin{aligned}\mathbf{I}_{ij}^{(n)}(\boldsymbol{\theta}) = E_X\left[\frac{\partial \ln p(\boldsymbol{x}|\boldsymbol{\theta})}{\partial\theta_i}\,\frac{\partial \ln p(\boldsymbol{x}|\boldsymbol{\theta})}{\partial\theta_j}\right] &= n\,E_X\left[\frac{\partial \ln p(x_1|\boldsymbol{\theta})}{\partial\theta_i}\,\frac{\partial \ln p(x_1|\boldsymbol{\theta})}{\partial\theta_j}\right]\\ &= n\,\mathbf{I}_{ij}^{(1)}(\boldsymbol{\theta});\end{aligned}$$

Example 4.4 Consider a random quantity $X \sim N(x|\mu,\sigma)$ with σ known and μ the parameter of interest with a prior density $\pi(\mu) = N(\mu|\mu_0,\sigma_0)$. Then

$$p(x,\mu|\mu_0,\sigma,\sigma_0) = N(x|\mu,\sigma)\,N(\mu|\mu_0,\sigma_0); \qquad p(x|\mu_0,\sigma,\sigma_0) = N(x|\mu_0,\sqrt{\sigma^2+\sigma_0^2})$$

so the amount of Expected Information from the experiment $e(1)$ on the parameter μ with respect to the knowledge contained in $\pi(\mu)$ will be

$$I(e(1)|\pi(\mu)) = \int_{-\infty}^{\infty} d\mu \int_{-\infty}^{\infty} dx\; p(x,\mu|\cdot)\,\ln\left(\frac{p(x,\mu|\cdot)}{\pi(\mu|\cdot)\,p(x|\cdot)}\right) = \ln\left(1+\frac{\sigma_0^2}{\sigma^2}\right)^{1/2}$$

Consider now the experiment $e(2)$ that provides two independent observations $\{x_1, x_2\}$. Then

$$p(x_1,x_2,\mu|\mu_0,\sigma,\sigma_0) = N(x_1|\mu,\sigma)\,N(x_2|\mu,\sigma)\,N(\mu|\mu_0,\sigma_0)$$

and

$$p(x_1,x_2|\mu_0,\sigma,\sigma_0) = N(x_1,x_2|\mu_0,\mu_0,\sqrt{\sigma^2+\sigma_0^2},\sqrt{\sigma^2+\sigma_0^2},\rho)$$

where $\rho = (1+\sigma^2/\sigma_0^2)^{-1}$; that is, the random quantities X_1 and X_2 are correlated through $\pi(\mu)$. In this case:

$$I(e(2)|\pi(\mu)) = \frac{1}{2}\ln\left(1 + 2\frac{\sigma_0^2}{\sigma^2}\right) = 2\,I(e(1)|\pi(\mu)) + \frac{1}{2}\ln(1-\rho^2)$$

so $D_{KL}[p(x_1, x_2|\cdot)\|p(x_1|\cdot)p(x_1|\cdot)] = -\frac{1}{2}\ln(1-\rho^2)$. In general, for n independent observations we have that

$$I(e(n)|\pi(\mu)) \;=\; \frac{1}{2}\ln\left(1 + n\frac{\sigma_0^2}{\sigma^2}\right)$$

that behaves asymptotically with n as $I(e(n)|\pi(\mu)) \sim \ln\sqrt{n}$.

- **In general, the grouping of observations implies a reduction of information.** Consider a partition of the sample space

$$\Omega_X = \bigcup\nolimits_{i=1}^{N} E_i; \qquad E_i \cap E_j = \emptyset \quad \forall i \neq j$$

and let's write the Mutual Information as:

$$\begin{aligned} I(X:\theta) &= \int_{\Omega_X} dx \int_{\Omega_\theta} d\theta\, p(x,\theta)\ln\left(\frac{p(x,\theta)}{p(x)\pi(\theta)}\right) = \\ &= \int_{\Omega_\theta} d\theta \int_{\Omega_X} dx\, p(x,\theta)\ln\left(\frac{p(x,\theta)}{p(x)}\right) - \int_{\Omega_\theta} d\theta\, p(\theta)\ln\pi(\theta) \end{aligned}$$

Introducing

$$\mu_1(E_i,\theta) = \int_{E_i} p(x,\theta)\,dx \qquad \text{y} \qquad \mu_2(E_i) = \int_{E_i} p(x)\,dx$$

we have for the first term that:

$$\begin{aligned} &\int_{\Omega_\theta} d\theta \sum_{i=1}^{N} \int_{E_i} dx\, p(x,\theta)\ln\left(\frac{p(x,\theta)}{p(x)}\right) = \\ &\quad = \int_{\Omega_\theta} d\theta \sum_{i=1}^{N} \int_{E_i} dx\, \frac{p(x,\theta)}{\mu_1(E_i,\theta)}\,\mu_1(E_i,\theta)\ln\left(\frac{p(x,\theta)/\mu_1(E_i,\theta)}{p(x)/\mu_2(E_i)}\,\frac{\mu_1(E_i,\theta)}{\mu_2(E_i)}\right) = \\ &\quad = \sum_{i=1}^{N} \int_{\Omega_\theta} d\theta\, \mu_1(E_i,\theta)\left\{\ln\frac{\mu_1(E_i,\theta)}{\mu_2(E_i)} + \int_{E_i} dx\, f_1(x,\theta)\ln\left(\frac{f_1(x,\theta)}{f_2(x)}\right)\right\} \end{aligned}$$

where

$$f_1(x,\theta) = \frac{p(x,\theta)}{\mu_1(E_i,\theta)} \geq 0 \qquad \text{and} \qquad f_2(x) = \frac{p(x)}{\mu_2(E_i)} \geq 0$$

The Mutual Information for the grouped data is

$$I_G(X:\theta) = \sum_{i=1}^{N} \int_{\Omega_\theta} d\theta\, \mu_1(E_i,\theta) \ln \frac{\mu_1(E_i,\theta)}{\mu_2(E_i)} - \int_{\Omega_\theta} d\theta\, p(\theta) \ln \pi(\theta)$$

and therefore

$$I(X:\theta) = I_G(X:\theta) + \int_{\Omega_\theta} d\theta \sum_{i=1}^{N} \mu_1(E_i,\theta) \int_{E_i} dx\, f_1(x,\theta) \ln \frac{f_1(x,\theta)}{f_2(x)}$$

Jensen's inequality for convex functions implies that

$$\int_{E_i} dx\, f_1(x,\theta) \ln \frac{f_1(x,\theta)}{f_2(x)} \geq 0$$

and, in consequence, $I(X:\theta) \geq I_G(X:\theta)$ being the equality satisfied if and only if

$$f_1(x,\theta) = f_2(x) \qquad \longrightarrow \qquad \frac{p(x,\theta)}{p(x)} = \frac{\mu_1(E_i,\theta)}{\mu_2(E_i)}$$

- **If sufficient statistics exist, using them instead of the whole sample does not reduce the information.**

Given a parametric model $p(x_1, x_2, \ldots, x_n|\boldsymbol{\theta})$, we know that the set of statistics $\boldsymbol{t} = \boldsymbol{t}(x_1, \ldots, x_n)$ is *sufficient* for $\boldsymbol{\theta}$ iff for all $n \geq 1$ and any prior distribution $\pi(\boldsymbol{\theta})$ it holds that

$$p(\boldsymbol{\theta}|x_1, x_2, \ldots, x_m) = p(\boldsymbol{\theta}|\mathbf{t})$$

Then, since $p(\boldsymbol{\theta}|\boldsymbol{t}) \propto p(\boldsymbol{t}|\boldsymbol{\theta})\, \pi(\boldsymbol{\theta})$ we have that

$$\begin{aligned} I(X:\boldsymbol{\theta}) &= \int_{\Omega_x} d\boldsymbol{x}\; p(\boldsymbol{x}) \int_{\Omega_\theta} d\theta p(\theta|\boldsymbol{x}) \ln \left(\frac{p(\theta|\boldsymbol{x})}{\pi(\boldsymbol{\theta})} \right) \\ &= \int_{\Omega_T} d\boldsymbol{t}\, p(\boldsymbol{t}) \int_{\Omega_\theta} d\theta p(\boldsymbol{\theta}|\boldsymbol{t}) \ln \left(\frac{p(\boldsymbol{\theta}|\boldsymbol{t})}{\pi(\boldsymbol{\theta})} \right) = I(T:\boldsymbol{\theta}) \end{aligned}$$

It is then clear that for inferences about $\boldsymbol{\theta}$, all the information provided by the data is contained in the set of sufficient statistics.

Example 4.5 Consider the random quantity $X \sim Ex(x|\lambda)$ and the experiment $e(n)$. Under independence of the sample $\boldsymbol{x} = \{x_1, x_2, \ldots, x_n\}$ we have that

$$p(x_1, \ldots, x_n|\lambda) = \lambda^n\, e^{-\lambda\,(x_1 + \cdots + x_n)}$$

We know already that there is a sufficient statistic $t = x_1 + \cdots + x_n$ with density $Ga(t|\lambda, n)$:

$$p(t|\lambda) = \frac{\lambda^n}{\Gamma(n)}\, e^{-\lambda t}\, t^{n-1}$$

Therefore, for a prior density $\pi(\lambda)$:

$$\begin{aligned} I(e(n)|\pi(\lambda)) &= \int_0^\infty d\lambda \int_0^\infty d\boldsymbol{x}\; p(\boldsymbol{x}, \lambda)\, \ln\left(\frac{p(\boldsymbol{x}, \lambda)}{\pi(\lambda)\, p(\boldsymbol{x})}\right) \\ &= \int_0^\infty d\lambda \int_0^\infty dt\; p(t, \lambda)\, \ln\left(\frac{p(t, \lambda)}{\pi(\lambda)\, p(t)}\right) \end{aligned}$$

In particular, for a conjugated prior $\pi(\lambda|a, b) = Ga(\lambda|a, b)$ we have that

$$I(e(n)|\pi(\lambda)) = \ln\frac{\Gamma(b)}{\Gamma(n+b)} - n + (n+b)\Psi(n+b) - b\Psi(b)$$

with $\Psi(x)$ the Digamma Function. This can be written as

$$\begin{aligned} I(e(n)|\pi(\lambda)) &= n\, I(e(1)|\pi(\lambda)) - \left\{ n\,(1 + \frac{1}{b}) + (n+1)\Psi(b) - (n+b)\,\Psi(n+b) \right. \\ &\qquad \left. - \ln\frac{b^n\, \Gamma(b)}{\Gamma(n+b)} \right\} \end{aligned}$$

where the last term in brackets the Kullback–Leibler discrepancy. Again, the asymptotic behaviour with n is non-linear: $I(e(n)|\pi(\lambda)) \sim \ln\sqrt{n}$.

4.7 Geometry and Information

This last section is somewhat marginal for the use of Information in the context that we have been interested in but, besides being an active field of research, illustrates a very interesting connection with geometry that almost surely will please most physicists. Consider the family of distributions $\mathcal{F}[p(x|\boldsymbol{\theta})]$ with $\boldsymbol{\theta} \in \Omega_\Theta \subseteq \mathcal{R}^n$. They all have the same functional form so their difference is determined solely by different values of the parameters. In fact, there is a one-to-one correspondence between each distribution $p(x|\boldsymbol{\theta}) \in \mathcal{F}$ and each point $\boldsymbol{\theta} \in \Omega_\Theta$ and the "*separation*" between them will be determined by the geometrical properties of this *parametric space*. Intuitively, we can already see that, in general, this space is a non-Euclidean Riemannian Manifold. Consider for instance the Normal density $N(x|\mu, \sigma)$ and two points (μ_1, σ_1) and (μ_2, σ_2) of the parametric space $\Omega_{\mu,\sigma} = \mathcal{R} \times \mathcal{R}^+$. For a real constant $a > 0$, if $\mu_2 = \mu_1 + a$ and $\sigma_1 = \sigma_2$ (same variance, different mean values) we have the Euclidean distance

$$d_E = \sqrt{(\mu_2 - \mu_1)^2 + (\sigma_2 - \sigma_1)^2} = a$$

We have the same Euclidean distance if $\mu_2 = \mu_1$ and $\sigma_2 = \sigma_1 + a$ (different variance, same mean values) but, intuitively, one would say that in this second case it will be more difficult to distinguish the two distributions. Even though the Euclidean distances are the same, we shall need more information to achieve the same level of "*separation*".

As we have seen (Problem 2.1), the Fisher's Matrix (from now on denoted by $g_{ij}(\boldsymbol{\theta})$)

$$g_{ij}(\boldsymbol{\theta}) = \int_{\Omega_X} p(\boldsymbol{x}|\boldsymbol{\theta}) \frac{\partial \ln p(\boldsymbol{x}|\boldsymbol{\theta})}{\partial \theta_i} \frac{\partial \ln p(\boldsymbol{x}|\boldsymbol{\theta})}{\partial \theta_j} d\boldsymbol{x}$$

behaves under a change of parameters $\boldsymbol{\phi} = \boldsymbol{\phi}(\boldsymbol{\theta})$ as a second order covariant tensor; that is:

$$\mathbf{g}_{ij}(\boldsymbol{\phi}) = \frac{\partial \theta_k}{\partial \phi_i} \frac{\partial \theta_l}{\partial \phi_j} \mathbf{g}_{kl}(\boldsymbol{\theta})$$

with the contravariant form $g^{ij}(\boldsymbol{\theta})$ satisfying $g^{ik}(\boldsymbol{\theta}) g_{kj}(\boldsymbol{\theta}) = \delta^i_{\ j}$. In the geometric context it is called the *Fisher-Rao metric tensor* and defines the geometry of the parameters' non-Euclidean Riemannian manifold. The differential distance is given by

$$(ds)^2 = g_{ij}(\boldsymbol{\theta})\, d\theta^i\, d\theta^j$$

and for any two pints $\boldsymbol{\theta}_1$ and $\boldsymbol{\theta}_2$ of the parametric space, the distance between them along the trajectory $\boldsymbol{\theta}(t)$ parametrized by $t \in [t_1, t_2]$ with end points $\boldsymbol{\theta}(t_1) = \boldsymbol{\theta}_1$ and $\boldsymbol{\theta}(t_2) = \boldsymbol{\theta}_2$ will be:

$$S(\boldsymbol{\theta}_1, \boldsymbol{\theta}_2) = \int_{\boldsymbol{\theta}_1}^{\boldsymbol{\theta}_2} ds = \int_{t_1}^{t_2} dt \frac{ds}{dt} = \int_{t_1}^{t_2} dt \left(g_{ij}(\boldsymbol{\theta}) \frac{d\theta^i}{dt} \frac{d\theta^j}{dt} \right)^{1/2}$$

that, for one parameter families, reduces to

$$S(\theta_1, \theta_2) = \int_{\theta_1}^{\theta_2} \sqrt{g(\theta)}\, d\theta$$

The path of shortest information distance between two points is always a geodesic curve determined by the second order differential equation

$$\frac{d^2\theta^i}{dt^2} + \Gamma^i_{jk}(\boldsymbol{\theta})\,\frac{d\theta^j}{dt}\,\frac{d\theta^k}{dt} = 0 \qquad \text{with} \qquad \Gamma^i_{jk}(\boldsymbol{\theta}) = \frac{1}{2}\,g^{im}\,\left(g_{jm,k} + g_{mk,j} - g_{jk,m}\right)$$

the Christoffel symbols and t the affine parameter along the geodesic. Each point of the manifold has associated a tensor (*Riemann tensor*) given, in its covariant representation, by

$$R_{iklm} = \frac{1}{2}\left(g_{im,kl} + g_{kl,im} - g_{il,km} - g_{km,il}\right) + g_{np}\left(\Gamma^n_{kl}\Gamma^p_{im} - \Gamma^n_{km}\Gamma^p_{il}\right)$$

that depends only the metric and provides a local measure of the curvature; that is, by how much the Fisher-Rao metric is not locally isometric to the Euclidean space. For a n-dimensional manifold it has n^4 components but due to the symmetries $R_{iklm} = R_{lmik} = -R_{kilm} = -R_{ikml}$ and $R_{iklm} + R_{imkl} + R_{ilmk} = 0$ they are considerably reduced. The only non-trivial contraction of indices of R_{iklm} is the *Ricci curvature tensor* $R_{ij} = g^{lm} R_{iljm}$ (symmetric) and its trace $R = g^{ij} R_{ij}$ is the *scalar curvature*. For two dimensional manifolds, the Gaussian curvature is $\kappa = R_{1212}/\det(g_{ij}) = R/2$. Last, note that the invariant differential volume element is $dV = \sqrt{|\det g(\boldsymbol{x})|}d\boldsymbol{x}$ and $\sqrt{|\det g(\boldsymbol{x})|}$ is the by now quite familiar Jeffrey's prior.

We have then that, in this geometrical context, the *Information* is the *source of curvature* of the parametric space and the curvature determines how the information *flows* from one point to another. The *geodesic distance* is an *intrinsic distance*, invariant under reparameterisations, and is related to the amount of information difference between two points; in other words, how easy is to *discern* between them: the larger the distance, the larger the *separation* and the easier will be to discern.

Let's start with a simple one-parameter case: the Binomial distribution $Bi(n|N, \theta)$. In this case

$$g_{11}(\theta) = \frac{1}{\theta(1-\theta)}; \qquad g^{11}(\theta) = \theta(1-\theta) \quad \text{and} \qquad \Gamma^1_{11}(\theta) = -\frac{1-2\theta}{2\theta(1-\theta)}$$

so the geodesic equation is:

$$\frac{d^2\theta}{dt^2} - \frac{1-2\theta}{2\theta(1-\theta)}\left(\frac{d\theta}{dt}\right)^2 = 0 \qquad \longrightarrow \qquad \theta(t) = \sin^2(at+b)$$

Then, for any two points $\theta_1, \theta_2 \in [0, 1]$ the geodesic distance will be:

$$S(\theta_1, \theta_2) = \int_{\theta_1}^{\theta_2} \sqrt{g_{11}(\theta)}\,d\theta = 2\,\left|\text{arc}\sin\sqrt{\theta}\right|_{\theta_1}^{\theta_2}$$

and for two infinitesimally close points $(\theta, \theta + \epsilon)$

$$S(\theta, \theta + \epsilon) \simeq \frac{\epsilon}{\sqrt{\theta(1-\theta)}} \propto \sqrt{g_{11}}$$

that is, Jeffrey's prior. We leave as an exercise to see that for the exponential family $Ex(x|1/\tau)$

$$g_{11}(\tau) = \frac{1}{\tau^2}; \qquad \Gamma^1_{11}(\tau) = -\frac{1}{\tau}; \qquad S(\tau_1, \tau_2) = \left|\log\frac{\tau_2}{\tau_1}\right|; \qquad S(\tau, \tau + \epsilon) \simeq \frac{\epsilon}{\tau} \propto \sqrt{g_{11}}$$

The Normal family $N(x|\mu, \sigma)$ is more interesting. The metric tensor for $\{\mu, \sigma\}$ is

$$g_{ij} = \begin{pmatrix} 1/\sigma^2 & 0 \\ 0 & 2/\sigma^2 \end{pmatrix}; \qquad \det(g_{ij}) = 2/\sigma^4; \qquad g^{ij} = \begin{pmatrix} \sigma^2 & 0 \\ 0 & \sigma^2/2 \end{pmatrix}$$

and therefore the non-zero Christoffel symbols are

$$\Gamma^1_{12} = \Gamma^1_{21} = -1/\sigma; \qquad \Gamma^2_{11} = 1/2\sigma; \qquad \Gamma^2_{22} = -1/\sigma$$

From them we can obtain the Gaussian and scalar curvatures:

$$\kappa = \frac{R_{1212}}{\det(g_{ij})} = -\frac{1}{2}; \qquad R = g^{ij} g^{lm} R_{iljm} = -1$$

and conclude that the parametric manifold (μ, σ) of the Normal family is hyperbolic and with constant curvature (thus, there is no complete isometric immersion in E^3). The geodesic equations are:

$$\frac{d^2\mu}{dt^2} - \frac{2}{\sigma}\frac{d\mu}{dt}\frac{d\sigma}{dt} = 0 \quad \text{and} \quad \frac{d^2\sigma}{dt^2} + \frac{1}{2\sigma}\left(\frac{d\mu}{dt}\right)^2 - \frac{1}{\sigma}\left(\frac{d\sigma}{dt}\right)^2 = 0$$

so writing

$$\frac{d\sigma}{dt} = \frac{d\sigma}{d\mu}\frac{d\mu}{dt} \quad \text{and} \quad \frac{d^2\sigma}{dt^2} = \frac{d^2\sigma}{d\mu^2}\left(\frac{d\mu}{dt}\right)^2 + \frac{d\sigma}{d\mu}\left(\frac{d^2\mu}{dt^2}\right)$$

we have that:

$$\sigma\frac{d^2\sigma}{d\mu^2} + \left(\frac{d\sigma}{d\mu}\right)^2 + \frac{1}{2} = \frac{d(\sigma\sigma')}{d\mu} + \frac{1}{2} = 0 \quad \longrightarrow \quad 2\,\sigma^2(\mu) = a - (\mu - b)^2$$

where $b - \sqrt{a} < \mu < b + \sqrt{a}$. For the points (μ_1, σ_1) and (μ_2, σ_2), the integration constants are

$$b = \frac{\mu_1 + \mu_2}{2} + \frac{\sigma_1^2 - \sigma_2^2}{\mu_1 - \mu_2} \quad \text{and} \quad a = \frac{[(\mu_1 - \mu_2)^2 + 2(\sigma_1 + \sigma_2)^2][(\mu_1 - \mu_2)^2 + 2(\sigma_1 - \sigma_2)^2]}{4(\mu_1 - \mu_2)^2}$$

and the geodesic distance is

$$S_{12} = \sqrt{a/2} \int_{\mu_1}^{\mu_2} \sigma(\mu)^{-2}\, d\mu = \sqrt{2}\, \ln\left\{\frac{1+\delta}{1-\delta}\right\}$$

with

$$\delta = \left(\frac{(\mu_1-\mu_2)^2 + 2(\sigma_1-\sigma_2)^2}{(\mu_1-\mu_2)^2 + 2(\sigma_1+\sigma_2)^2}\right)^{1/2} \in [0, 1)$$

Suppose for instance that (μ_1, σ_1) are given. Then, under the hypothesis $H_0 : \{\mu = \mu_0\}$:

$$\inf_\sigma S(\mu_1, \sigma_1; \mu_0, \sigma) \qquad \longrightarrow \qquad \sigma^2 = \sigma_1^2 + (\mu_1 - \mu_0)^2/2$$

and, under $H_0 : \{\sigma = \sigma_0\}$:

$$\inf_\mu S(\mu_1, \sigma_1; \mu, \sigma_0) \qquad \longrightarrow \qquad \mu = \mu_1$$

Problem 4.1 Show that for the Cauchy distribution

$$p(x|\alpha, \beta) = \frac{\beta}{\pi}\left[1 + \beta^2(x-\alpha)^2\right]^{-1}; \qquad x \in \mathcal{R}; \qquad (\alpha, \beta) \in \mathcal{R} \times \mathcal{R}^+$$

we have for $\{\alpha, \beta\}$:

$$g_{ij} = \begin{pmatrix} \beta^2/2 & 0 \\ 0 & 1/(2\beta^2) \end{pmatrix};$$
$$\Gamma^2_{11} = -\beta^2; \quad \Gamma^1_{12} = -\Gamma^2_{22} = \beta^{-1}; \quad R_{1212} = -1/2; \quad R = 2\kappa = -4$$

and, as for the Normal family, the geodesic equation has a parabolic dependence $\beta^{-2} = a - (\alpha - b)^2$. It may help to consider that

$$\int_{-\infty}^{\infty} p(x|\alpha, \beta)\left[1 + \beta^2(x-\alpha)^2\right]^{-n} = \frac{\Gamma(n+1/2)}{\Gamma(1/2)\Gamma(n+1)}; \qquad n \in \mathcal{N}$$

As for the Normal family, the parametric manifold for the Cauchy family of distributions is also hyperbolic with constant curvature. Show that this is the general case for distributions with location and scale parameters $p(x|\alpha, \beta) = \beta f[\beta(x-\alpha)]$ where $x \in \mathcal{R}$ and $(\alpha, \beta) \in \mathcal{R} \times \mathcal{R}^+$.

Problem 4.2 Show that for the metric of the Example 2.21 (ratio of Poisson parameters)

$$g_{ij}(\theta, \mu) = \begin{pmatrix} \mu/\theta & 1 \\ 1 & (1+\theta)/\mu \end{pmatrix}$$

the Riemann tensor is zero and therefore the manifold for $\{\theta, \mu\}$ is locally isometric to the two dimensional Euclidean space. Show that the geodesics are given by $\theta(\mu) = (b_0 + b_1\mu^{-1/2})^2$ and, for the affine parameter t, $\mu(t) = (a_0 + a_1 t)^2$. If we define the new parameters $\phi_1 = 2(\theta\mu)^{1/2}$ and $\phi_2 = 2\mu^{1/2}$, What do you expect to get for the manifold $\mathcal{M}\{\phi_1, \phi_2\}$?

As we move along the information geodesic, there are some quantities that remain invariant. The *Killing vectors* ζ_μ are given by the first order differential equation $\zeta_{\mu,\nu} + \zeta_{\nu,\mu} - 2\Gamma^{\rho}_{\mu\nu}\zeta_\rho = 0$ and if we denote by $u^\mu = dx^\mu/dt$ the tangent vector to the geodesic, with t the affine parameter, one has that $\zeta_\mu u^\mu =$ constant. For n-dimensional spaces with constant curvature there are $n(n+1)/2$ of them and clearly and linear combination with constant coefficients will be also a Killing vector. In the case of the Normal family we have that:

$$\zeta_\mu = c_1\left(\frac{\mu^2 - 2\sigma^2}{4\sigma^2}, \frac{\mu}{\sigma}\right) + c_2\left(\frac{\mu}{2\sigma^2}, \frac{1}{\sigma}\right) + c_3\left(\frac{1}{\sigma^2}, 0\right)$$

with $c_{1,2,3}$ integration constants and setting $c_i = 1, c_{j\neq i} = 0$ for $i, j = 1, 2, 3$ we get the 3 independent vectors. Therefore, along the information geodesic:

$$\frac{\mu^2 - 2\sigma^2}{4\sigma^2}\frac{d\mu}{dt} + \frac{\mu}{\sigma}\frac{d\sigma}{dt}, \qquad \frac{\mu}{2\sigma^2}\frac{d\mu}{dt} + \frac{1}{\sigma}\frac{d\sigma}{dt} \qquad \text{and} \qquad \frac{1}{\sigma^2}\frac{d\mu}{dt}$$

will remain constant. In fact, it is easier to derive from these first order differential equations the expression of the geodesic as function of the affine parameter t:

$$\mu(t) = b + \sqrt{a}\tanh(c_1 t + c_0) \qquad \text{and} \qquad \sigma(t)^{-1} = \sqrt{\frac{2}{a}}\cosh(c_1 t + c_0)$$

where for $t \in [0, 1]$, $(\mu_1, \sigma_1)_{t=0}$ and $(\mu_2, \sigma_2)_{t=1}$:

$$c_0 = \tanh\left(\frac{\mu_1 - b}{\sqrt{a}}\right) \qquad \text{and} \qquad c_0 - c_1 = \tanh\left(\frac{\mu_2 - b}{\sqrt{a}}\right)$$

With respect to the standardized Normal $N(x|0, 1)$, all points of the (μ, σ) of the manifold $\mathcal{R} \times \mathcal{R}^+$ with the same geodesic distance $S \geq 0$ are given by

$$\mu = \pm\left(4\sigma\cosh(S/\sqrt{2}) - 2(1 + \sigma^2)\right)^{1/2}$$

Figure 4.1(left) shows the curves in the (μ, σ) plane whose points have the same geodesic distance with respect to the Normal density $N(x|0, 1)$. The inner set corresponds to a distance of $d_G = 0.1$, the outer one to $d_G = 1.5$ and those in between to increasing steps of 0.2.

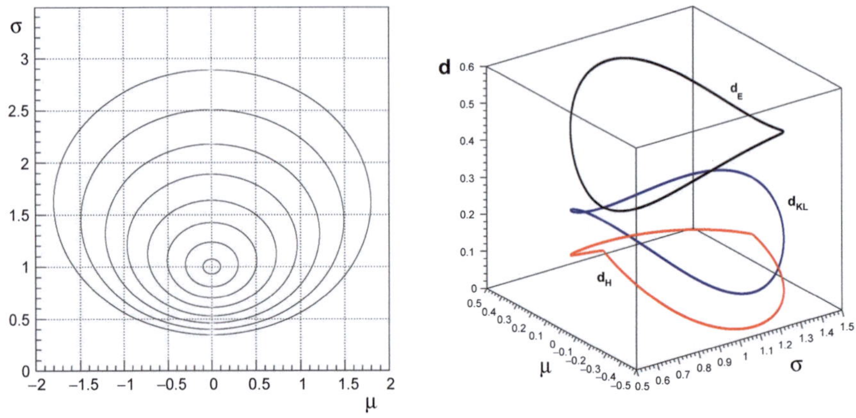

Fig. 4.1 *Left* Set of points in the (μ, σ) plane have the same geodesic distance with respect to the Normal density $N(x|0, 1)$ points in the (μ, σ). The inner set corresponds to a distance of $d_G = 0.1$, the outer one to $d_G = 1.5$ and those in between to increasing steps of 0.2. *Right* Euclidean (d_E), symmetrized Kullback–Leibler discrepancy (d_{KL}) and the Hellinger distance (d_H) for the set of (μ, σ) points that have the same geodesic distance $d_G = 0.5$

For two normal densities, the symmetrized Kullback–Leibler discrepancy is

$$J[p_1; p_2] = D(p_1 \| p_2) + D(p_2 \| p_1) = \frac{(\mu_1 - \mu_2)^2 + (\sigma_1^2 - \sigma_2^2)^2}{2\sigma_1^2 \sigma_2^2}$$

and the Hellinger distance

$$d_H[p_1; p_2] = 1 - \sqrt{\frac{2\sigma_1\sigma_2}{\sigma_1^2 + \sigma_2^2}} \exp\left\{ -\frac{(\mu_1 - \mu_2)^2}{4(\sigma_1^2 + \sigma_2^2)} \right\}$$

Figure 4.1(right) shows the euclidean (d_E), the symmetrized Kullback–Leibler (d_{KL}) and the Hellinger distances (d_H) for the set of points that have the same geodesic distance $d_G = 0.5$.

References

1. C. Shannon, A mathematical theory of communication. Bell Syst. Tech. J. **27** (1948)
2. S. Kullback, *Information Theory and Statistics* (Dover, New York, 1968)
3. J.M. Bernardo, A.F.M. Smith, *Bayesian Theory* (Wiley, New York, 1994)
4. J.M. Bernardo, J. R. Stat. Soc. Ser. B **41**, 113–147 (1979)